Push your Career Publish your Thesis

Science should be accessible to everybody. Share the knowledge, the ideas, and the passion about your research. Give your part of the infinite amount of scientific research possibilities a finite frame.

Publish your examination paper, diploma thesis, bachelor thesis, master thesis, dissertation, or habilitation treatises in form of a book.

A finite frame by infinite science.

Infinite Science
Publishing

A University Press Imprint of
Infinite Science GmbH
MFC 1 | Technikzentrum Lübeck
BioMedTec Wissenschaftscampus
Maria-Goeppert-Straße 1
23562 Lübeck
book@infinite-science.de
www.infinite-science.de

6th International Workshop on

Magnetic Particle Imaging
IWMPI 2016

March 16–18, 2016 | Lübeck, Germany

Book of Abstracts

T. M. Buzug, J. Borgert, and T. Knopp (Eds.)

Infinite Science
Publishing

© 2016   Infinite Science Publishing
University Press and
Academic Printing

Imprint of Infinite Science GmbH,
MFC 1 | BioMedTec Wissenschaftscampus
Maria-Goeppert-Straße 1
23562 Lübeck, Germany

Cover Design and Illustration: Uli Schmidts, metonym
Editorial: Universität zu Lübeck

Publisher: Infinite Science GmbH, Lübeck, www.infinite-science.de
Printed in Germany, BoD, Norderstedt

ISBN Paperback: 978-3-945954-19-5

Bibliografische Information der Deutschen Nationalbibliothek:
Die Deutsche Nationalbibliothek verzeichnet diese Publikation in der Deutschen Nationalbibliografie; detaillierte bibliografische Daten sind im Internet über http://dnb.d-nb.de abrufbar.

Scientific Committees

Workshop Chairs

Thorsten M. Buzug	University of Lübeck	Germany
Jörn Borgert	Philips GmbH Innovative Technologies, Hamburg	Germany
Tobias Knopp	University Medical Center Hamburg-Eppendorf (UKE)	Germany

Program Committee

Gerhard Adam	University Medical Center Hamburg-Eppendorf (UKE)	Germany
Christoph Alexiou	University Medical Center Erlangen	Germany
Meltem Asilturk Akdeniz	University, Antalya	Turkey
Jörg Barkhausen	UKSH Lübeck	Germany
Volker Behr	University of Würzburg	Germany
Ayhan Bingolbali	Yildiz Technical University, Istanbul	Turkey
Jeff Bulte	John Hopkins University, Baltimore	USA
Thorsten M. Buzug	University of Lübeck	Germany
Steven M. Conolly	University of California, Berkeley	USA
Nurcan Dogan	Gebze Institute of Technology, Kocaeli	Turkey
Silvio Dutz	Technical University of Ilmenau	Germany
Matthew Ferguson	LodeSpin Labs, Seattle	USA
Dominique Finas	Evangelisches Krankenhaus Bielefeld	Germany
Patrick W. Goodwill	University of California, Berkeley	USA
Mark Griswold	Case Western Reserve University, Cleveland	USA
Urs Häfeli	University of British Columbia, Vancouver	Canada
Jens Haueisen	Technical University of Ilmenau	Germany
Ulrich Heinen	Bruker BioSpin MRI GmbH, Ettlingen	Germany
Yasutoshi Ishihara	Meiji University, Tokyo	Japan
Peter Jakob	University of Würzburg	Germany
Tobias Knopp	University Medical Center Hamburg-Eppendorf (UKE)	Germany
Kannan Krishnan	University of Washington, Seattle	USA
Frank Ludwig	Technical University of Braunschweig	Germany
Mauro Magnani	University of Urbino	Italy
Jan Niehaus	CAN Center für Applied Nanotechnology, Hamburg	Germany
Stefan Odenbach	Technical University of Dresden	Germany
Ulrich Pison	Charité Universitätsmedizin Mitte, Berlin	Germany
Jürgen Rahmer	Philips GmbH Innovative Technologies, Hamburg	Germany
Anna Cristina Samia	Case Western Reserve University, Cleveland	USA
Emine Ulku Saritas	Bilkent University, Bilkent/Ankara	Turkey
Meinhard Schilling	Technical University of Braunschweig	Germany
Ingo Schmale	Philips GmbH Innovative Technologies, Hamburg	Germany
Jörg Schnorr	Charité Universitätsmedizin Mitte, Berlin	Germany
Gunnar Schütz	Bayer HealthCare Pharmaceuticals, Berlin	Germany
Bennie ten Haken	University of Twente, Enschede	Germany
Lutz Trahms	PTB Physikalisch-Technische Bundesanstalt, Berlin	Germany
John B. Weaver	Dartmouth-Hitchcock Medical Center, Lebanon	USA
Jürgen Weizenecker	University of Applied Sciences, Karlsruhe	Germany
Frank Wiekhorst	PTB Physikalisch-Technische Bundesanstalt, Berlin	Germany
Barbara Wollenberg	University Medical Center Schleswig-Holstein, Lübeck	Germany

Preface and Acknowledgements

Dear Colleagues,

we are very pleased to host the 6th International Workshop on Magnetic Partical Imaging again in Lübeck where the IWMPI history started six years ago and would like to thank Emine U. Sarias and her team for the outstanding meeting IWMPI2015 in exciting Istanbul. We are proud to announce a record number of contributions of participants from 14 different countries presenting 45 talks, 96 posters with combined elevator speech and 4 keynote speeches from all fields of MPI namely Instrumentation, Application, Methodology and Tracer Materials.

Since the first workshop in 2010, the International Workshop on MPI (IWMPI) has been the premier forum for researchers working in the MPI field. The workshop aims at covering the status and recent developments of both the instrumentation and the tracer material, as they are equally important in designing a well performing MPI system. The main topics presented at the workshop include hardware developments, image reconstruction and systems theory, nanoparticle physics and theory, nanoparticle synthesis, spectroscopy, patient safety, and medical/research applications of MPI.

We encourage you and your colleagues to contribute your research and results to IWMPI, where you will have an opportunity to interact and collaborate with the greater MPI community, and to take steps in advancing the field of MPI. The workshop will provide a great opportunity to present your research results, as well as to learn more about the technical aspects and clinical potential of MPI.

In 2015, as new publication platform the International Journal on Magnetic Particle Imaging (IJMPI) has been launched as a future format for publishing high quality research articles on MPI (journal.iwmpi.org). The scope of the IJMPI ranges from imaging sequences and reconstruction over scanner instrumentation as well as particle synthesis and particle physics to pre-clinical and potential future clinical applications. Journal articles will be published online with open access under a Creative Commons License. In order to share ideas and experiences with a focused audience, we encourage submission of research papers to the new journal. IJMPI will publish research articles that can be submitted at any time.

As chairs of the workshop we would like to thank the members of the program committee for their exceptional service for the MPI community: G. Adam, University Medical Center Hamburg-Eppendorf (UKE); C. Alexiou, University Medical Center Erlangen; M. Asilturk, Akdeniz University, Antalya; J. Barkhausen, UKSH Lübeck; V. Behr, University of Würzburg; A. Bingolbali, Yildiz Technical University, Istanbul; J. Bulte, John Hopkins University, Baltimore; T. M. Buzug, University of Lübeck; S. M. Conolly, University of California, Berkeley; N. Dogan, Gebze Institute of Technology, Kocaeli; S. Dutz, Technical University of Ilmenau; M. Ferguson, LodeSpin Labs, Seattle; D. Finas, Evangelisches Krankenhaus Bielefeld; P. W. Goodwill, University of California, Berkeley; M. Griswold, Case Western Reserve University, Cleveland; U. Häfeli, University of British Columbia, Vancouver; J. Haueisen, Technical University of Ilmenau; U. Heinen, Bruker BioSpin, Ettlingen; Y. Ishihara, Meiji University, Tokyo; P. Jakob, University of Würzburg; T. Knopp, University Medical Center Hamburg-Eppendorf; K. Krishnan, University of Washington, Seattle; F. Ludwig, Technical University of Braunschweig; M. Magnani, University of Urbino; J. Niehaus, CAN Center for Applied Nanotechnology, Hamburg; S. Odenbach, Technical University of Dresden; U. Pison, Charité, Berlin; J. Rahmer, Philips GmbH Innovative Technologies, Hamburg; A. C. Samia, Case Western Reserve University, Cleveland; E. U. Saritas, Bilkent University, Bilkent/Ankara; M. Schilling, Technical University of Braunschweig; I. Schmale, Philips GmbH Innovative Technologies, Hamburg; J. Schnorr, Charité, Berlin; G. Schütz, Bayer HealthCare, Berlin; B. ten Haken, University of Twente, Enschede; L. Trahms, PTB, Berlin; J. B. Weaver, Dartmouth-Hitchcock Medical Center, Lebanon; J. Weizenecker, University of Applied Sciences, Karlsruhe; F. Wiekhorst, PTB, Berlin; B. Wollenberg, UKSH Lübeck.

Most importantly, we would like to thank our partners for their support and cooperation: Bruker BioSpin, Eurocomp Elektronik GmbH, Philips Healthcare, magnetic INSIGHT, nanoPET Pharma GmbH, Pure Devices GmbH, PACK LitzWire and would also like to extend our gratitude to the members of the local organization teams for their outstanding efforts and work.

We wish all of us an inspiring workshop and are already looking forward to IWMPI2017 that will be held March 22-24, 2017 and hosted in China at HUST.

Thorsten M. Buzug, Jörn Borgert and Tobias Knopp
Lübeck and Hamburg, March 2016

Contents

Session 1: Methodology 1

Keynote: Magnetic Nanoparticle Dynamics in Spectroscopy and its Applications
Dr. John B. Weaver
3

Spatial Resolution in MPI: The Role of Harmonic Number
H. Bagheri and M.E. Hayden
5

Sparse source reconstruction for nanomagnetic relaxometry
Sara Loupot, Wolfgang Stefan, Reza Medankan, Kelsey Mathieu, David Fuentes, and John D. Hazle
6

Symmetries in Nanoparticle Dynamics Found From The Buckingham Pi Theorem Improve Sensing Strategies
Yipeng Shi, John B. Weaver
7

Functional Magnetic Particle Imaging in measurement and simulation
Thilo Viereck, Christian Kuhlmann, Sebastian Draack, Frank Ludwig, Meinhard Schilling
8

Experimental Distinction of Different Viscosities using Multispectral Magnetic Particle Imaging
Martin Hofmann, Jan Dieckhoff, Harald Ittrich, Tobias Knopp
9

Calibration-Free Color MPI
Yavuz Muslu, Mustafa Ütkür, Omer Burak Demirel, Emine Ulku Saritas
10

Elevator speeches 1: Instrumentation

An Ultra-Low Noise Preamplifier Design for Magnetic Particle Imaging
Beliz Gunel, Bo Zheng, Steven Conolly
13

Suppress Direct Feedthrough Induced by Excitation Magnetic Field
Tao Jiang, Shiqiang Pi, and Wenzhong Liu
14

Switching Power Amplifier for Magnetic Particle Imaging
Timo F. Sattel, Christian Vollertsen, Jan Gressmann, Oliver Woywode
15

A novel compensation technique for gradiometer receive coils in MPI/MPS
Florian Fidler, Karl-Heinz Hiller, Peter M. Jakob
16

A Tunable Gradiometer Receive Coil for Magnetic Particle Imaging
Steffen Bruns, Matthias Weber, Thorsten M. Buzug
17

MPS and ACS with an atomic magnetometer
Simone Colombo, Victor Lebedev, Zoran D. Grujić, Vladimir Dolgovskiy, Antoine Weis.
18

Development of a K-Rb Hybrid Atomic Magnetometer toward MPI
Yosuke Ito, Tetsuo Kobayashi
19

MPS test measurments with phase angle detection
Przemysław Wróblewski, Waldemar Smolik
20

An Arbitrary Excitation Waveform Relaxometer
Zhi Wei Tay, Daniel W. Hensley, Laura A. Taylor, Beliz Gunel, Patrick W. Goodwill, Bo Zheng, Steven M. Conolly
21

Effects of Viscosity on Nanoparticle Relaxation Time Constant
Mustafa Ütkür, Yavuz Muslu, Ahmet Alacaoglu, Ali Alper Ozaslan, Emine Ulku Saritas
22

Reconstruction of a 2D Phantom Recorded with a Single-Sided MPI Device
Ksenija Gräfe, Anselm von Gladiss, Mandy Ahlborg, Gael Bringout, Thorsten M. Buzug
23

Single-sided FFL-based MPI Device with Depth Encoding
Alexey Tonyushkin
24

High Resolution Tomographic MPI with a Field Free Line Electromagnet
ELAINE YU, PATRICK W. GOODWILL, ZHI WEI TAY, PAUL KESELMAN, XINYI ZHOU, RYAN ORENDORFF, DANIEL W. HENSLEY, MATT FERGUSON, BO ZHENG, STEVEN M. CONOLLY ... 25

MPI Cube – fully 3D field free line scanner
P. VOGEL, M.A. RÜCKERT, V.C. BEHR ... 26

Self-Shielded, High-Resolution, and High-Sensitivity MPI FFL Imager
PATRICK GOODWILL, JUSTIN KONKLE, STEVEN SUDDARTH, ANNA CHRISTENSEN ... 27

Modular mobility MPI system
SEBASTIAN DRAACK, CHRISTIAN KUHLMANN, THILO VIERECK, FRANK LUDWIG, MEINHARD SCHILLING ... 28

Bimodal TWMPI-MRI hybrid scanner – first MRI results
PETER KLAUER, EBERHARD ROMMEL, PATRICK VOGEL, MARTIN A. RÜCKERT, VOLKER C. BEHR ... 29

Studies on the Optimization of Efficient Selection and Focus Field Coil Configurations
JULIA MRONGOWIUS, CHRISTIAN KAETHNER, THORSTEN M. BUZUG ... 30

Magnetic Particle Imaging by Using Multichannel Coil Arrays
SHU-HSIEN LIAO, JEN-JIE CHIEH, HERNG-ER HORNG, HONG-CHANG YANG, SABURO TANAKA ... 31

Designing coils to minimize the maximal induced electrical field amplitude in a patient
GAEL BRINGOUT, JOHAN LÖFBERG, PATRICIA ULLOA, MARTIN A. KOCH, THORSTEN M. BUZUG ... 32

Novel Selection Coils Design for 3D FFL-based MPI
ALEXEY TONYUSHKIN ... 33

Evaluation of the spatial confidence and dual modal FOV-center conformity of a highly integrated MPI-MRI hybrid system
JOCHEN FRANKE, ULRICH HEINEN, ALEXANDER WEBER, HEINRICH LEHR, MICHAEL HEIDENREICH, WOLFGANG RUHM, VOLKMAR SCHULZ ... 34

Metallic artefact suppression in intraoperative magnetometers
SEBASTIAAN WAANDERS, ROGIER WILDEBOER, ERIK KROOSHOOP, BENNIE TEN HAKEN ... 35

Systematic Background Estimation
MARCEL STRAUB, BERNHARD GLEICH, JÜRGEN RAHMER, VOLKMAR SCHULZ ... 36

Controlling the Position of the Field-Free-Point in Magnetic Particle Imaging
A. WEBER, J. WEIZENECKER, R. PIETIG, U. HEINEN, T.M. BUZUG ... 37

Force analysis device for magnetic manipulation
DAVID WELLER, THORSTEN M. BUZUG AND THOMAS FRIEDRICH ... 38

Study of temperature measurement on pn-junction of light-emitting diodes using magnetic nanothermometer
ZHONGZHOU DU, KAI WEI, RIJIAN SU, YONG GAN, AND WENZHONG LIU ... 39

Elevator speeches 1: Applications

In vitro MPI iron quantification of labeled cells for a metastasis-tracking study
VERA PAEFGEN, MARCEL STRAUB, FABIAN KIEßLING, VOLKMAR SCHULZ ... 43

MPI-Detection of Multicore Iron Oxide Nanoparticles dedicated for Magnetic Drug Targeting
STEFAN LYER, TOBIAS KNOPP, FRANZISKA WERNER, LUTZ TRAHMS, FRANK WIEKHORST, TOBIAS STRUFFERT, TOBIAS ENGELHORN, ARNDT DÖRFLER, TOBIAS BÄUERLE, MICHAEL UDER, CHRISTOPH ALEXIOU ... 44

Different Behavior of MPI Signals from Magnetic Nanoparticles Internalized by Macrophages and Colon Cancer Cells
HISAAKI SUZUKA, ATSUSHI MIMURA, YOSHIMI INAOKA, KOHEI NISHIMOTO, NATSUO BANURA, KENYA MURASE ... 45

Processing of SPIO in macrophages and tumor tissue for MPI lymph node imaging in breast cancer
DOMINIQUE FINAS, JANINE STEGMANN-FREHSE J, BENJAMIN SAUER, GEREON HÜTTMANN, ACIM RODY, THORSTEN BUZUG,
KERSTIN LÜDTKE-BUZUG   46

Magnetic Particle Spectrometer for the Analysis of Magnetic Particle Heating Applications
ANDRÉ BEHRENDS, THORSTEN M. BUZUG, ALEXANDER NEUMANN   47

Visualization and Quantification of the Intratumoral Distribution and Time-Dependent Change of
Magnetic Nanoparticles in Magnetic Hyperthermia Using Magnetic Particle Imaging
TOMOMI KUBOYABU, ISAMU YABATA, MARINA AOKI, AKIKO OHKI, MIKIKO YAMAWAKI, YOSHIMI INAOKA, KAZUKI SHIMADA,
KENYA MURASE   48

Towards Simultaneous MFH and Temperature Monitoring with MPI
CAGLA DENIZ BAHADIR, MUSTAFA ÜTKÜR, EMINE ULKU SARITAS   49

First Results: Phantoms for MPI and Ultrasound Therapy
ANKIT MALHOTRA, CORINNA STEGELMEIER, THOMAS FRIEDRICH, KERSTIN LÜDTKE- BUZUG AND THORSTEN M. BUZUG   50

Magnetic particle imaging in a mouse model of acute ischemic stroke
PETER LUDEWIG, NADINE GDANIECJAN SEDLACIK, SARAH BEHR, SCOTT J. KEMP, R. MATTHEW FERGUSON,
AMIT P. KHANDHAR, KANNAN M. KRISHNAN, JENS FIEHLER, CHRISTIAN GERLOFF, TOBIAS KNOPP, TIM MAGNUS   51

First in-vivo Perfusion Imaging with MPI
RYAN ORENDORFF, PAUL KESELMAN, STEVEN M. CONOLLY   52

Long term in vivo biodistribution and clearance of tailored MPI tracers
PAUL KESELMAN, BO ZHENG, PATRICK W. GOODWILL, AND STEVEN M. CONOLLY   53

Stem cell tracking potential of Magnetic Particle Imaging compared with 19F Magnetic Resonance
Imaging
FRISO G. HESLINGA, STEFFEN BRUNS, ELAINE YU, PAUL KESELMAN, XINYI Y. ZHOU, BO ZHENG, SEBASTIAAN WAANDERS,
PATRICK W. GOODWILL, M. WENDLAND, BENNIE TEN HAKEN, STEVEN M. CONOLLY   54

Growth inhibition of Pseudomonas Aeruginosa by extremely low frequency Pulsed Magnetic Field
(PMF)
FADEL M.ALI, NERMEEN.SERAG, A. M. KHALIL   55

Compression of FFP System Matrix with a Special Sampling Rate on the Lisssajous Trajectory
MARCO MAASS, KLAAS BENTE, MANDY AHLBORG, HANNE MEDIMAGH, HUY PHAN, THORSTEN M. BUZUG, AND
ALFRED MERTINS   56

Investigation and Removal of Artifacts Due to Particles Located Outside the Field-Free-Point Trajectory
A. WEBER, F. WERNER, J. WEIZENECKER, T.M. BUZUG, T. KNOPP   57

MMSE MPI Reconstruction Using Background Identification
HANNA SIEBERT, MARCO MAASS, MANDY AHLBORG, THORSTEN M. BUZUG, AND ALFRED MERTINS   58

Optimizing the Coil Setup for a Three-Dimensional Magnetic Particle Spectrometer
XIN CHEN, ANDRÉ BEHRENDS, MATTHIAS GRAESER, ALEXANDER NEUMANN, THORSTEN M. BUZUG   59

Development and Testing of Magnetic Nanoparticle-Gel Materials for Magnetic Particle Imaging
Phantoms
R. SANDIG, A. MATTERN, D. BAUMGARTEN, O. KOSCH, F. WIEKHORST, A. WEIDNER, S. DUTZ   60

Dynamic Magnetization of Immobilized Magnetic Nanoparticles for Cases with Aligned and Randomly
Oriented Easy Axes
TAKASHI YOSHIDA, THILO VIERECK, TERUYOSHI SASAYAMA, KEIJI ENPUKU, MEINHARD SCHILLING, AND FRANK LUDWIG   61

Elevator speeches 2: Methodology

A Novel Magnetic Particle Imaging Scanner with Lower Amplitude of an Excitation Field
XINGMING ZHANG, TUẤN ANH LÊ, AND JUNGWON YOON — 63

RDS Toolbox – Simulation of 3D Rotational Drift
A. VILTER, M. A. RÜCKERT, T. KAMPF, V. J. F. STURM, V. C. BEHR — 64

Magnetic signal detection method based on active vibration of magnetic nanoparticles
AKIHIRO MATSUHISA, TOMOKI HATSUDA, TOMOYUKI TAKAGI, MASAHIRO ARAYAMA, YASUTOSHI ISHIHARA — 65

The Influence of Discretization of DC Field on Magnetic Nanothermometer
LE HE, SHIQIANG PI, QINGGUO XIE, WENZHONG LIU — 66

The effect of dc field strength on the performance of a magnetic nanothermometer
JING ZHONG, FRANK LUDWIG, MEINHARD SCHILLING — 67

Magnetic signal separation using independent component analysis
MASAHIRO ARAYAMA, TOMOYUKI TAKAGI, TOMOKI HATSUDA, AKIHIRO MATSUHISA, HIROKI TSUCHIYA, YASUTOSHI ISHIHARA — 68

MPI meets MRI – simultaneous measurement of MPI and MRI signals
P. VOGEL, T. KAMPF, M.A. RÜCKERT, A. VILTER, P.M. JAKOB, V.C. BEHR — 69

Effects of Safety Limits on Image Quality in MPI
ECEM BOZKURT, OMER BURAK DEMIREL, DAMLA SARICA, YAVUZ MUSLU, EMINE ULKU SARITAS — 70

Lissajous Node Points for a Sytem Matrix based MPI Image Reconstruction Approach
CHRISTIAN KAETHNER, MANDY AHLBORG, WOLFGANG ERB, THORSTEN M. BUZUG — 71

Basic Study of Image Reconstruction Method Using Neural Networks with Additional Learning for Magnetic Particle Imaging
TOMOKI HATSUDA, TOMOYUKI TAKAGI, AKIHIRO MATSUHISA, MASAHIRO ARAYAMA, HIROKI TSUCHIYA, YASUTOSHI ISHIHARA — 72

A new 3D MPI model using realistic magnetic field topologies for algebraic reconstruction
WOLFGANG ERB, GAEL BRINGOUT, JÜRGEN FRIKEL, THORSTEN M. BUZUG — 73

Nonlinear Scanning in X-Space MPI
AHMET ALACAOGLU, ALI ALPER OZASLAN, OMER BURAK DEMIREL, EMINE ULKU SARITAS — 74

X-Space and Chebyshev Reconstruction in Magnetic Particle Imaging: A First Experimental Comparison
TOBIAS KNOPP, CHRISITAN KAETHNER, MANDY AHLBORG AND THORSTEN M. BUZUG — 75

Self Calibration for Relaxation- and System-Induced Delays in X-space MPI
BATURALP BUYUKATES, DAMLA SARICA, EMINE ULKU SARITAS — 76

Spatial Resolution in MPI: Modeling the Role of Harmonic Number
H. BAGHERI AND M.E. HAYDEN — 77

Rapid Scanning in X-Space MPI: Impacts on Image Quality
OMER BURAK DEMIREL, DAMLA SARICA, EMINE ULKU SARITAS — 78

Influence of Particle Size Distribution of Magnetic Nanoparticles on the Spatial Resolution of Magnetic Particle Imaging
XIUYING WANG, SHIQIANG PI, AND WENZHONG LIU — 79

DC Shift Imaging for X-Space MPI Reconstruction
DAMLA SARICA, OMER BURAK DEMIREL, YAVUZ MUSLU, EMINE ULKU SARITAS — 80

Limitations of Magnetic Particle Imaging Resolving Large Contrasts
NADINE GDANIEC, MARTIN HOFMANN, TOBIAS KNOPP — 81

Deconvolving Relaxation Effects in Multi-Dimensional X-space MPI
GAMZE ONUKER, OMER BURAK DEMIREL, DAMLA SARICA, YAVUZ MUSLU, EMINE ULKU SARITAS — 82

Enhancing the sensitivity in Magnetic Particle Imaging by Background Subtraction
K. THEM, M. G. KAUL, C. JUNG, M. HOFMANN, T. MUMMERT, F. WERNER, T. KNOPP   83

Correction of Blurring due to a Difference in Scanning Direction of Field-Free Line in Projection-Based Magnetic Particle Imaging
KENYA MURASE, KAZUKI SHIMADA, NATSUO BANURA   84

Sensitivity enhancement for stem cell monitoring in Magnetic Particle Imaging
KOLJA THEM, J. SALAMON, M. G. KAUL, CLAUDIA LANGE, H. ITTRICH, TOBIAS KNOPP   85

Towards the Characterization of Distortion Artifacts in Elongated Trajectory MPI
ANNIKA HÄNSCH, CHRISTIAN KAETHNER, AILEEN CORDES, THORSTEN M. BUZUG   86

Fiducial-Based Geometry Planning and Image Registration for Magnetic Particle Imaging
F. WERNER, C. JUNG, M. HOFMANN, R. WERNER, J. SALAMON, D. SÄRING, M. G. KAUL, K. THEM, O. M. WEBER, T. MUMMERT, G. ADAM, H. ITTRICH, T. KNOPP   87

Predicting 2D MPI imaging performance using a conventionally acquired or a hybrid 2D system function
HANNE MEDIMAGH, THORSTEN M. BUZUG   88

Experimental Results on 3D Real-Time Magnetic Particle Imaging of Large Fields-of-View
JÜRGEN RAHMER, BERNHARD GLEICH, CLAAS BONTUS, INGO SCHMALE, JOACHIM SCHMIDT, OLIVER WOYWODE, AND JÖRN BORGERT   89

Rotational Drift Spectroscopy (RDS): Measuring Fast Relaxing Magnetic Nanoparticle Ensembles
M.A. RÜCKERT, A. VILTER, P. VOGEL, V.C. BEHR   90

Dependence of Brownian and Néel Time Constants on Magnetic Field
FRANK LUDWIG, JAN DIECKHOFF, DIETMAR EBERBECK   91

Harmonic phases of the nanoparticle magnetization and their variation with temperature.
ENEKO GARAIO, JUAN-MARI COLLANTES, JOSE ANGEL GARCIA, FERNANDO PLAZAOLA, IRATI RODRIGO AND OLIVIER SANDRE   92

Heat Transfer Simulation for Optimization and Treatment Planning of Magnetic Hyperthermia Using Magnetic Particle Imaging
NATSUO BANURA, ATSUSHI MIMURA, KOHEI NISHIMOTO, KENYA MURASE   93

Magnetic Nanoparticle Temperature Estimation Using Dual-Frequency Magnetic Filed
KAI WEI, SHIQIANG PI, WENZHONG LIU   94

3D-GUI Simulation Environment for MPI
P. VOGEL, M.A. RÜCKERT, V.C. BEHR   95

Elevator speeches 2: Tracer Materials

Biocompatible Magnetite Nanoparticles as Tracer Material for Magnetic Particle Imaging
CORINNA STEGELMEIER, ANKIT MALHOTRA, KERSTIN LÜDTKE-BUZUG   99

Continuous Synthesis of Single-Core Iron Oxide Nanoparticles for Biomedical Applications
ABDULKADER BAKI, REGINA BLEUL, CHRISTOPH BANTZ, RAPHAEL THIERMANN, MICHAEL MASKOS   100

Diffusion-Controlled Synthesis of Magnetic Nanoparticles
DAVID HEINKE, NICOLE GEHRKE, DANIEL SCHMIDT, UWE STEINHOFF, THILO VIERECK, HILKE REMMER, FRANK LUDWIG, ANDREAS BRIEL   101

Development and Physicochemical Characterization of Continuously Manufactured Single-Core Iron Oxide Nanoparticles
CHRISTOPH BANTZ, REGINA BLEUL, ABDULKADER BAKI, RAPHAEL THIERMANN, NORBERT LÖWA, DIETMAR EBERBECK, LUTZ TRAHMS, MICHAEL MASKOS   102

Formation of a Protein Corona on Magnetic Nanoparticles Affects Nanoparticle-Cell Interactions
A. WEIDNER, C. GRÄFE, M. V.D. LÜHE, C. BERGEMANN, J.H. CLEMENT, F.H. SCHACHER, S. DUTZ   103

Development of Magnetic Nanocarriers Based on Thermosensitive Liposomes and Their Visualization Using Magnetic Particle Imaging

SHUKI MARUYAMA, KOHEI ENMEIJI, KAZUKI SHIMADA, KENYA MURASE
104

Quantitative biodistribution studies of optimized MPI tracers radiolabeled for multimodal SPECT/CT imaging

HAMED ARAMI, KATHAYOUN SAATCHI, ERIC TEEMAN, ALYSSA TROKSA, HAYDIN BRADSHAW, URS O. HÄFELI, AND KANNAN M. KRISHNAN
105

Magnetic Separation to Extract Suitable Cells for MPI Cell Tracking

ANGELA ARIZA DE SCHELLENBERGER, NORBERT LÖWA, JÖRG SCHNORR, HARALD KRATZ, MATTHIAS TAUPITZ, FRANK WIEKHORST
106

Evaluation of harmonic signal from blood-pooling magnetic nanoparticles for magnetic particle imaging

SATOSHI OTA, RYUJI TAKEDA, TSUTOMU YAMADA, YASUSHI TAKEMURA
107

Does a highly concentrated sample generate a better system function?

OLAF KOSCH, NORBERT LÖWA, FRANK WIEKHORST, LUTZ TRAHMS
108

In vivo measurement und comparison of two Magnetic Particle Imaging tracer: LS-008 and Resovist

MICHAEL G. KAUL, CAROLINE JUNG, JOHANNES SALAMON, TOBIAS MUMMERT, MARTIN HOFMANN, SCOTT J. KEMP, R. MATTHEW FERGUSON, AMIT P. KHANDHAR, KANNAN M. KRISHNAN, HARALD ITTRICH, GERHARD ADAM, TOBIAS KNOPP
109

Correlation of MPS with Colorimetric Iron Content Measurements

LISA WENDT, KERSTIN LÜDTKE-BUZUG
110

Magnetic Particle Spectrometry of LS-008 driven at 153 kHz, 15 mT/μ_0

R. MATTHEW FERGUSON, AMIT P. KHANDHAR, SCOTT J. KEMP, AND KANNAN M KRISHNAN
111

MPS study on new MPI tracer material

CHRISTINA DEBBELER, CATHRINE FRANDSEN, NICOLE GEHRKE, CORDULA GRÜTTNER, DAVID HEINKE, CHRISTER JOHANSSON, ANJA JOHL, MARÍA DEL PUERTO MORALES, MIRIAM VARÓN, KERSTIN LÜDTKE-BUZUG
112

Imaging Characterization of MPI Tracers Employing Offset Measurements in a two Dimensional Magnetic Particle Spectrometer

DANIEL SCHMIDT, MATTHIAS GRAESER, ANSELM VON GLADISS, THORSTEN M. BUZUG, UWE STEINHOFF
113

The Particle Response of Blended Nanoparticles in MPI

ANSELM VON GLADISS, MATTHIAS GRAESER, R. MATTHEW FERGUSON, AMIT P. KHANDHAR, SCOTT J. KEMP, KANNAN M. KRISHNAN, THORSTEN M. BUZUG
114

Determining magnetic impurities and nonspecific magnetic nanoparticle adhesion of MPI phantom materials

PATRICIA RADON, NORBERT LÖWA, FELIX PTACH, DIRK GUTKELCH, FRANK WIEKHORST
115

Session 2: Application 1

Keynote: Potential Clinical Applications of MPI

DR. MED. HARALD ITTRICH AND DR. MED. JOHANNES SALAMON
119

Color MPI for Cardiovascular Interventions

JULIAN HAEGELE, SARAH VAALMA, NIKOLAOS PANAGIOTOPOULOS, JÖRG BARKHAUSEN, FLORIAN M. VOGT, JÖRN BORGERT, JÜRGEN RAHMER
121

The next step towards interventional MPI: Real Time 3D MPI-guided treatment of a vessel stenosis using a blood pool agent and MRI Road Map approach

JOHANNES SALAMON; MARTIN HOFMANN; CAROLINE JUNG; MICHAEL GERHARD KAUL, RUDOLPH REIMER; ANNIKA VOM SCHEIDT; GERHARD ADAM; TOBIAS KNOPP; HARALD ITTRICH
122

Quantification of Vascular Stenosis Phantoms using Traveling Wave MPI

S. HERZ, P. VOGEL, V.C. BEHR, T.A. BLEY
123

Session 3: Methodology 2

Resolution Improvement for X-Space MPI having Low Gradient Field
HAMED JABBARI ASL, JUNGWON YOON   127

X-space Deconvolution for Multidimensional Lissajous-based Data-Acquisition Schemes
AILEEN CORDES, CHRISTIAN KAETHNER, MANDY AHLBORG, THORSTEN M. BUZUG   128

Flexible reconstruction method for Traveling Wave MPI
T. KAMPF, P. VOGEL, M.A. RÜCKERT, V.C. BEHR   129

Reconstruction of Experimental 2D MPI Data using a Hybrid System Matrix
MATTHIAS GRAESER, ANSELM VON GLADISS, PATRYK SZWARGULSKI, MANDY AHLBORG, TOBIAS KNOPP, THORSTEN M. BUZUG   130

Artefact Suppression in Time-resolved Magnetic Particle Imaging
ALEXANDER WEBER, JOCHEN FRANKE, HEINRICH LEHR, WOLFGANG RUHM, MICHAEL HEIDENREICH, THORSTEN M. BUZUG
ULRICH HEINEN   131

Fused Lasso Regularization for Magnetic Particle Imaging
MARTIN STORATH, CHRISTINA BRANDT, MARTIN HOFMANN, TOBIAS KNOPP, ALEXANDER WEBER, ANDREAS WEINMANN   132

Session 4: Instrumentation 1

Keynote: Safety Limits in MPI and Implications for Image Quality
DR. EMINE ULKU SARITAS   135

Signal path for a 10 kHz and 25 kHz mobility MPI System
CHRISTIAN KUHLMANN, SEBASTIAN DRAACK, THILO VIERECK, FRANK LUDWIG, MEINHARD SCHILLING   137

First Spectrum Measurements with a Rabbit-Sized FFL-Scanner
JAN STELZNER, GAEL BRINGOUT, ANSELM VON GLADISS, HANNE MEDIMAGH, MANDY AHLBORG, TIMO F. SATTEL,
THORSTEN M. BUZUG   138

Micro Traveling Wave MPI – initial results with optimized tracer LS-008
P. VOGEL, M.A. RÜCKERT, S.J. KEMP, A.P. KHANDHAR, R.M. FERGUSON, A. VILTER, P. KLAUER, K.M. KRISHNAN, V.C. BEHR   139

M(H) dependence and size distribution of SPIONs measured by atomic magnetometry
SIMONE COLOMBO, VICTOR LEBEDEV, ZORAN D. GRUJIĆ, VLADIMIR DOLGOVSKIY, ANTOINE WEIS.   140

The Design of Magnetic Particle Imaging Gradient Magnetic Field Generator using Finite Element Method
SHIQIANG PI, JINGJING CHENG, WENZHONG LIU   141

A 1.4 T/m Field Free Line Magnetic Particle Imaging Device
MATTHIAS WEBER, KLAAS BENTE, STEFFEN BRUNS, ANSELM VON GLADISS, MATTHIAS GRAESER, THORSTEN M. BUZUG   142

Session 5: Application 2

Assessing flow dynamics in a 3D printed aneurysm model by magnetic particle imaging
JAN SEDLACIK, ANDREAS FRÖLICH, JOHANNA SPALLEK, NILS D. FORKERT, TOBIAS D. FAIZY, FRANZISKA WERNER,
TOBIAS KNOPP, DIETER KRAUSE, JENS FIEHLER, JAN-HENDRIK BUHK   145

Differential pick-up coils in magnetic particle spectrometry to detect low concentration SPIO nanoparticle tracers
BHARADWAJ MURALIDHARAN, THOMAS E. MILNER AND CHUN HUH   146

Devices for remote magnetic operation in an MPI scanner
CHRISTIAN STEHNING, PETER MAZURKEWITZ, BERNHARD GLEICH, JÜRGEN RAHMER   147

First Murine *in vivo* Cancer Imaging with MPI
ELAINE YU, MINDY BISHOP, PATRICK W. GOODWILL, BO ZHENG, MATT FERGUSON, KANNAN M. KRISHNAN,
STEVEN M. CONOLLY   148

In-vivo Measurements with UW-tracers in a harmonic 5.5 T/m MPI
MARCEL STRAUB, VERA PÄFGEN, ERIC TEEMAN, KANNAN M. KRISHNAN, FABIAN KIEßLING, VOLKMAR SCHULZ   149

Multi-patch MPI allows whole body imaging of mice using a long circulating blood pool tracer

C. Jung, J. Salamon, P. Szwargulski, N. Gdaniec, M. Hofmann, M.G. Kaul, G. Adam, S.J. Kemp, R.M. Ferguson, A.P. Khandhar and K.M. Krishnan, T. Knopp, H. Ittrich ... 150

Preliminary results of a hybrid cardio vascular *in vivo* study using a highly integrated hybrid MPI-MRI system

Jochen Franke, Nicoleta Baxan, Ulrich Heinen, Alexander Weber, Heinrich Lehr, Martin Ilg, Michael Heidenreich, Wolfgang Ruhm and Volkmar Schulz ... 151

Systemic Real-time Cell Tracking with Magnetic Particle Imaging

Bo Zheng, Marc P. von See, Elaine Yu, Beliz Gunel, Kuan Lu, Tandis Vazin, David V. Schaffer, Patrick W. Goodwill, Steven M. Conolly ... 153

Session 6: Tracer Materials 1

Keynote: High Resolution Temperature Estimation by using Magnetic Nanoparticles

Prof. Wenzhong Liu ... 155

Localization of magnetic nanoparticles and its effect on magnetic relaxation evaluated by dynamic magnetization measurement for magnetic particle imaging

Yasushi Takemura* and Satoshi Ota ... 157

In vivo velocity determination in the inferior vena cava in mice by Magnetic Particle Imaging and Magnetic Resonance Imaging

Michael G. Kaul, Tobias Mummert, Johannes Salamon, Martin Hofmann, Harald Ittrich, Gerhard Adam, Tobias Knopp, Caroline Jung ... 158

Imaging brain cancer xenografts *in vivo* using tailored nanoparticles functionalized for glioma tumor targeting and MPI-NIRF contrast

Hamed Arami, Eric Teeman, Alyssa Troksa, Haydin Bradshaw, Denny Liggitt, and Kannan M. Krishnan ... 159

Study on the *in vivo* survival of murine Ferucarbotran-loaded RBCs for their use as new MPI contrast agents

Antonella Antonelli, Carla Sfara, Ulrich Pison, Oliver Weber and Mauro Magnani ... 160

Session 7: Tracer Materials 2

Blood half-life of a long-circulating MPI tracer (LS-008)

Amit P Khandhar, Paul Keselman, Scott J Kemp, R Matthew Ferguson, Bo Zheng, Patrick W Goodwill, Steven M Conolly and Kannan M Krishnan ... 163

Concentration Dependent MPI Tracer Performance

Norbert Löwa, Patricia Radon, Olaf Kosch, Frank Wiekhorst ... 164

MPI Analysis of Metal Doped and Anisotropic Nanoparticles

Lisa M. Bauer, Shu F. Situ, Mark A. Griswold, Anna Cristina S. Samia ... 165

Oncogenic protease detection using magnetic particle spectrometry

Sonu Gandhi, Hamed Arami and Kannan M. Krishnan ... 166

Session 8: Instrumentation 2 / Methodology 3

Imaging and Localized Nanoparticle Heating with MPI

Daniel Hensley, Patrick Goodwill, Rohan Dhavalikar, Zhi Wei Tay, Bo Zheng, Carlos Rinaldi, Steven Conolly ... 171

Device manipulation in an MPI-Scanner

Daniel Wirtz, Claas Bontus, Jürgen Rahmer, Peter Mazurkewitz, Christian Stehning and Bernhard Gleich ... 172

Magnetic particle detection based on non-linear response to magnetic susecptibilty changes

Florian Fidler, Karl-Heinz Hiller, Peter M. Jakob ... 173

MPI system matrix reconstruction: making assumptions on the imaging device rather than on the tracer spatial distribution

GAEL BRINGOUT, KSENIJA GRÄFE, THORSTEN M. BUZUG   174

The Influence of Trajectory and System Matrix Overlap on Image Reconstruction Results in Magnetic Particle Imaging

M. AHLBORG, C. KAETHNER, T. KNOPP, P. SZWARGULSKI AND T.M. BUZUG   175

Fast Implicit Reconstruction of Focus Field Data in MPI

P. SZWARGULSKI, M. HOFMANN, N. GDANIEC, AND T. KNOPP   176

Interactive Positioning and Sizing of the Imaging Volume in Real-Time Magnetic Particle Imaging

JÜRGEN RAHMER, CLAAS BONTUS, JÖRN BORGERT   177

Session 1:

Methodology 1

Keynote:

Magnetic Nanoparticle Dynamics in Spectroscopy and its Applications

Dr. John B. Weaver

Dartmouth-Hitchcock Medical Center, Lebanon, USA

ABSTRACT MPI provides information about how many nanoparticles (NPs) collect in a given location. That information is very useful in delineating vessel shape and location as well as the position of cells loaded with NPs. In contrast, we have maximized the effects of the local microenvironment by using larger NPs that relax using the Brownian mechanism instead of smaller Neel relaxing NPs. Brownian relaxation is affected by the temperature and the viscosity of the media the NPs are in as well as chemical binding. Further, if the binding is well characterized, the rigidity of the matrix to which the NPs are bound also impacts the relaxation. We are using spectroscopy to characterize the local microenvironment of the magnetic NPs. We showed that the spectroscopic signal from magnetic NPs depends on temperature, viscosity and binding. We developed scaling methods of quantitating those parameters.

The amplitude of the applied field can be scaled to estimate temperature at low frequencies. The frequency of the applied field can be scaled to estimate the relaxation time. More complex combinations of the amplitude and frequency can also be scaled. The importance of the relaxation time is that it reflects chemical binding resulting in a method of estimating the concentration of a free molecule. By decorating the NPs so they form a sandwich around the biomarker molecule very low concentrations can be measured in vivo. For example, we have measured cytokines resulting from infection in mice. Many diseases have significant immune components that can be evaluated using cytokine concentrations. We will talk about two, immunotherapy monitoring and surgical site infection detection, but there are many more ranging from Multiple Sclerosis to heart disease.

Spatial Resolution in MPI: The Role of Harmonic Number

H. Bagheri[*] and M.E. Hayden

Department of Physics, Simon Fraser Univeristy, Burnaby BC Canada V5A 1S6
[*] Corresponding author; email: hbagheri@sfu.ca

INTRODUCTION Previous studies of spatial resolution in Magnetic Particle Imaging (MPI) have focused on diverse parameters, ranging from the strength of imposed oscillating magnetic fields and magnetic field gradients to particle size, distribution, and susceptibility [1]. Here we examine the role of harmonic number in relation to image resolution, for a fixed selection (gradient) field and particle type. We observe a strong correlation between harmonic number n and resolution that is ultimately limited by factors such as signal-to-noise ratio (SNR) at high harmonics.

MATERIAL AND METHODS Our experiments are performed using an MPI scanner in which the Field Free Point (FFP) is defined by a 9 T/m radial (18 T/m axial) magnetic field gradient. It operates in a manner wherein the processes of FFP manipulation (over a two-dimensional axially-symmetric Field of View) and particle excitation (via an axial oscillating drive field $\mathbf{H}_1$; leading to harmonic generation) are entirely decoupled from one another [2]. We employ phantoms consisting of dense agglomerations of 10 nm mean diameter iron oxide particles [3] confined to squat cylindrical voids (1 mm dia. × 0.3 mm) in an acrylic substrate. Particle excitation occurs at 76 kHz.

RESULTS Figure 1a shows images of two coplanar particle-filled voids separated by 1.8 mm (centre-to-centre) and positioned to lie within the FFP plane. An improvement in resolution is evident as the harmonic number is increased, under otherwise identical conditions. Insight into this trend is provided by Fig. 1b, which shows the maximum signal amplitude extracted from images of a single particle-filled void as the plane in which the void is situated is displaced axially from the FFP plane. A factor of three improvement in resolution is observed at $n = 9$ relative to $n = 3$ if the first zero crossings of these curves are interpreted as crude measures of minimum resolvable feature sizes.

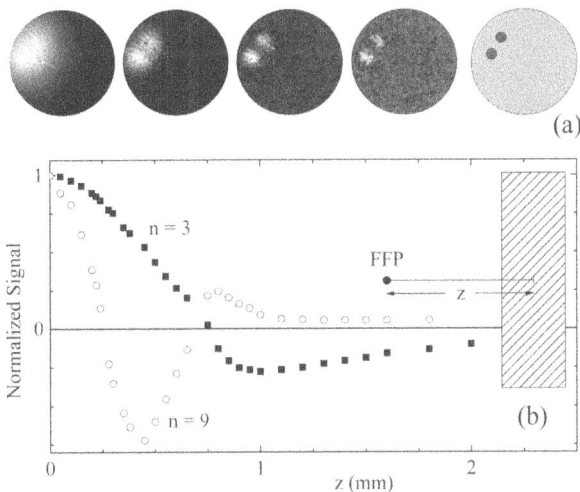

(a)

(b)

Figure 1: (a) Left-to-right: Images of a phantom acquired at the 3rd, 5th, 7th, and 9th harmonics, along with an outline of the phantom geometry. (b) Normalized maximum signal amplitude as the phantom is displaced axially from the FFP plane (parallel to $\mathbf{H}$). Note that phase sensitive detection is employed.

Figure 2 shows data illustrating the manner in which maximum signal amplitudes ε_n extracted from images scale with excitation level and harmonic number when a single particle-filled void is placed in the FFP plane. Over the range of H_1 considered, we observe that ε_n increases approximately in proportion to H_1^α, with $\alpha = (n + 1)/2$. And, we observe that it falls off rapidly as n is increased. Insofar as image resolution is concerned, the latter implies that resolution cannot be improved indefinitely by increasing n. Ultimately noise in the detection chain (or other considerations) will limit the minimum resolvable feature size. And, to some extent, the increase in the exponent α with n decreases the dynamic range of excitation amplitudes over which image contrast can be generated.

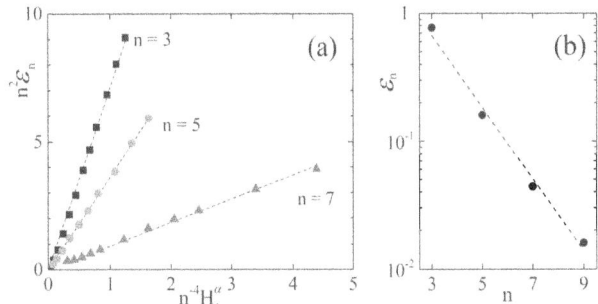

Figure 2: (a) Maximum signal amplitudes ε_n (in mV) as a function of excitation field amplitude H_1 (in kA/m) and harmonic number. The factors of n^2 and n^{-4} used here simply provide convenient scaling for display purposes. (b) Maximum signal amplitudes as a function of harmonic number at a fixed excitation field amplitude $H_1 = 9$ kA/m. The particle-filled void is in the FFP plane. Lines are intended as guides for the eye.

CONCLUSION The data presented here contribute to the understanding of a feature of MPI that has not previously received much direct attention. Image resolution increases as higher harmonics of the excitation frequency are employed. At the same time, signal amplitudes decrease rapidly as the harmonic number is increased, and hence SNR (or other considerations) ultimately limit the maximum achievable resolution. Interpretation of these data is reinforced in an accompanying contribution focused on modeling [4]. Note also that aspects of this work are complementary to Magnetic Particle Spectroscopy studies of particle magnetization dynamics [5].

ACKNOWLEDGEMENTS This work is funded by the Natural Sciences and Engineering Research Council of Canada.

REFERENCES
[1] e.g. Vogel et al., *IEEE Trans. Magn.* **51**, 6502104 (2015); Knopp et al. *IEEE Trans. Med. Imag.* **30**, 1284 (2011); Goodwill and Conolly, *IEEE Trans. Magn.* **29**, 1851 (2010); or Weaver et al., *Med. Phys.* **35**, 1988 (2008); and references therein.
[2] Bagheri et al., *IEEE* doi: 10.1109/IWMPI.2015.7107089
[3] Ferrofluidics AO5; FerroTec Corp., Santa Clara CA USA.
[4] Bagheri and Hayden, these proceedings; 6th *IWMPI* (2016).
[5] e.g. Shah et al., *J. Appl. Phys.* **116**, 163910 (2014).

Sparse source reconstruction for nanomagnetic relaxometry

Sara Loupot[a,b*], Wolfgang Stefan[b], Reza Medankan[b], Kelsey Mathieu[b], David Fuentes[b], and John D. Hazle[b]

[a] The University of Texas Graduate School of Biomedical Sciences at Houston;
[b] Department of Imaging Physics, The University of Texas MD Anderson Cancer Center, Houston, Texas 77030
[*] Corresponding author, email: slloupot@mdanderson.org

INTRODUCTION: Superparamagnetic Relaxometry (SPMR) is an emerging technique for nanoparticle detection with the unique benefit of being able to distinguish biologically bound from unbound particles.[1,2] Our lab has been working to validate this technology in several pre-clinical cancer models. The relaxometer (second generation MRX™ system, Senior Scientific, Albuquerque, NM) uses six superconducting quantum interference devices (SQUIDs) to detect the remnant magnetization from a nanoparticle sample after a magnetization pulse is applied with a pair of Helmholtz coils. Reconstructing the location and number of bound particles is challenging because the magnetic inverse problem is ill posed. In this work, we present an approximate l0-norm minimization approach to source localization, and demonstrate its feasibility on distributions with multiple sources.

By defining a discretized field of view with respect to the sensor locations, the magnetic field detected by the detectors can be described by the linear system Ax=b+v, where b is the signal at each of the detectors, v is the uncertainty in this signal, A is a matrix of geometric factors between each sensor and each voxel in the field of view, and x is the contribution of each voxel to the total moment of the system. In the application of early detection, the vector x can be assumed to be sparse, since only a few voxels will contain a substantial number of bound particles. This allows the use of sparse sensing methods to distinguish the true source distribution from an infinite amount of source distributions that satisfy the linear system. We approximated a minimum l0-norm solution by iterating over the minimization problem:[3]

$$\min \|x\|_1 \; s.t. \; \|Ax - b\|_2 < \epsilon$$

MATERIAL AND METHODS We performed a proof-of-concept study with two point source phantoms to demonstrate the feasibility of the sparse reconstruction approach. We used a self dual minimization method to solve the minimization problem.[4] Point source phantoms were constructed by drying a known amount of 25 nm PrecisionMRX™ nanoparticles, provided by Senior Scientific, LLC, onto Q-tips in zero field. Two such phantoms with equal amounts of nanoparticles were placed as shown in Figure 1.

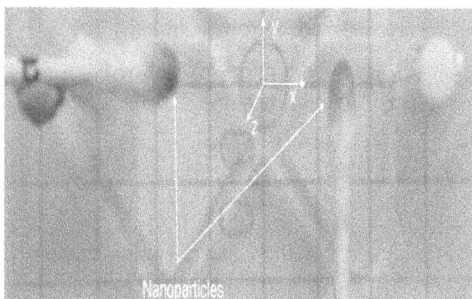

Figure 1: Phantom setup.

Relaxation curves were measured in our MRX™ system. The relaxation curves were used to determine the magnitude of the magnetic field at each detector immediately after the field was removed. The initial field values were used for the source reconstruction. The reconstruction was done on a 10 cm x 10 cm x 4.5 cm field of view discretized into 25x25x25 pixels.

RESULTS The algorithm was able to distinguish the two sources. The reconstructed sources were within 1 cm the actual location of the phantoms, as shown in Figure 2.

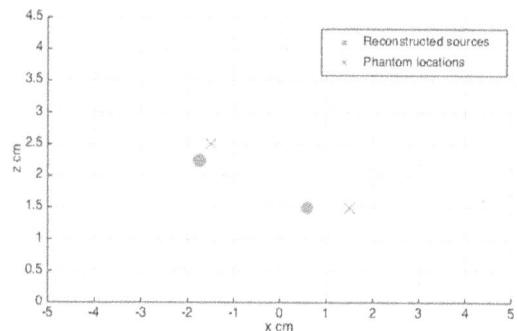

Figure 2: Reconstructed sources (circles) and phantom locations (x).

CONCLUSION This proof-of-concept study demonstrates the ability of the sparse reconstruction algorithm to successfully distinguish between two sources. Future work will focus on improving the robustness of the algorithm and determining the ultimate limits of resolution.

ACKNOWLEDGEMENTS The authors would like to thank Ms. Caterina Kaffes for collecting the data used in this study. This work was performed, in part, at the Center for Integrated Nanotechnologies, an Office of Science User Facility operated for the U.S. Department of Energy (DOE) Office of Science by Los Alamos National Laboratory (Contract DE-AC52-06NA25396) and Sandia National Laboratories (Contract DE-AC04-94AL85000).

REFERENCES
[1] Frank Wiekhorst, et al. *Pharm Res*, 29.5 (2012): 1189-1202. doi: 10.1007/s11095-011-0630-3
[2] Flynn and Bryant *PMB* 50 (2005): 1273-1293. doi:10.1088/0031-9155/50/6/016
[3] Irina F Gorodnitsky, et al. *Electroencephalography and clinical Neurophysiology* 95.4 (1995): 231-251. doi:10.1016/0013-4694(95)00107-A
[4] Michael Grant and Stephen Boyd. (2013) http://cvxr.com/cvx
[5] Leyma P. De Haro, et al. *Biomed. Eng. Biomedical Tech.*, (2015): doi: 10.1515/bmt-2015-0053

Symmetries in Nanoparticle Dynamics Found From The Buckingham Pi Theorem Improve Sensing Strategies

Yipeng Shi [a,*], John B. Weaver [a,b]

[a] Department of Physics & Astronomy, Dartmouth College
[b] Dartmouth-Hitchcock Medical Center
[*] Corresponding author, email: yipeng.shi.GR@dartmouth.edu

INTRODUCTION Magnetic nanoparticles (NP) are used in a variety of diagnostic and therapeutic applications. In diagnostic applications, the microenvironment can be characterized by measuring the magnetization induced by the NPs in an applied alternating magnetic field. Methods such as Magnetic Spectroscopy of Brownian motion (MSB) has been used to measure many properties including viscosity, binding, temperature and cell uptake. [1-6]

However, the form of the expression for the magnetization is generally quite complex. In general, to estimate the parameters characterizing the microenvironment from the magnetization, a special relationship between different variables has to be found. For example, MSB temperature estimation exploits either the symmetric relationship between amplitude of the applied field and the temperature of MNPs [1] or the symmetry between frequency and relaxation time [7]. However, there is no theoretical framework to guide the selection of these important relationships and it generally involves some guesswork. We present the general theoretical framework to systematically identify these crucial relationships to design more accurate NP sensing techniques.

MATERIAL AND METHODS Dimensional analysis using the Buckingham Pi theorem describes a physical parameter such as magnetization using different dimensionless collections. Further, all possible combinations of variables are provided. Through Buckingham Pi theorem, it can be shown that the critical relationships previously exploited in MSB results from dimensionless collections such as $\omega\tau$ and $\frac{M_s V_c H}{kT}$. Using this mathematical theory, new useful relationships can be found systematically. Furthermore, the general form of other physical parameters such as harmonic phase angle can be found using this theory without resorting to special models or theoretical approximations. The results can be used in designing new sensing methods.

RESULTS The dependence of $\omega\tau$ predicted by the Buckingham Pi theorem for harmonic phase angle has been experimentally shown. Method previously used in MSB is applied to harmonic phase angle to measure relaxation time. Compared to previous methods, in average a 50% improvement of measurement accuracy is acquired using harmonic phase angle (from 0.95% mean error using the older method to 0.48%)

CONCLUSION Buckingham Pi Theorem provides invaluable insight into the symmetries inherent in nanoparticle behavior. The symmetries can be used to provide improved sensing methods like those provided using the phase angle of the harmonics as an alternative to the ratio of the harmonics.

Acknowledgements Department of Radiology, DHMC.

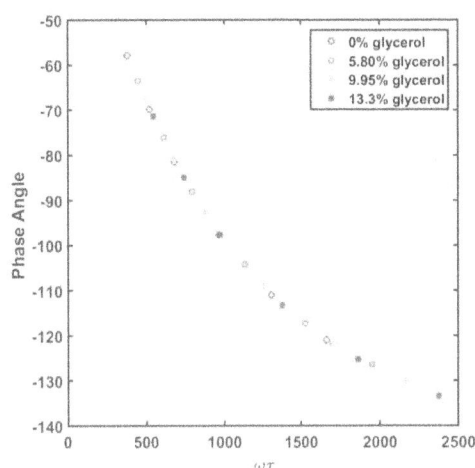

Figure 1: The dependence of phase angle on $\omega\tau$ for the second harmonic of the NP magnetization predicted by Buckingham Pi theorem is experimentally shown. NPs in different glycerol solution (different viscosity) exhibits the same variable dependence.

References

[1] J.B. Weaver, A.M Rauwerdink, E.W. Hansen, "Magnetic nanoparticle temperature estimation". Med. Phys. 36, 5 (2009).

[2] Rauwerdink AM, Weaver JB. Measurement of molecular binding using the Brownian motion of magnetic nanoparticle probes. Applied Physics Letters 96(3), 033702 (2010).

[3] Zhang X, Reeves DB, Perreard IM, Kett WC, Griswold KE, Gimi B, Weaver JB. Molecular sensing with magnetic nanoparticles using magnetic spectroscopy of nanoparticle Brownian motion. Biosensors and Bioelectronics 50(0), 441-446 (2013).

[4] Tu L, Wu K, Klein T, Wang J-P. Magnetic nanoparticles colourization by a mixing-frequency method. Journal of Physics D: Applied Physics 47(15), 155001 (2014).

[5] Remmer H, Dieckhoff J, Schilling M, Ludwig F. Suitability of magnetic single-and multi-core nanoparticles to detect protein binding with dynamic magnetic measurement techniques. Journal of Magnetism and Magnetic Materials 380: 236-240 (2015).

[6] A.J. Giustini, I Perreard, A.M. Rauwerdink, P.J. Hoopes, J.B. Weaver. Noninvasive assessment of magnetic nanoparticle–cancer cell interactions. Integrative Biology 4(10) 1283-1288 (2012).

[7] I. Perreard, D.B. Reeves, X. Zhang, E. Kuehlert, E. Forauer, J.B. Weaver. Temperature of the magnetic nanoparticle microenvironment: estimation from relaxation times. PMB 59(5), 1109 (2014).

Functional Magnetic Particle Imaging in measurement and simulation

Thilo Viereck [a,*], Christian Kuhlmann[a], Sebastian Draack[a], Frank Ludwig[a], Meinhard Schilling[a]

[a] Institut fuer Elektrische Messtechnik und Grundlagen der Elektrotechnik, TU Braunschweig, Germany
[*] Corresponding author, email: t.viereck@tu-bs.de

INTRODUCTION Magnetic Particle Imaging (MPI) has been acknowledged as a fast imaging modality with high spatial resolution and high contrast [1]. Recently, with the introduction of Mobility MPI [2] and the Color MPI reconstruction [3] approach, potential has been identified for functional imaging in MPI. In that sense, MPI enables the discrimination of different viscosities or binding states in the field-of-view, in additional to a concentration-weighted image.

Our group has developed various ideas for obtaining access to viscosity information in MPI. The 'best' approach is subject to the imaging parameters and especially the available signal-noise-ratio of the acquired data. In this contribution, we show examples of functional MPI in phantom measurement and simulation. We focus on viscosity estimation from a single drive field frequency in comparison with a newly implemented dual-frequency mMPI acquisition protocol.

MATERIAL AND METHODS At IWMPI 2015, we presented our mMPI approach, which estimates the viscosity of a single sort of particles, i.e. FeraSpin™ R, in the imaging volume at a drive field frequency of 10 kHz. As a next step, we utilize calibration data obtained on a viscosity series (water-glycerin mixtures with adjustable viscosity) to analyze the objective image quality and the estimation error for concentration and viscosity from mMPI and cMPI reconstruction and we introduce the dual-frequency mMPI approach (initially presented in simulation form at IWMPI 2012) as a reference method for functional MPI.

Imaging experiments were performed with our custom-built MPI scanner [4] in 1-D mode (with a single fast-scanning axis) and with a new dual-frequency MPI scanner operating at 10 kHz and 25 kHz alternately.

An imaging phantom covering six different viscosities in the range of 1 – 100 mPas and identical iron concentration was prepared (Fig. 1). The images were reconstructed a) using standard F-space reconstruction with mMPI post-processing for the viscosity estimation, b) using the Color MPI (cMPI) reconstruction to obtain concentration and viscosity images simultaneously and c) via the dual-frequency mMPI protocol (delta image from both frequencies).

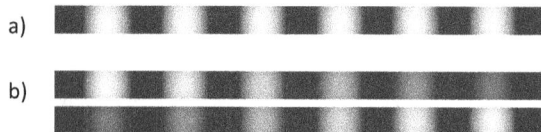

Figure 1: a) Imaging phantom with six spots of FeraSpin™ R with identical concentration embedded in a viscous water-glycerin matrix (viscosity range 1 – 100 mPas, from left to right). b) Expected cMPI reconstruction (simulated from acquired calibration data) using the lowest (top) and highest (bottom) viscosity calibration sample.

RESULTS The estimation error for concentration and viscosity is closely related to the SNR of the acquired image and calibration data. In general, a good SNR gives a good viscosity estimation. In that case, the Color MPI (cMPI) reconstruction yields the smallest viscosity deviation from reference values. Under poor SNR conditions, cMPI reconstruction suffers from large errors and fails at reconstructing a large number of different viscosities simultaneously. Mobility MPI (mMPI) processing provides a reasonable viscosity estimation and, being a time-domain method, excels at small particle concentrations. Figure 2 shows the minimum SNR requirements for the different reconstruction strategies.

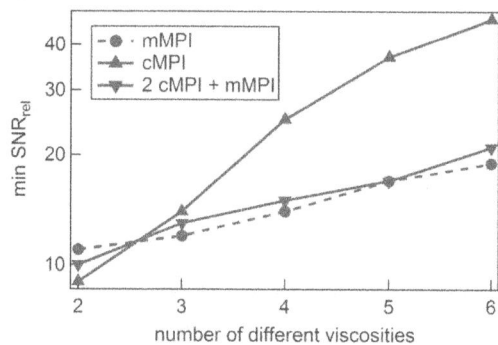

Figure 2: Minimal SNR required for reconstructing a variable number (2 – 6) of different viscosities in the FOV for Mobility MPI (mMPI) and Color MPI (cMPI) reconstruction strategies.

CONCLUSION Color MPI reconstruction is a good instrument to obtain viscosity- and concentration-weighted images from a single MPI experiment. While mMPI delivers viscosity estimations with larger errors, the approach is superior to cMPI under low SNR conditions, such as diluted samples with low iron concentration in practical application scenarios. Experimental validation and evaluation of dual-frequency mMPI data and 2-dimensional MPI data is ongoing.

ACKNOWLEDGEMENTS Financial support by the German Research Foundation DFG (LU 800/5-1) is acknowledged.

REFERENCES
[1] B. Gleich and J. Weizenecker. *Nature*, 435(7046):1217—1217, 2005. doi: 10.1038/nature03808.
[2] T. Wawrzik, et. al. *IEEE Xplore* 6528372, 2013. doi: 10.1109/IWMPI.2013.6528372.
[3] J. Rahmer, et. al. *Phys. Med. Biol.* 60:1775—1791, 2015. doi: 10.1088/0031-9155/60/5/1775.
[4] M. Schilling, et. al. *Biomed. Tech.* 58(6):557—563, 2013. doi: 10.1515/bmt-2013-0014.

Experimental Distinction of Different Viscosities using Multispectral Magnetic Particle Imaging

Martin Hofmann [a,b,*], Jan Dieckhoff [c], Harald Ittrich [c], Tobias Knopp [a,b]

a Section for Biomedical Imaging, University Medical Center Hamburg-Eppendorf, Hamburg
b Institute for Biomedical Imaging, Hamburg University of Technology, Hamburg
c Diagnostic and Interventional Radiology Department and Clinic, University Medical Center Hamburg-Eppendorf, Hamburg
* Corresponding author, email: m.hofmann@uke.de

INTRODUCTION Magnetic Particle Imaging (MPI) is a tracer-based imaging technique. The particles and their physical environment influence the measurement signal, which can be used to separate different particle types and aggregation states [1]. Using magnetic particle spectroscopy even a quantification of changes in the particle environment like for instance the viscosity is possible [2]. The purpose of this work is to investigate whether it is possible to discriminate viscosity differences with spatial resolution using the multi-spectral reconstruction technique developed in [1].

MATERIAL AND METHODS A series of Resovist-water-glycerin mixtures were prepared with an iron concentration of 40 mmol/L, a volume of 50 µL, and glycerin volume fraction of 80% (G80), 70% (G70), 60% (G60), 50% (G50), and 0% (G00). At a sample temperature of 25°C following literature values are found for the water-glycerin viscosity: 70 mPas (G80), 27 mPas (G70), 13 mPas (G60), 7 mPas (G50), and 0.9 mPas (G00).
The samples were measured pairwise with a center to center distance of 1 cm along the bore of the scanner. Experiments were carried out with a preclinical MPI system (Bruker/Philips) operated in 3D mode with 1 T/m selection field gradient and 14 mT drive field amplitudes.
For reconstruction a G80 and G00 system matrix were measured using 3×3×3 mm³ delta samples on a 26×26×13 grid. For each measured dataset three images were reconstructed: Two with the separate monospectral reconstruction using the G80 and G00 system matrices and the third with the combined multispectral approach [1].

RESULTS In the monospectral images both samples are visible but the sample with the lower glycerin concentration has a stronger intensity independent of the system matrix used for reconstruction (see fig. 1 for the sample pair G80 and G00). Using the multispectral reconstruction the signal of the sample with the higher glycerin volume fraction is assigned mainly to the G80 part of the image and the signal of the other sample is assigned predominantly to the G00 part (see figs. 1 and 2). If the samples do match the samples used for acquisition of the system matrix, the best separation is achieved. If there is a mismatch between sample and system function each sample is also visible in the mismatching part of the multispectral reconstruction as can be seen in fig. 2. In any case, a clear separation of the different viscosities is possible.

CONCLUSION In summary, multispectral magnetic particle imaging can be used to discriminate between spatially separated samples with different viscosities.

ACKNOWLEDGEMENTS We gratefully acknowledge funding and support of the German Research Foundation (DFG, grant number AD 125/5 - 1) and (DFG, SFB 841).

References

[1] J. Rahmer, A. Halkola, B. Gleich, I. Schmale and J. Borgert, Phys. Med. Biol. 60(5):1775-1791, 2015. doi: 10.1088/0031-9155/60/5/1775.
[2] A. M. Rauwerdink and J. B. Weaver, J. Magn. Magn. Mater. 322(6):609-613, 2010. doi:10.1016/j.jmmm.2009.10.024.

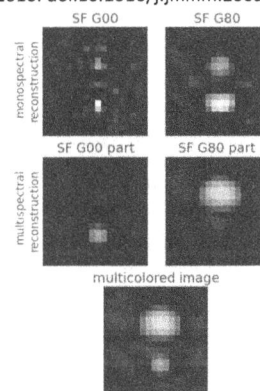

Figure 1: Comparison of monospectral (top row) and multispectral (middle row) reconstructed images of G80 (upper sample) and G00 (lower sample), with the color-coded combined image of the multispectral reconstruction shown in the bottom row.

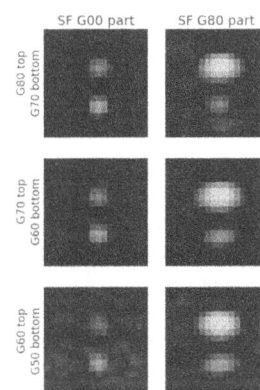

Figure 2: Multispectral reconstructed images for each of the pairs G80 and G70 (top row), G70 and G60 (middle row), and G60 and G50 (bottom row). The sample with higher glycerin content is located at the top in each image.

Calibration-Free Color MPI

Yavuz Muslu[a,b], Mustafa Ütkür[a,b], Omer Burak Demirel[a,b], Emine Ulku Saritas[a,b*]

[a] Department of Electrical and Electronics Engineering, Bilkent University, Ankara, Turkey
[b] National Magnetic Resonance Research Center (UMRAM), Bilkent University, Ankara, Turkey
[*] saritas@ee.bilkent.edu.tr

INTRODUCTION Nanoparticle relaxation reduces image resolution in x-space magnetic particle imaging (MPI) by blurring the images along the scan direction [1]. When estimated correctly, deconvolution with the relaxation kernel can increase the resolution, if the signal-to-noise ratio (SNR) is sufficiently high. One can also take advantage of the relaxation effects to identify different magnetic nanoparticles, thus colorizing the MPI image [2-3]. Here, we introduce a technique that reconstructs a relaxation map from image acquisition at only one drive field amplitude, without any need for prior calibration. We provide a proof-of-concept validation for this technique with experimental results from a custom magnetic particle spectrometer (MPS).

MATERIAL AND METHODS Previous studies modeled the MPI signal as a convolution of the nanoparticle's Langevin response with an exponential function [1]. Using the mirror symmetry of the positive and negative half cycles for the original Langevin response, we developed an algorithm to estimate the relaxation time constant, τ, directly from the MPI signal [4]. Here we adopt the same algorithm to generate 2D relaxation maps for calibration-free color MPI: we scan through the entire field-of-view (FOV), calculate τ for each partial field-of-view (pFOV), then generate a 2D relaxation map. Using the MPI image, we apply a threshold to suppress the background noise. At this phase of the study, we demonstrated the proposed technique using a custom MPI simulation toolbox developed in MATLAB. Simulations were done according to real-life conditions: (3,-6,3) T/m selection field in x-y-z directions, 15mT(peak) drive field at 10kHz, 20nm particle diameter, first harmonic filtered out [5], noise added. According to experimental results reported in, we utilized $\tau = 2.9\mu s$ and $\tau = 1.1\mu s$ for Resovist and UW33 nanoparticles, respectively [6]. For the experimental validation on our custom MPS setup (also known as an MPI relaxometer), signal from nanomag-MIP nanoparticles (Micromod Gmbh, Germany) at 10.2kHz, and 15mT(peak) drive field were used.

RESULTS Figure 1 shows the particle distributions (a,d), corresponding MPI images (b,e) and relaxation maps (c,f) for two different separations between nanoparticle distributions. For both cases, the estimated maps give mean τ values with less than 4% error. For Fig.1f, there is a slight interference effect in regions where the distributions are closest to each other. Figure 2 shows the experimental validation results from the MPS setup. Relaxation time constant was estimated as 4.46μs for the nanomag-MIP nanoparticle.

CONCLUSION Here we have demonstrated a technique to map the relaxation time constant of nanoparticles, without any calibration. The proposed technique can also be extended to probe viscosity, particle concentration in a mixture, and binding state of the particles. Future work includes the improvement and imaging demonstration of the mapping technique.

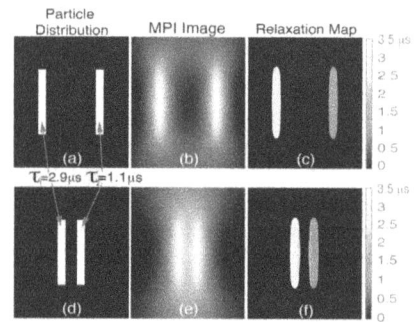

Figure 1: Results of 2D simulations for calibration-free color MPI, for two nanoparticle distributions with different relaxation time constants (2.9μs vs. 1.1μs). [Top] For 30 mm separation between the distributions, mean estimated time constants were 2.90μs vs. 1.06μs, respectively. [Bottom] For 10 mm separation, mean estimated time constants were 2.92μs vs. 1.14μs. Imaging Parameters: 6x6cm^2 FOV; 1 cm pFOV; 80% pFOV overlap.

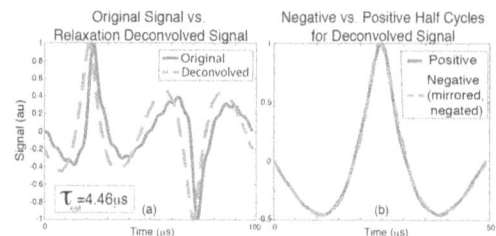

Figure 2: Experimental validation of the proposed technique on a custom MPS setup. For nanomag-MIP at 10.2 kHz and 15 mT drive field, estimated τ was 4.46μs. (a) Comparison of the original and relaxation-deconvolved signals. (b) Positive and negative half cycles of the deconvolved signal show mirror symmetry, validating the accuracy of the estimation scheme.

ACKNOWLEDGEMENTS This work was supported by the Scientific and Technological Research Council of Turkey through a TUBITAK 3501 Grant (114E167), by the European Commission through an FP7 Marie Curie Career Integration Grant (PCIG13-GA-2013-618834), and by the Turkish Academy of Sciences through TUBA-GEBIP 2015 program.

REFERENCES
[1] L.R. Croft *et al.*, *IEEE Transactions on Medical Imaging*, *31*(12), 2335–2342, 2012. doi: 10.1109/TMI.2012.2217979.
[2] J. Rahmer *et al.*, *Phys Med Biol*, vol. 60, no. 5, pp.1775 -91 2015. doi: doi:10.1186/1471-2342-9-4.
[3] D. Hensley *et al.*, *Proc of International Workshop on Magnetic Particle Imaging*. doi: 10.1109/IWMPI.2015.7106993
[4] G. Onuker *et al.*, *Proc of International Workshop on Magnetic Particle Imaging*. doi: 10.1109/IWMPI.2015.7107042
[5] K. Lu *et al.*, *IEEE Transactions on Medical Imaging*, 32(9):1565-1575,2013. doi: 10.1109/TMI.2013.2257177.
[6] L.R. Croft *et al.*, "Low Drive Field Amplitude for Improved Image Resolution in Magnetic Particle Imaging", *Medical Physics*, In press

Elevator speeches 1:

Instrumentation

An Ultra-Low Noise Preamplifier Design for Magnetic Particle Imaging

Beliz Gunel[1], Bo Zheng[*2], Steven Conolly[1,2]

[1]Department of Electrical Engineering and Computer Science, UC Berkeley
[2]Department of Bioengineering, UC Berkeley
[*] Corresponding author, email: bozheng@berkeley.edu

INTRODUCTION Magnetic Particle Imaging (MPI) is linearly quantitative and can measure as few as 200 cells *in vivo* with no background signal and no penetration depth limitations. This combination is unique across existing medical imaging modalities. The gold-standard molecular imaging modality, Positron Emission Tomography, has a detection limit of 10,000 cells.[4] Hence, MPI shows great promise for applications requiring high sensitivity. However, the true physical sensitivity limits have not yet been achieved for MPI. Importantly, the preamplifier stage of an MPI receiver chain must be optimized to achieve low noise figures for the reactive inductor sensors used in MPI [3], and the usable bandwidth should be increased by reducing the preamplifier input capacitance.

In this work, we present a novel ultra-low noise preamplifier design using a common source JFET amplifier with cascode connection and feedback circuitry, which will enable up to 2-fold improved detection sensitivity and signal linearity over existing preamplifier architectures.

MATERIAL AND METHODS

MPI receiver coil is assumed to have 1Ω and $300\ \mu H$ impedance, and 100 pF of self-capacitance. We designed a custom ultra-low noise preamplifier for noise matching to that coil in order to improve the sensitivity of the scanner. In X-space MPI, the bandwidth is limited by the resonant frequency between the receiver coil and the amplifier input capacitance due to amplitude and phase effects from the LC resonance. Therefore, the input capacitance plays a key role in the overall usable bandwidth. Also, linearity of gain is often a problem while working with JFETs, since the drain current is quadratic in the gate drive for operation above subthreshold. Hence, we designed a two-stage amplifier, where we used common source BF862 JFET amplifiers with cascode connection at the first stage to reduce Miller capacitance, and used LM6171A operational amplifier with negative feedback at the second stage to achieve high signal linearity.

RESULTS

Our preamplifier demonstrated a high gain of 34 dB, low voltage noise (1 nV/√Hz), and low input capacitance (44 pF). Also, linear gain was achieved across different frequencies. We expect to have a noise figure at 3 dB and a bandwidth of 500 kHz using a standard MPI receiver coil and a 1:8 noise-matching transformer. Importantly, the amplified MPI signal is extremely linear: the second and third-order harmonic distortion was found to be less than -100 dB relative to a test signal of 10 mVpp at 20 kHz.

Figure 1. Custom Berkeley two-stage preamplifier with cascode and feedback for linearizing gain.

CONCLUSION We designed and built a preamplifier with BF862 JFETs and LM6171A operational amplifiers. In this design, there is a trade-off between signal linearity and low noise figure, since the voltage noise is limited by the second-stage operational amplifier that achieves the signal linearity through negative feedback. We plan to explore different operational amplifiers to further optimize the voltage noise while maintaining the signal linearity. Also, we plan to optimize the board layout to reduce input capacitance in order to improve the overall bandwidth.

ACKNOWLEDGEMENTS The authors gratefully acknowledge support from these research grants: NIH R01: 5R01EB013689-03, CIRM RT2:RT2-01893, Keck Foundation: 034317, NIH R24:1R24MH106053-01, NIH R01: 1R01EB019458-01, ACTG: 037829.

REFERENCES
[1] Y. Netzer, "The design of low-noise amplifiers," Proc. IEEE, vol. 69, no. 6, pp. 728–741, 1981.
[2] Wencong Zhang; Bo Zheng; Goodwill, P.; Conolly, S., "A custom low-noise preamplifier for Magnetic Particle Imaging," in Magnetic Particle Imaging (IWMPI), 2015 5th International Workshop on , vol., no., pp.1-1, 26-28 March 2015
[3] I. Schmale, B. Gleich, J. Borgert, and H. Karlsruhe, Proc. IWMPI 2010, Lvbeck, Germany
[4] P. K. Nguyen, J. Riegler, and J. C. Wu, Cell Stem Cell, vol. 14, no. 4, pp. 431–44, Apr. 2014.
[5] B. Zheng, T. Vazin, P. Goodwill, D. Schaffer, S. Conolly, Proc. IWMPI 2014, Berlin, Germany
[6] Horowitz, Paul,Hill, Winfield.*The Art Of Electronics*. Cambridge [England] : Cambridge University Press, 1989.

Suppress Direct Feedthrough Induced by Excitation Magnetic Field

Tao Jiang[a], Shiqiang Pi [a], and Wenzhong Liu[a,*]

[a] School of Automation, Huazhong University of Science and Technology, Wuhan 430074, China
* Corresponding author, email: lwz7410@hust.edu.cn

INTRODUCTION As a novel modality of imaging, magnetic particle imaging (MPI) is featured with high sensitivity and spatial resolution. It provides a theoretical support for applying MPI to medical imaging that the magnetic susceptibilities of bone, muscle, and internal organs are far less than that of superparamagnetic iron oxide (SPIO). This imaging modality can be implemented in many applications, such as disease diagnosis [1] and cardiovascular imaging [1]. So far, many scientific workers have been dedicated to the researches of MPI and many great breakthroughs have been made.

There is a common problem in MPI that output signal is induced not only by SPIO magnetization but also by excitation magnetic field [2]. The former kind of induced voltage is called sample response and the latter is called direct feedthrough. The response of SPIOs is always far less than direct feedthrough and merged by it [3]. To suppress direct feedthrough, many researchers used a high-pass filter. Here we introduce a new way to suppress direct feedthrough.

MATERIAL AND METHODS In our work, we tried to suppress the direct feedthrough induced by excitation magnetic field for a clearer magnetic particle image.

Two compensate coils and the receive coil were wrapped on a cylindrical plastic framework with wires linked reversely. The receive coil was put within field of view (FOV) while compensate coils were far away from it (see green dotted box in Fig. 1). Turns and cross-section area of compensate coils could be changed to match the receive coil. In our research, we made many attempts at the receive coil of different turns.

Afterwards, amplitude and phase adjustors were added for better suppression effectiveness. The amplitude adjustors were made up of variable gain amplifiers whereas the phase adjustors were realized by all-pass filters. Matched coils and another coil with a few turns were linked to amplitude and phase adjustors, respectively (see blue dotted box in Fig. 1). We suppressed the direct feedthrough approximately to the noise level by adjusting amplitude and phase of voltage on each channel (see Fig. 2).

RESULTS For a receive coil with only 10 turns, the output voltage was suppressed to -60dB with compensate coils used solely. When turn number came to 100, the output voltage was only weakened to -40dB. What was even worse, the output signal would be no less than -20dB if turn number was up to 1000.

However, when amplitude and phase adjustors were added, output voltage was suppressed to about -85dB for a receive coil with 1000 turns (see Fig. 2).

CONCLUSION As phase of induced voltage in each coil is affected by inductance, resistance, and distributed capacitance, it is not effective to suppress direct feedthrough with compensate coils used solely, especially for coils with large turn number.

However, the amplitude and phase adjustors work well with the output signal suppressed to -85dB, indicating the effectiveness of the method presented herein.

Some work needs improvements. As time goes on, optimized signal tends to rise. Therefore, robustness is needed for a long-term imaging process. Optimizing hardware structure is a way to reduce the drift effects.

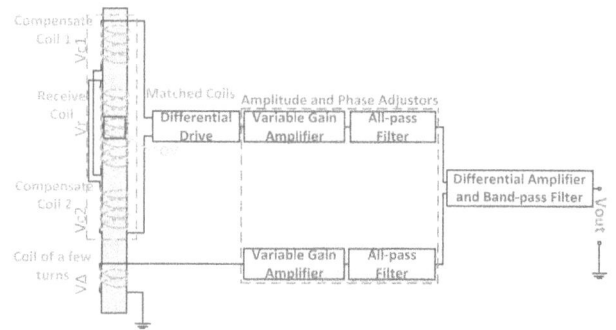

Figure 1: Hardware structures of the detection module and signal-processing module. The receive coil was matched with two compensate coils. The amplitude adjustors are made up of variable gain amplifiers whereas the phase adjustors are realized by all-pass filters.

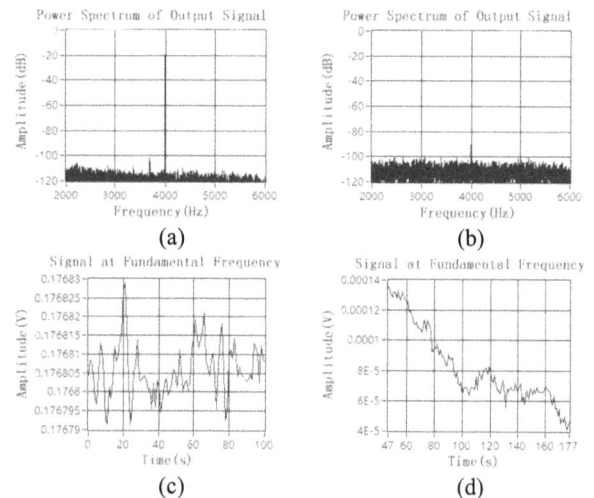

Figure 2: (a) and (c) show output voltage suppressed by compensate coils solely. (b) and (d) show output voltage suppressed by compensate coils and amplitude and phase adjustors.

ACKNOWLEDGEMENTS This work was supported by 61571199 (NSFC) and Hubei Provincial project of 2014AEA048.

REFERENCES

[1] H. Arami, A. P. Khandhar, A. Tomitaka, E. Yu, P. W. Goodwill, S. M. Conolly, and K. M. Krishnan. *Biomaterials*, 52, 251—261, 2015. doi: 10.1016/j.biomaterials.2015.02.040.

[2] M. Zhou, J. Zhong, W. Liu, Z. Du, Z. Huang, M. Yang, and P. C. Morais. *IEEE Transaction on Magnetics*, 51(9), 6101006, 2015. doi: 10.1109/TMAG.2015.2434322.

[3] M. Graeser, T.Knopp, M. Gruttner, T. F. Sattel and T. M. Buzug. *Medical Physics*, 40(4), 042303, 2013. doi: 10.1118/1.4794482

Switching Power Amplifier for Magnetic Particle Imaging

Timo F. Sattel[a,*], Christian Vollertsen[a], Jan Gressmann[a], Oliver Woywode[a]

[a] Philips Medical Systems DMC GmbH, Hamburg, Germany
* Corresponding author, email: timo.frederik.sattel@philips.com

INTRODUCTION The magnetic particle imaging (MPI) method applies a strong magnetic gradient field, providing a field free point (FFP) or, alternatively, a field free line (FFL) [1, 2]. During the imaging process, the FFP or the FFL is moved through the field of view by superimposed drive fields. The drive fields are generated by electromagnetic coils carrying oscillating currents. To provide the required power to the coils, so far, analogue power amplifiers have been used. As MPI systems are scaled-up to fit larger objects and enable human full-body imaging, the drive field coils are designed larger, requiring more electric power. Due to peripheral nerve stimulation and specific absorption rate considerations, it is reasonable to increase the applied drive field frequency from about 25 kHz, typical for lab sized systems, to about 150 kHz for full-body systems [3]. At these frequencies, the requirements quickly exceed the power limits of commercially available analogue amplifiers.

To increase the power capability, several amplifiers can be combined by a transformer network. To realize an MPI signal chain as an intermediate step to a full-body demonstrator system [4], three analogue amplifiers have been combined, increasing the total output power to 4 kW at 150 kHz.

Considering a full-body system, the drive field power requirements may exceed 20 kW per channel. Even though, it is technically feasible to realize such a system using combined analogue amplifiers, they would require much floor space and be cost intensive. Furthermore, due to their poor efficiency, an additional active cooling and considerably higher mains power would be required.

To overcome these drawbacks, switching power amplifier concepts can be applied. In [5], a pulse width modulated (PWM) driven class-D power amplifier concept is proposed to generate a sinusoidal output signal at 25 kHz. To generate output signals at 150 kHz with reasonable spectral purity, the PWM switching frequency needs to be increased accordingly.

There are commercial class-D amplifiers running at PWM frequencies above 1 MHz. However, due to their low power and low voltage capabilities, massive paralleling would be required. An implementation of the PWM concept with high voltage switches is not feasible because of their limited switching frequency.

In this contribution, a simple switching power amplifier concept for MPI is introduced. Instead of applying a pulse density modulation or a PWM, the amplifier periodically switches rectangular pulses at a fixed frequency, which is the output signal drive field frequency.

MATERIAL AND METHODS Figure 1 shows the functional units of a switching power amplifier for MPI. It consists of a controller, an inverter, and an output filter stage.

The full-bridge inverter runs at a fixed frequency of 150 kHz. Hard commutation is avoided. The fixed duty-ratio of 50 % results in a square-wave output. The output signal contains a rich spectrum of odd harmonics. Since MPI relies on the spectral purity of the drive field signal, an analogue filter with a high quality factor is required before feeding the signal into the drive field chain.

The amplitude of the inverter output signal is adjusted by controlling its DC-link voltage. To compensate for phase drifts of the load impedance, the width of single pulses can be adapted.

To achieve a stable drive field coil current in terms of amplitude and phase, these parameters are measured as feedback and compared to the settings of the magnetic field sequence by a system control unit. For compensation of signal deviations, adapted parameters are given to the switching power amplifier. In this way, a control loop is realized.

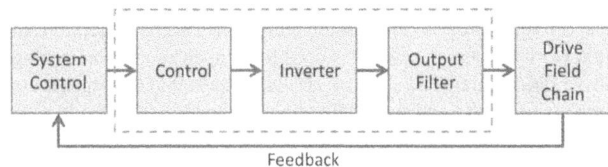

Figure 1: Functional units of a switching power amplifier for MPI

RESULTS The presented concept was successfully implemented and tested with an MPI signal chain at 150 kHz. It meets the requirements and gives great results in terms of overall efficiency (about 90 %) and spectral signal quality.

CONCLUSION Power consumption increases with the size of MPI scanner systems. Aside from this fact, the power requirements depend on various parameters, which are not yet established for clinical systems. Nevertheless, the presented switching power amplifier concept cover even high power requirements while overcoming the drawbacks of analogue and typical class-D amplifiers.

ACKNOWLEDGEMENTS This work was supported by the German Federal Ministry of Education and Research (BMBF grant FKZ 13N11086).

REFERENCES
[1] B. Gleich and J. Weizenecker. *Nature*, 435(7046):1217—1217, 2005. doi: 10.1038/nature03808.
[2] J. Weizenecker, B. Gleich and J. Borgert. *J. Phys. D: Appl. Phys.*, 41, 105009, 2008. doi: 10.1088/0022-3727/41/10/105009
[3] I. Schmale, B. Gleich, J. Schmidt, J. Rahmer, C. Bontus, R. Eckart, B. David, M. Heinrich, O. Mende, O. Woywode, J. Jockram and J. Borgert. *IWMPI, International Workshop on*, 1-1, 23-24, 2013. doi: 10.1109/IWMPI.2013.6528346.
[4] T. F. Sattel, O. Woywode, J. Weizenecker, J. Rahmer, B. Gleich and J. Borgert. *Magnetics, IEEE Transactions on*, 51, 2, 1-3, 2015. doi: 10.1109/TMAG.2014.2326256
[5] C. Loef, P. Luerkens and O. Woywode. *IWMPI, International Workshop on*, 2010. doi: 0.1142/9789814324687_0019

A novel compensation technique for gradiometer receive coils in MPI/MPS

Florian Fidler[a,*], Karl-Heinz Hiller[a], Peter M. Jakob[a]

[a] Research Center Magnetic-Resonance-Bavaria (MRB), Am Hubland, D-97074 Würzburg, Germany
[*] Corresponding author, email: fidler@mr-bavaria.de

INTRODUCTION A major challenge in Magnetic Particle Imaging (MPI) [1] and Spectroscopy (MPS) is the signal reception while the drive field is on. Generated Voltage in the receive coil induced by the drive field is substantially higher than the signal from the particles, and for the first harmonic the frequency is equal [2]. Filtering the drive field is a widely spread technique, but end in the lack of the first harmonic signal which leads to missing signal in spectroscopic data. A gradiometer design of the receive coil offers the possibility to eliminate the drive field signal passively, but still the remaining signal depends on the accuracy of the gradiometer itself as well as on the homogeneity of the drive field. A minimized remaining signal is necessary to use the full dynamic range of the receiver for signal detection. In this work a passive network is used to compensate the remaining drive field signal. The network is easily adjustable to compensate phase shifts and amplitude differences of the individual coils used as a gradiometer. This allows detecting even the first harmonic using the maximized dynamic range of the receiver. Despite improving the dynamic range, in MPS spectroscopy the first harmonic contains valuable information and is necessary to reconstruct the magnetic moment time curve. In MPI imaging it is uncommon to use information from the first harmonic, but as shown in [3] spatial encoding is not based on the detection of higher harmonics, but in the detection of a non-linear behavior of the magnetic moment in an alternating field. It was shown, that the non-linear behavior could be detected by only using first harmonic signal during a power sweep, therefore spatial encoding is also possible with the first harmonic signal as stated. Moreover this technique can reduce the requirements for receive filters drastically.

MATERIAL AND METHODS The passive network consist of a low impedance resistive network (A1,A2) attached in series and a high impedance network attached parallel (P1,P2) to each gradiometer coil, afterwards signal is combined to cancel out (Fig. 1). The low impedance network reduces the amplitude of a single coil. As only one of the gradiometer coil amplitudes has to be corrected, it is clear, that this network for the coil with the lower amplitude is set to zero. The achievable induced voltage is the voltage induced in this coil, keeping in mind that these voltages are quite similar by construction. The high resistive network shifts the phase of each coil as it forms a combined complex impedance with an altered real part. The direction of the phase shift is given and the same for each coil, so only one of the network is used, which is depending on the actual coil. If the impedance of the phase shift network is high compared to the impedance of the coil, it does virtually not affect the achievable signal-to-noise ratio, even though lossy elements are used.

This compensation mechanism was build for a gradiometer coil with 2 times 140 turns and a diameter of 7 mm. The impedance of the amplitude network was adjustable in a range of 0 Ohm to 50 Ohm, the phase shift network was adjustable in range from 0 to 20kOhm.

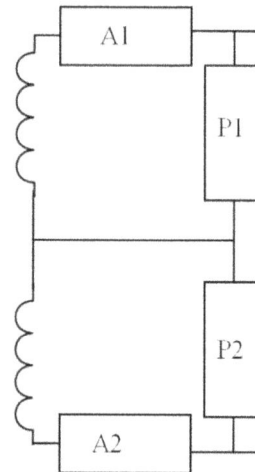

Figure 1: Schematic of the compensation networks.

RESULTS The presented network was able to reduce the remaining signal from the build gradiometer coil from 6V to less than 50µV in an excitation field of 40 mT at 20kHz. It works well without the need for readjustment in a wide frequency range from 5 to 40 kHz with a remaining voltage less than 2 mV, which is sufficient to use the dynamic range of the receiver electronics in our system. It is possible to manage even phase shifts of less than 0.0001° and is easy adjustable if necessary.

CONCLUSION The presented technique provides an easy and robust method for operating gradiometer receive coils in Magnetic Particle Imaging (MPI) and Spectroscopy (MPS). Using this technique reduces the requirements for receive filters drastically and offers the opportunity to measure the valuable first harmonic signal.

ACKNOWLEDGEMENTS This work was supported by the EU FP7 HEALTH program IDEA – "Identification, homing and monitoring of therapeutic cells for regenerative medicine – Identify, Enrich, Accelerate" under grant agreement no 279288.

REFERENCES
[1] B. Gleich and J. Weizenecker. *Nature*, 435(7046):1217—1217, 2005. doi: 10.1038/nature03808.
[2] T. Knopp and T. M. Buzug. Springer, Berlin/Heidelberg, 2012. doi: 10.1007/978-3-642-04199-0.
[3] F. Fidler et al. P28, Proc. IWMPI, 2015.

A Tunable Gradiometer Receive Coil for Magnetic Particle Imaging

Steffen Bruns[a,*], Matthias Weber[a], Thorsten M. Buzug[a]

[a] Institute of Medical Engineering, University of Luebeck
[*] Corresponding author, email: {bruns,buzug}@imt.uni-luebeck.de

INTRODUCTION An issue in Magnetic Particle Imaging (MPI) is the direct coupling of the drive field coil signal into the receive chain. Therefore, mechanisms to decouple the drive field signal and particle signal from each other were proposed [1]. So far, only rigid gradiometer coils were used and tested [1, 2]. For the first time, we built and experimentally tested a tunable gradiometer coil to even better remove direct coupling between transmit and receive chain and to suppress coupled interferences.

MATERIAL AND METHODS For comparison purposes, we measured the induced voltages with and without Resovist® (Bayer Schering Pharma AG, Leverkusen, Germany) particles in three different receive coils in a Magnetic Particle Spectrometer (MPS). The first coil is a cylindrical receive coil (length 36 mm, radius 8 mm) with 80 copper wire windings. The second coil is a gradiometer coil with same radius, but it is surrounded by two 18 mm parts being wound in the opposite direction on both ends with 40 windings each. The third one is our tunable receive coil in gradiometer configuration. The coil consists of a fixed part, a movable part, and a closure, which were designed and 3D-printed (see fig. 1). Its movable part is 23 mm long and has 51 windings. When the closure is removed, one can adjust the distance from the main part to the movable part with an M3 plastic screw inside the cylinder. After the adjustment, the closure part can be stuck into the movable end of the coil and turned around to inhibit further movements.

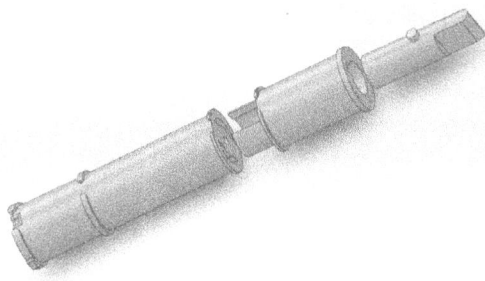

Figure 1: Tunable gradiometer receive coil model consisting of three parts: fixed part, movable part, and closure.

An arbitrary waveform generator (RIGOL, Beijing, China) generates 25 kHz signals from 1 up to 6 V_{pp}, which are 20 times amplified by a power amplifier (AE Techron 2105, Elkhart, USA). This signal is lowpass-filtered with a first-order Butterworth filter to suppress higher harmonics in the transmit chain. The receive signal from our coils is then bandstop-filtered with a fourth-order Butterworth filter and amplified by a custom-built low-noise amplifier. For evaluation, the induced and amplified voltage at 75 kHz is measured by using a signal analyzer (Rohde & Schwarz, Munich, Germany). Consequently, the three different receive coils are put into the drive field coil and measurements of the 75 kHz amplitude are processed for the different output voltages both with and without particles in the receive coil. To adjust our tunable gradiometer coil perfectly, the distance of the movable part to the fixed part is varied in small steps and the received voltage is measured at 75 kHz without particles in the coil.

RESULTS Figure 2 shows the signal-to-interference-and-noise-ratio (SINR) for all coils at six different output voltages with

$$SINR = \frac{V(particles) - V(empty)}{V(empty)}.$$

An exponential decay with increasing output voltage due to particle saturation effects can be observed for all coils. However, the SINR for the rigid gradiometer coil is almost an order of magnitude higher than the one for the regular coil at all six output voltages. The SINR for our new tunable gradiometer coil is even another factor of 1.5 higher.

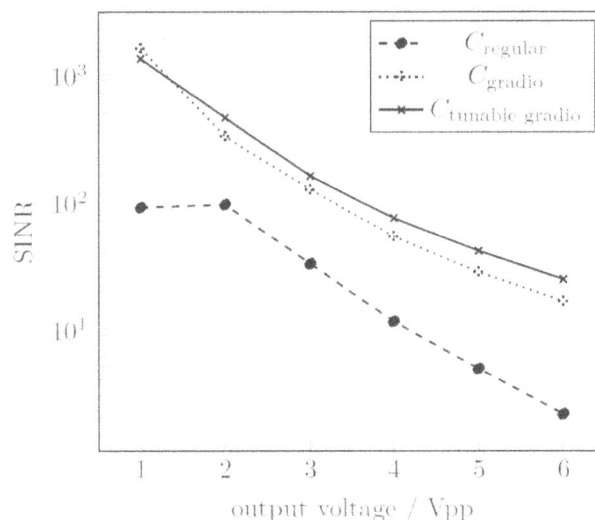

Figure 2: Signal-to-interference-and-noise-ratio (SINR) for the regular receive coil, the rigid gradiometer coil, and the tunable gradiometer coil.

CONCLUSION Using gradiometer coils for receiving MPI signals is a valuable method to suppress both directly coupled signals from the transmit chain and interferences from outside the system. With our novel concept of a tunable gradiometer coil, we were able to further improve this suppression and therefore, to significantly enhance the overall SNR possible for a given system.

ACKNOWLEDGEMENTS This work has been supported by the German Research Foundation under Grant number BU 1436/10-1.

REFERENCES
[1] B. Gleich and J. Weizenecker. *Nature*, 435(7046):1214-1217, 2005. doi: 10.1038/nature03808.
[2] P. W. Goodwill and S. M. Conolly. *IEEE TMI*, 29(11):1851-1859, 2010. doi: 10.1109/TMI.2010.2052284.

MPS and ACS with an atomic magnetometer

Simone Colombo[a], Victor Lebedev[a,*], Zoran D. Grujić[a], Vladimir Dolgovskiy[a], Antoine Weis[a].

[a] Département de Physique, Université de Fribourg, Chemin du Musée 3, 1700 Fribourg, Switzerland
[*] Corresponding author, email: victor.lebedev@unifr.ch

INTRODUCTION The technique of Magnetic Particle Imaging (MPI), following its introduction in 2005 [1], has evolved along two major pathways, viz., frequency-space MPI and X-space MPI, both approaches having their respective merits and drawbacks.

Most MPI approaches developed so far share the common feature that the detected signal is recorded by induction coil(s).

Because of Faraday's induction law signal/noise considerations call for large drive frequencies.

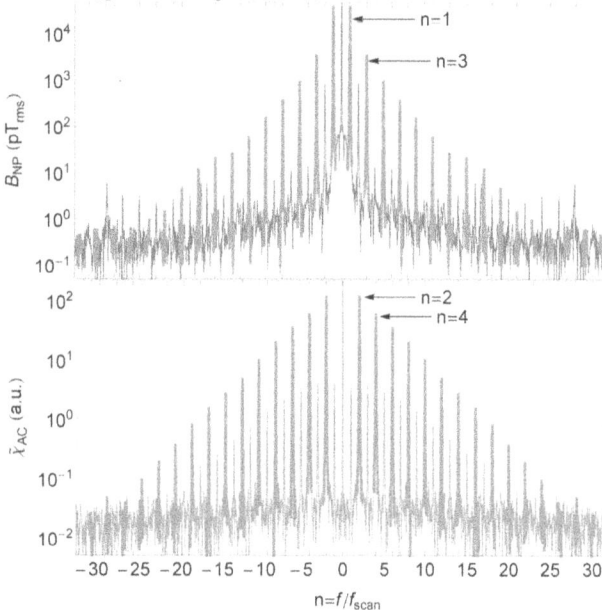

Figure 1: Fourier spectra of nanoparticle produced induction B_{NP} (top) and AC susceptibility $\chi_{AC}=dM/dH$ (bottom).

Here we report on the detection of the anharmonic magnetic response of MNPs by a high sensitivity optically pumped atomic magnetometer (OPM), and demonstrate that an OPM can be used for Magnetic Particle Spectroscopy (MPS) and AC susceptometry (ACS).

Laser-driven OPMs are compact and versatile instruments, mostly operating at room temperature that can detect magnetic field changes in the femto- or even sub-femto-Tesla range [2].

MPS is often referred to as zero-dimensional MPI, and any high-sensitivity MPS method is thus a necessary prerequisite for developing an MPI system.

MATERIAL AND METHODS The OPM used in our apparatus is based on optically detected magnetic resonance in spin-polarized Cs vapor [3]. The recordings were done on a Ferrotec-EMG707 sample containing 3.4 mg of iron. For MPS measurements we excite the sample by a field $H_{scan}(t)$ of amplitude up to 15 mT$_{pp}$/μ_0 that sinusoidally oscillates at a frequency f_{scan} of 600 mHz. For ACS measurements we add to H_{scan} a 1 mT$_{pp}$/μ_0 amplitude modulation field that oscillates at $\approx$800 Hz.

We record time series (sampled at a rate of 320 S/s) of the drive current $I_{scan}(t)$ and the corresponding induced induction signals $B_{NP}(t)$. Performing the Fourier transform of data from 30 consecutive scan cycles yields the harmonics spectrum.

We have extended the method discussed above for measuring $M(H)$ curves towards the direct and simultaneous recording of the derivative, i.e., $\chi_{AC}(H)=dM/dH(H)$-dependence.

RESULTS Figure 1 (top) shows the MPS spectrum. When rescaled to unity bandwidth, the noise floor in the figure represent a power spectral density of 4 pT/Hz$^{1/2}$ which limits the detectable number of harmonics in the $M(H)$ signals to 23. While the direct $M(H)$ method is sensitive to drift and low frequency noise of the background field at the sensor (noise pedestal under the upper spectrum in Fig. 1), the derivative spectrum (Fig. 1, bottom) is insensitive to low-frequency perturbations and allows extraction of a larger number of harmonics

CONCLUSION We have demonstrated that an atomic magnetometer in an unshielded environment can be used for a direct quantitative measurement of MPS and ACS spectra of magnetic nanoparticles in the sub-kHz frequency range. Comparison of the corresponding Fourier spectra reveals the superior power of modulation spectroscopy.

We believe that because of their high sensitivity and large bandwidth (DC up to hundreds of kHz), OPMs, when combined with a variant of X-space MPI, have the potential to yield a complementary, low-frequency MPI technique.

At current stage, our method allows absolute iron content determinations at a sub-μg level. Based on the excellent signal quality demonstrated above we now pursue the goal of designing a 2D X-space MPI scanner based on induction field detection by atomic magnetometers.

ACKNOWLEDGEMENTS Work supported by Swiss National Science Foundation Grant No. 200021_149542.

REFERENCES
[1] B. Gleich and J. Weizenecker. *Nature*, 435(7046):1217—1217, 2005. doi: 10.1038/nature03808.
[2] D. Budker and D.F. Jackson Kimball. Optical Magnetometry. Cambridge University Press, 2013.
[3] G. Bison, N. Castagna, A. Hofer, P. Knowles, J. L. Schenker, M. Kasprzak, H. Saudan, and A. Weis. *Appl.Phys. Lett.*, 95:173701, 2009. doi: 10.1063/1.3255041.

Development of a K-Rb Hybrid Atomic Magnetometer toward MPI

Yosuke Ito[*], Tetsuo Kobayashi

Graduate School of Engineering, Kyoto University, Japan
[*] Corresponding author, email: yito@kuee.kyoto-u.ac.jp

INTRODUCTION Optically pumped atomic magnetometers (OPAMs) have been the subject of many research papers as sensors for biomagnetic measurements due to their extremely-high sensitivity to magnetic fields without cryogenic cooling [1, 2]. Since the sensitivity of the OPAMs does not depend on the frequency, it is expected to operate MPI scanners at low driving frequency by using the OPAM as a receiving sensor instead of an induction coil. This is an important advantage to scaling up the scanner to human-size. In a previous study, we proposed a K-Rb hybrid atomic magnetometer, which has spatially homogeneous sensor properties in all over the sensor cell, so that we can integrate the sensing area in the cell [3]. It leads to enlarging the sensing area of MPI. In this study, we demonstrate the ultra-high sensitivity K-Rb hybrid atomic magnetometer with multiple sensing regions.

MATERIAL AND METHODS Figure 1 shows the experimental setup. We applied a charge-coupled device (CCD) sensor, which has 64×256 pixels, to detector of the K-Rb hybrid atomic magnetometer. The expanded probe beam, which was slightly-detuned from the resonant wavelength of Rb atoms, brought information of the magnetic fields at different locations along the pump beam, which was tuned to the resonant wavelength of K, and was detected with the CCD sensor. The size of the sensor cell is $5 \times 5 \times 5$ cm^3. It contains K and Rb with He and N$_2$ as buffer and quenching gases. By measuring magnetic fields from a test loop coil with a diameter of 10 mm, we examined the performance of the hybrid OPAM.

RESULTS Figure 2 shows the magnetic field distributions measured with our hybrid OPAM and calculated by Biot-Savart law. Although the measured distribution agrees relatively well with calculated results, the intensities of measured results far from the test coil were larger than those of calculated ones. This might be caused by smear and/or blooming, which are specific problem to CCD sensors. The sensitivity of each channel was about 10 pT/Hz$^{1/2}$ at 40 Hz, which was worse comparing to highest sensitivity so far [4]. The issues are originated from the detector. Therefore, we plan to improve the sensitivity and the sensing area by application of a photodiode array, which has lower noise level and larger sensor area as the detector.

CONCLUSION We showed the ultra-high sensitivity K-Rb hybrid atomic magnetometer with multiple sensing regions. The measured magnetic field distributions originated from a test coil showed good agreement with calculated results. These results demonstrate the feasibility of the hybrid OPAM as a sensor toward MPI scanners.

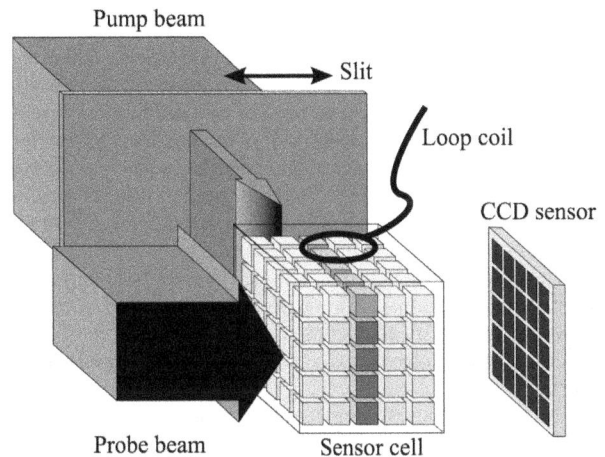

Figure 1: Experimental setup.

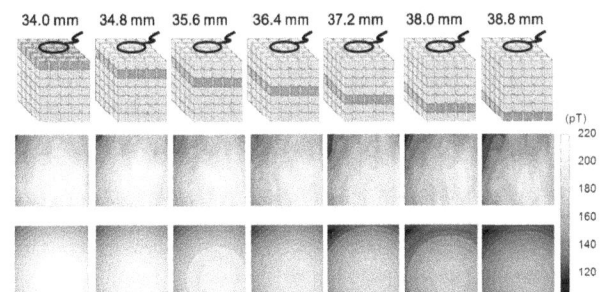

Figure 2: Magnetic field distributions measured with a CCD sensor. Top row illustrates sensing area, middle and bottom rows show measured and calculated maps, respectively.

ACKNOWLEDGEMENTS This work was supported in part by the Innovative Techno-Hub for Integrated Medical Bio-imaging of the Project for Developing Innovation Systems, Grant-in-Aid for Scientific Research (A) (15H01813), Grant-in-Aid for Scientific Research (C) (15K06106) and Grant-in-Aid for Challenging Exploratory Research(2650466), all from the Ministry of Education, Culture, Sports, Science, and Technology (MEXT), Japan.

REFERENCES
[1] G. Lembke, S. N. Erné, H. Nowak, B. Menhorn and A. Pasquarelli. *Biomed. Opt. Express*, 5(3):876-881, 2014. doi: 10.1364/BOE.5.000876.
[2] O. Alem, A. M. Benison, D. S. Barth, J. Kitching and S. Knappe. *J. Neurosci.*, 34(43):14324-14327, 2014. doi: 10.1523/JNEUROSCI.3495-14.2014.
[3] Y. Ito, H. Ohnishi, K. Kamada, and T. Kobayashi. *IEEE Trans. Magn.*, 48(11):3715-3718, 2012. doi: 10.1109/TMAG.2012.2199966.
[4] H. B. Dang, A. C. Maloof and M. V. Romalis. *Appl. Phys. Lett.*, 97(15):151110, 2010. doi: 10.1063/1.3491215.

MPS test measurments with phase angle detection

Przemysław Wróblewski[*], Waldemar Smolik

[a] Division of Nuclear and Medical Electronics, Institute of Radioelectronics and Multimedia Technology, Warsaw University of Technology, Poland
[*] Corresponding author, email: P.Wroblewski@ire.pw.edu.pl

INTRODUCTION Magnetic Particles Imaging (MPI) is a tomographic imaging technique that measures the spatial distribution of superparamagnetic nanoparticles. Quality and value of those measurement is highly dependent on the type and parameters of the nanoparticles used as well as the properties of the environment that they are in. One of measurement techniques used to characterized magnetic nanoparticles is Magnetic Particles Spectrometry (MPS). It can be used to directly acquire amplitude and phase spectrum of nanoparticles magnetization signal. For example, amplitude spectrum allows to estimate the particles size distribution of the tracer, while phase angle can allow research the impact of various parameters on nanoparticle's Brownian motion. By combining information from both of those sources valuable information about the particle's environment can be gathered.

This work presents test measurements of amplitude and phase spectrums of MPS signal using laboratory setup including SR865 DSP Lock-in Amplifier.

MATERIAL AND METHODS Magnetic particles spectrometer developed in our Division was used to conduct test measurements of amplitude and phase spectrums of MPS signal. As a sample 10 µL of water solution of nanoPET's FeraSpin™ XL (50-60 nm) nanoparticles were used. FeraSpin™ XL has a concentration of 10 mmol(Fe)/L. Particles were excited using 10 kHz sinusoidal signal with amplitude of 8 [mT/µ0]. In the receiving channel of our measurement setup excitation signal was attenuated by use of a reference coil connected in opposite direction. This way over 60 dB damping of excitation signal can be achieved. After this procedure the signal was fed to SR865 Lock-in Amplifier, which allows measurements of both amplitude and phase of up to 99 harmonics of fundamental frequency. First, a signal from the background was measured in our measurement method. Then the values of spectral magnetic moment and phase angle of harmonics were calculated by subtraction of this background noise signal from signal acquired in the presence of the sample.

RESULTS Using the setup described above we were able to measure amplitude and phase spectrum up to 25 harmonic only. Unfortunately the harmonics over 250 kHz were distorted due to the interference with a peek at 490 kHz induced by audio amplifier used in excitation channel. Only odd harmonics were measured.

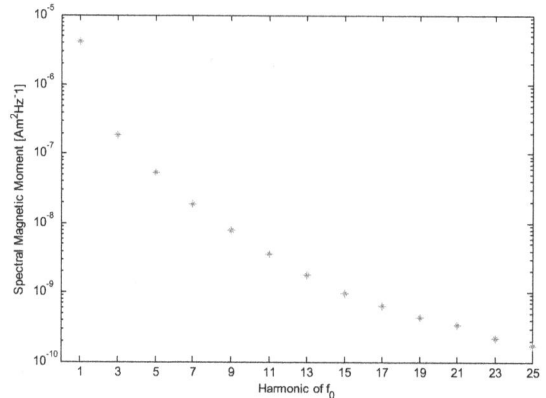

Figure 1: Magnetic moment spectrum of FeraSpin™ XL at a field strength of 8 mT/µ0.

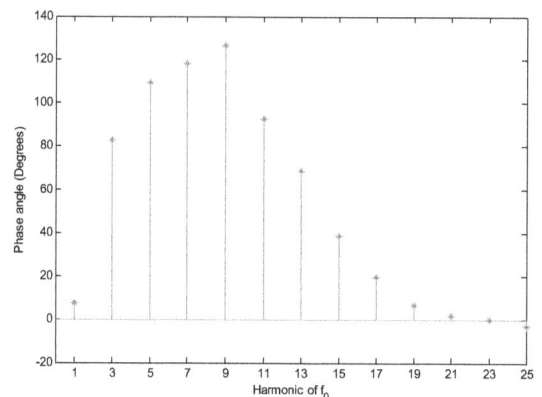

Figure 2: Phase angle spectrum of FeraSpin™ XL at a field strength of 8 mT/µ0.

CONCLUSION By introducing the SR865 Lock-in amplifier to our measurement setup we are now able to measure both amplitude and phase spectrum simultaneously. The measurements showed that the applied audio amplifier was the main source of distortion at high frequencies and that it is necessary to use an amplifier with improved performance. In the future we are planning to conduct further MPS measurements using this setup with particles of different diameters and dispersed in various environments.

REFERENCES
[1] Biederer, S., et al. "Magnetization response spectroscopy of superparamagnetic nanoparticles for magnetic particle imaging." *Journal of Physics D: Applied Physics* 42.20 (2009): 205007.
[2] Rauwerdink, Adam M. and Weaver, John B., "Harmonic phase angle as a concentration-independent measure of nanoparticle dynamics", *Medical Physics*, 37, 2587-2592 (2010)

An Arbitrary Excitation Waveform Relaxometer

Zhi Wei Tay[a,*], Daniel W. Hensley[a], Laura A. Taylor[a], Beliz Gunel[a], Patrick W. Goodwill, Bo Zheng[a], Steven M. Conolly[a]

[a] Department of Bioengineering and EECS, University of California Berkeley, USA
[*] Corresponding author, email: zwtay@berkeley.edu

INTRODUCTION The response of Magnetic Nanoparticles to a magnetic drive-field determines critical characteristics such as resolution and signal-to-noise ratio in Magnetic Particle Imaging (MPI) [1-4]. Recently, many groups have demonstrated that MPI tracer performance is significantly affected by both the drive-field frequency and amplitude [5-7]. However, the excitation (drive-field) waveform used in these studies is limited completely to sinusoids. Here, we expand upon our prior work of an untuned relaxometer [8] to construct an arbitrary waveform relaxometer and utilize its wide-bandwidth excitation capability for the first experimental demonstration of arbitrary excitation waveforms (triangle wave) for characterizing MPI tracers.

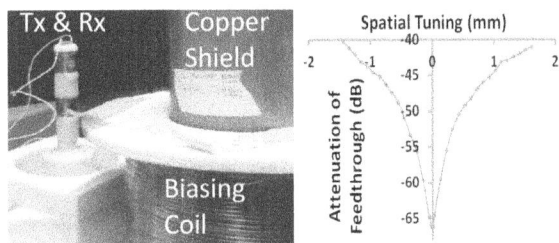

Figure 2a: Experimental trace of a triangle drive waveform and the experimental response from a 25 nm SPIO. No 1st harmonic filtering was applied. **2b:** Experimental reconstructed PSF for Triangle and Sine drive wave showing only minor differences.

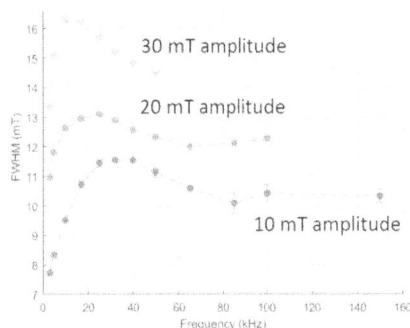

Figure 1: (left) Arbitrary Excitation Waveform Relaxometer. (right) Spatially-tunable gradiometer with 25μm increments allows for -67 dB attenuation of feedthrough

MATERIAL AND METHODS The Tx (ID=10 mm, L = 7 μH, N=20 turns) surrounds a gradiometric Rx (ID = 6 mm, N=30 turns) that fits a 200μL PCR tube. No analog transmit or receive filters were used. To deal with the now wide-bandwidth direct feedthrough, the gradiometric Rx is spatially tunable to achieve wide-bandwidth feedthrough attenuation of -67 dB. The signal is amplified by a low noise preamplifier (Stanford Research Systems SR560). The Tx is driven with a DC-coupled power amplifier (AE Techron7224) with pulsed power of up to 1.2 kW. Waveform shape was verified by a Rogowski coil (Pemuk LFR 6/60). 150μg of 25 or 27.4 nm nanoparticles (Senior Scientific PrecisionMRX carboxylic acid functionalized iron oxide nanoparticles) was used. Both sine and triangle drive waveforms are 20 mT amplitude and 25 kHz. X-space reconstruction is used [3] for both sine and triangle excitations, thus signal is directly gridded to the instantaneous applied field.

RESULTS Triangular excitation waveform was verified by the Rogowski coil demonstrating the arbitrary waveform capability of the relaxometer (Fig 2a). To assess the performance of the SPIO under a triangle wave excitation, we tested the same 25 nm SPIO sample with a sinusoidal and triangular excitation. Both excitation waves have exactly the same frequency and amplitude, differing only in their shape. Our preliminary results (Fig 2b) show modest improvements in resolution with triangle wave. One important benefit of the triangle wave is that it does not need velocity-compensation during reconstruction therefore reducing computational load and reconstruction error. The initial results suggest that triangle wave performs at least as well as a sinusoidal drive waveform but with an easier reconstruction.

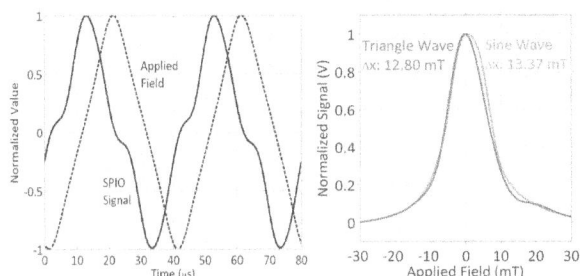

Figure 3: 27.4 nm SPIO PSF FWHM across 12 frequencies from 3 kHz to 150 kHz (sinusoidal) and 3 excitation amplitudes

CONCLUSION: Our improved gradiometer allows us to dispense with tuned circuit elements, making for simpler construction with no bulky capacitor banks and allowing convenient arbitrary frequency switching. The only tuning needed is one-time gradiometer tuning and we observe comparable SNR to our previous relaxometers. In conclusion, we demonstrate a novel relaxometer capable of testing arbitrary excitation waveforms and any arbitrary frequency <400kHz. This enables investigating non-sinusoidal excitation waveforms and optimizing sinusoidal excitation waveforms towards better MPI performance.

ACKNOWLEDGEMENTS We gratefully acknowledge support from NIH 5R01EB013689-03, CIRM RT2-01893, Keck Foundation 034317, NIH 1R24MH106053-01, NIH 1R01EB019458-01, ACTG: 037829 and the Agency of Science, Technology and Research, Singapore (fellowship).

REFERENCES
[1] B. Gleich and J. Weizenecker. *Nature*, 435(7046):1217—1217, 2005.
[2] T. Knopp and T. M. Buzug. Springer, Berlin/Heidelberg, 2012.
[3] P. Goodwill and S. Conolly. *IEEE Trans Med Imag 29 (11): 1851–59*. (2010)
[4] L.R. Croft, P.W. Goodwill, and S. Conolly. *IEEE Trans Med Imag*, 31: 2335–42. (2012)
[5] S. Shah, R. M. Ferguson, and K. M. Krishnan. *Journal of Applied Physics* 116 (16): 163910 (2014)
[6] A. Tomitaka, R. M. Ferguson, A. P. Khandhar, S. J. Kemp, K. Nakamura, Y. Takemura, S. Ota, and K. M. Krishnan. *IEEE Trans Magnetics* 51 (2). (2015)
[7] C. Kuhlmann, A. P. Khandhar, R. M. Ferguson, S. Kemp, T. Wawrzik, M. Schilling, K. M. Krishnan, and F. Ludwig. *IEEE Trans Magnetics* 51 (2). 2015..
[8] Z.W. Tay, P. W. Goodwill, D. W. Hensley, and S. Conolly. *In Magnetic Particle Imaging (IWMPI), 2015 5th International Workshop* 1–1.(2015)

Effects of Viscosity on Nanoparticle Relaxation Time Constant

Mustafa Ütkür [a,b], Yavuz Muslu [a,b], Ahmet Alacaoglu [a,b], Ali Alper Ozaslan [a,b], Emine Ulku Saritas [a,b,*]

[a] Department of Electrical and Electronics Engineering, Bilkent University, Ankara, Turkey
[b] National Magnetic Resonance Research Center (UMRAM), Bilkent University, Ankara, Turkey
[*] saritas@ee.bilkent.edu.tr

INTRODUCTION Magnetic particle imaging (MPI) shows promise as a functional imaging modality, with the potential to probe viscosity *in vivo*. Through Brownian relaxation, the nanoparticle's magnetization response is altered by external viscosity. Previous work has looked at the effects of viscosity on the harmonic ratio of the nanoparticle signal [1-2]. Here, the effects are analyzed via the relaxation time constant of the nanoparticle for the first time. Initial results show great promise as the Brownian relaxation becomes dominant at lower frequencies, and viscosity affects the time constant significantly.

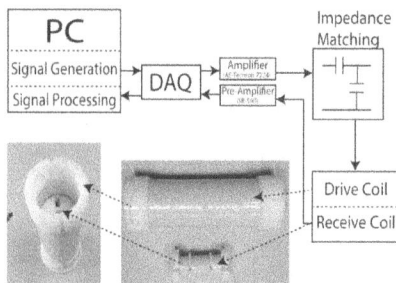

Figure 1: Experimental MPS setup. Arrows show the workflow of the transmit/receive chain. Photos of the drive/receive coils in the experimental configuration are given on the leftmost photo.

MATERIAL AND METHODS Undiluted nanomag-MIP (Micromod GmbH, Germany) particles with 89mM Fe/L concentration were used in this experiment. 11 tubes were prepared starting with 50µL of undiluted nanoparticle solution, then mixed with varying amounts of water and glycerol so that the viscosities ranged from 0.89cP to 15.33cP at 25°C (0.89cP tube for nanoparticle in water only). The total volume was kept constant at 200µL to avoid any Fe concentration bias in the results.

Our custom magnetic particle spectrometer (MPS) setup (also known as an MPI relaxometer) utilizes the drive coil design previously described in [3]. The receive coil was a three-section gradiometer-type solenoid with 7 layers having 62 turns per layer, allowing for a test tube of 1.5 cm in length and 0.8 cm diameter. Figure 1 shows the workflow of the transmit/receive chain, as well as photos of the coils.

The prepared tubes were tested at 4 different frequencies (250 Hz to 10.8 kHz) at 30 mT-pp drive field amplitude. Signal acquisition was repeated three times for each tube at each frequency. Using the exponential relaxation time constant (τ) model presented in [4], we previously developed a method to estimate τ from the MPI signal [5]. This technique is directly applicable to the MPS data. In addition, we looked at the fifth-to-third harmonic ratio for comparison with previous work [1-2].

RESULTS Figure 2 shows example MPS data at 550 Hz at two different viscosities. The signal at high viscosity (15.3cP) is considerably narrower than the low viscosity (0.89cP, i.e., water) signal. Figure 3 shows the results at all four tested frequencies.

As the frequency is reduced, the viscosity of the environment affects the SPIO signal more extensively. It should be noted that the relaxation time constant decreases monotonically with viscosity at 550 Hz and higher frequencies. On the contrary, the trend at 250 Hz is non-monotonic, which may pose a problem for viscosity mapping purposes.

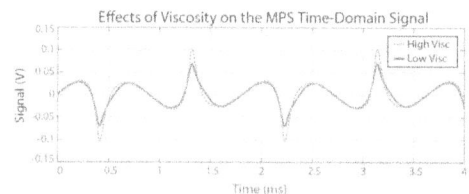

Figure 2: Measured data from our custom MPS setup at 550 Hz, for high viscosity (15.3cP) and low viscosity (0.89cP). $\tau_{est} = 50\mu s$ and $\tau_{est} = 84\mu s$, respectively.

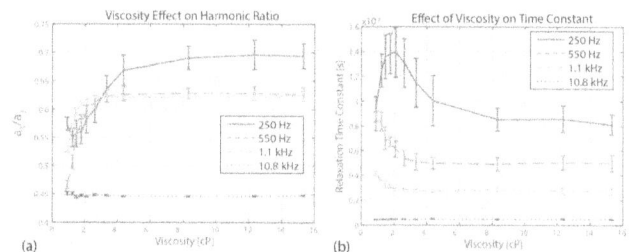

Figure 3: Effects of viscosity on (a) harmonic ratio and (b) relaxation time constants. Mean and std of 3 repetitions are shown at 4 different frequencies.

CONCLUSION In this work, viscosity effects on both signal harmonics and relaxation time constant were investigated experimentally. In general, the estimated time constant increases as the viscosity of the external environment decreases. Increase in error at lower frequencies was due to inherently lower SNR of those frequencies. We plan to improve our results with further enhancements on the transmit/receive chains.

ACKNOWLEDGEMENTS This work was supported by the Scientific and Technological Research Council of Turkey through a TUBITAK 3501 Grant (114E167), by the European Commission through an FP7 Marie Curie Career Integration Grant (PCIG13-GA-2013-618834), and by the Turkish Academy of Sciences through TUBA-GEBIP 2015 program.

REFERENCES
[1] A.M. Rauwerdink and J.B. Weaver, *J of Magnetism and Magnetic Materials*, 322(6):609-613, 2010. doi: 10.1016/j.jmmm.2009.10.024.
[2] T. Wawrzik et al., *Proc of International Workshop on Magnetic Particle Imaging*, 2013. doi: 10.1109/IWMPI.2013.6528371.
[3] M. Ütkür and E.U. Saritas, *Proc of International Workshop on Magnetic Particle Imaging*, 2015. doi: 10.1109/IWMPI.2015.7107082.
[4] L. R. Croft et al., *IEEE Trans Med Imag*, 31(12):2335-2342, 2012. doi: 10.1109/TMI.2012.2217979.
[5] G. Onuker et al., *Proc of International Workshop on Magnetic Particle Imaging*, 2015. doi: 10.1109/IWMPI.2015.7107042.

Reconstruction of a 2D Phantom Recorded with a Single-Sided MPI Device

Ksenija Gräfe [a,*], Anselm von Gladiss [a], Mandy Ahlborg [a], Gael Bringout [a], Thorsten M. Buzug [a,*]

[a] Institute of Medical Engineering, University of Lübeck
[*] Corresponding author, email: {graefe, buzug}@imt.uni-luebeck.de

INTRODUCTION There are different topologies for magnetic particle imaging (MPI) scanning devices, one of those can be used for a single-sided imaging device. Thanks to its coil arrangement, the object size is not limited.

In [1], the first reconstruction results of different single dot phantoms are presented which have been measured with the single-sided scanner. In order to realize 2D imaging, a pair of D-shaped coils has been added to the two circular coils, which are used in the prototype [2, 3]. The circular coils are arranged concentrically. Direct currents on the circular coils, which flow in opposite directions, generate the selection field. An additional alternating current on the inner coil enables the movement of the field free point (FFP) vertically to the coil setup (x-direction). Another alternating current on the D-shaped coil pair moves the FFP parallel to the scanner setup (y-direction). Applying the excitation frequencies $f_x = \frac{2.5}{99} MHz$ and $f_y = \frac{2.5}{96} MHz$, the FFP moves on a 2D Lissajous trajectory with the size of 30 x 30 mm². Table 1 lists the dimensions of the different coils and the applied current values.

Table 1: Dimensions and parameters of different coils used for the single-sided MPI device [4].

	Inner Circular Coil	Outer Circular Coil	D-shaped Coil Pair
Outer Diameter	29 mm	140 mm	140 mm
Inner Diameter	19 mm	107.5 mm	114 mm
Direct Current	65 A	55 A	–
Alternating Current	42 A	–	80 A
Power Loss	97 W	157 W	70 W

MATERIAL AND METHODS To be able to reconstruct images, a system matrix measurement $(\hat{S})$ and a phantom measurement $(\hat{u})$ have been performed. Undiluted Resovist (Bayer Schering Pharma, Berlin, Germany) is used as tracer material. The first 10 periods of the receive signal have been ignored and the 500 following periods have been averaged from. Only the frequency components above a SNR threshold of 3.56 are used for reconstruction. The distribution of the particle concentration (c) is calculated by solving the equation

$$(\widehat{S^*}W\hat{S} + \lambda I)c = \widehat{S^*}W\hat{u}$$

with $\widehat{S^*}$ the conjugate transpose of the truncated system matrix, a weighting matrix W, the identity matrix I and the regularization parameter λ by a modified Kaczmarz algorithm [5].

RESULTS Fig. 2 shows a model of the phantom and the reconstruction result. The penetration depth of the scanner is about 10 mm. It is possible to distinguish tracer material with a distance of 2 mm in x-direction and 4 mm in y-direction.

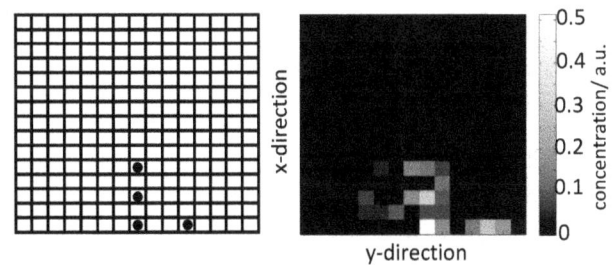

Figure 2: A four-dot phantom (left) and the reconstruction result (right). Each pixel of 2 x 2 mm².

CONCLUSION The result is promising for medical applications. Tracer material can be detected and its spatial distribution can be approximated, especially near to the scanner surface.

The reconstruction may be further improved by using dedicated reconstruction methods. Additionally, the hardware setup may be optimized to improve the penetration depth.

ACKNOWLEDGEMENTS This work has been supported by the German Federal Ministry of Education and Research under the grant numbers 01EZ0912, 13N11090 and 13GW0069A and the European Union and the state Schleswig-Holstein (Program for the Future Economy) under the grant number 122-10-004.

REFERENCES
[1] K. Gräfe, A. von Gladiss, G. Bringout, M. Ahlborg, and T. M. Buzug, "2D imaging with a single-sided MPI device," in 2015 International Workshop on Magnetic Particle Imaging (IWMPI). IEEE Xplore, 2015. doi: 10.1109/IWMPI.2015.7107024
[2] T. F. Sattel, T. Knopp, S. Biederer, B. Gleich, J. Weizenecker, J. Borgert, and T. M. Buzug, "Single-sided device for magnetic particle imaging," Journal of Physics D: Applied Physics, vol. 42, no. 2, pp. 1–5, 2009.
[3] K. Gräfe, G. Bringout, M. Graeser, T. F. Sattel, and T. M. Buzug, "System matrix recording and phantom measurements with a single-sided magnetic particle imaging device," IEEE Transactions on Magnetics, vol. 51, no. 2, pp. 6502303, 2015. doi: 10.1109/TMAG.2014. 2330371
[4] T. F. Sattel, M. Erbe, S. Biederer, T. Knopp, D. Finas, K. Diedrich, K. Lüdtke-Buzug, J. Borgert, and T. M. Buzug, "Single-sided magnetic particle imaging device for the sentinel lymph node biopsy scenario," in Proceedings of SPIE, vol. 83170, 2012. doi: 10.1117/12.912733
[5] T. Knopp, J. Rahmer, T. F. Sattel, S. Biederer, J. Weizenecker, B. Gleich, J. Borgert, and T. M. Buzug, "Weighted iterative reconstruction for magnetic particle imaging," Physics in Medicine and Biology, vol. 55, pp. 1577–1589, 2010.

23

Single-sided FFL-based MPI Device with Depth Encoding

Alexey Tonyushkin

Department of Physics, University of Massachussettss Boston, Boston, Massachusetts 02125, USA
email: alexey.tonyushkin@umb.edu

INTRODUCTION A Magnetic Particle Imaging (MPI) device has yet to be introduced to the clinical practice. The major challenge is to generate sufficiently large gradient field that penetrates whole body. An open geometry scanner will be also highly desirable. One way to make such a practical MPI device is to use a single-sided geometry [1,2]. The single-sided device has all the hardware on one side from the imaging volume and therefore could be used on a whole body as well as a local volume imaging. Here we propose a novel hardware design for single-sided device that is capable of 3D imaging. Moreover, different from recent field-free-point (FFP) based developments [3,4], in our geometry we utilize a potentially more sensitive field-free-line (FFL) configuration.

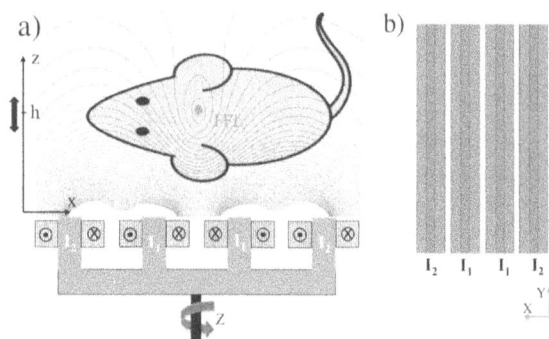

Figure 1: a) Single-sided selection coils with FFL (contour plot) at height *h* above the coils; b) four elements coil structure (top view).

MATERIAL AND METHODS The apparatus and topology of selection coils are shown in Fig. 1a. The selection coils consist of four straight parallel coils (see Fig. 1b). The generation of FFL works as following: two inner coil elements with equal DC current with amplitudes I_1 create field gradient at certain height from the surface, while two outer coils with current amplitudes I_2 create oscillating bias field that moves the height of FFL. Thus by switching the relative current between the inner and outer pairs of coils FFL could oscillate along z-axis encoding the depth inside the subject while the gradient strength could be kept relatively constant by dynamically adjusting the amplitude of I_1. The base height (depth) of FFL is defined by the width of each coil element and the gap between the coils. The in-plane (x-y) encoding is done through projection imaging by mechanical rotation of the device (or subject) around z-axis (up to 180°). The two inner coils could also be replaced with permanent magnets with alternating poles thus reducing power consumption of such device. More outer coil elements could be utilized for additional fine control of homogeneity (shimming) of the field.

RESULTS We carried out mathematical simulations using Mathematica (Wolfram) software and Radia package (ESRF, France) as well as experimentally measured properties of a limited capability two-element prototype coils [5].

Figure 2 shows simulations of the magnetic field pattern produced by four-element coil structures with dimensions: 14cm wide (x-axis) by 18cm long (y-axis). An example shows the DC currents of I_1=100A and I_2=0A that produce FFL at h=1.4cm with a field gradient of ~2 T/m. At DC currents of I_1=200A and I_2=200A the height of FFL moves to h=1.9cm and field gradient becomes ~1.5T/m. The usable FFL field flatness of a ~18cm-long device limits 2D FOV (x-y) to up to ⌀=4 cm.

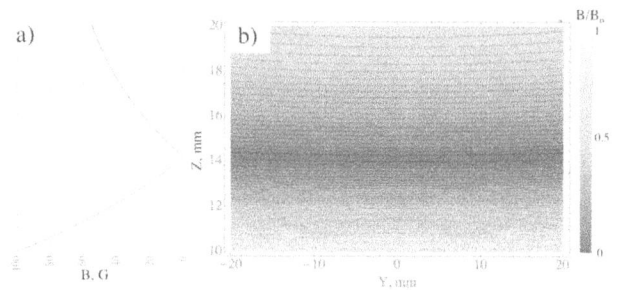

Figure 2: a) Magnetic field and b) normalized field contour plot show FFL at z=14mm.

CONCLUSION We presented a novel design of a single-sided FFL-based device which is capable of 3D imaging. With sufficient current amplitudes FFL could encode the whole volume of a small animal or penetrate deep enough into the human organs such as vascular system. An MPI device based on the proposed selection scheme could be a compact and robust alternative to the state-of-the-art FFP-based MPI scanners.

REFERENCES
[1] T. F. Sattel, T. Knopp, S. Biederer, B. Gleich, J. Weizenecker, J. Borgert, T. M. Buzug, *J. Phys D: Appl. Phys.*, 42(2), 1-5, 2009. ☐
[2] K. Gräfe, M. Grüttner, T. F. Sattel, M. Graeser, T. M. Buzug, *Proc. of SPIE*, vol. 8672, 6pp., 2013.
[3] B. Gleich and J. Weizenecker. *Nature*, 435(7046):1217—1217, 2005.
[4] K. Gräfe, A. von Gladiß, G. Bringout, M. Ahlborg, T. M. Buzug, *5th International Workshop on Magnetic Particle Imaging*, 2015.
[5] A. Tonyushkin and M. Prentiss. *J. Appl. Phys.* 108, 094904, 2010.

High Resolution Tomographic MPI with a Field Free Line Electromagnet

Elaine Yu[a*], Patrick W. Goodwill[a, c], Zhi Wei Tay[a], Paul Keselman[a], Xinyi Zhou[a], Ryan Orendorff[a], Daniel W. Hensley[a], Matt Ferguson[d], Bo Zheng[a], Steven M. Conolly[a, b]

[a] Department of Bioengineering, University of California, Berkeley, CA, USA
[b] Department of Electrical Engineering and Computer Science, University of California, Berkeley, CA, USA
[c] Magnetic Insight, Inc., Newark, CA, USA
[d] LodeSpin Labs, PO Box 95632, Seattle, WA, USA
[*] Corresponding author, email: elaineyu@berkeley.edu

INTRODUCTION While magnetic particle imaging (MPI) shows extraordinary promise for biomedical applications, one major improvement required is in its native spatial resolution, currently at 1 mm [1, 2]. Here we have designed and constructed a laminated iron-core Field-Free Line (FFL) MPI magnet with a 5.5 T/m gradient strength. The improved resolution of high gradient 3D projection reconstruction will enable future medical applications such as angiography and cancer screening.

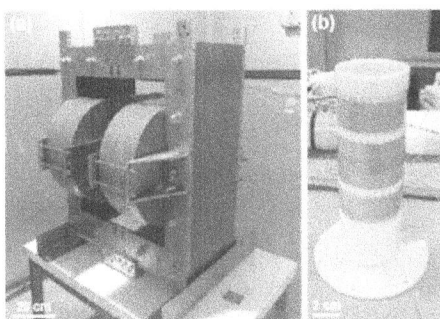

Figure 1: (a) High-resolution water-cooled FFL MPI electromagnet produces a 5.5 T/m Field Free Line selection field using 26 kW of continuous power. Magnet free bore: 13 cm. (b) Litz wire receive coil assembly with (-1, 2, -1) gradiometer configuration. Imaging free bore: 5 cm.

MATERIAL AND METHODS Unlike our prior selection field magnets, which were permanent magnets, here we use water-cooled electromagnets with a laminated iron-core to both create and shift the selection field. The iron-core was manufactured with C5-coated, and epoxy-bonded 16-gauge cold-rolled steel laminations to maximize power efficiency. A water-cooled copper bore provides electromagnetic shielding. The electromagnets are "racetrack coils" with parallel water cooling circuits. The full assembly weighs ~900 kg.

A water-cooled solenoid generates a drive field at 20.225 kHz. A receiver coil was wound with litz wire (100/44 SPNSN) on a 3D printed former in a gradiometer configuration to minimize direct feedthrough (30 dB); see Fig. 1(b). A 90 dB passive notch filter at 20.225 kHz rejects first harmonic direct feed through. A commercial low noise preamplifier (Stanford Research Systems, SR560) was matched to the receive coil with a transformer. MATLAB was used in conjunction with a National Instruments DAQ module for signal generation and collection as well as image reconstruction.

A spiral phantom was created using LodeSpin SPIO particles and we used x-space MPI and Filtered Back Projection to reconstruct the 3D image (35 projection angles) [2, 3].

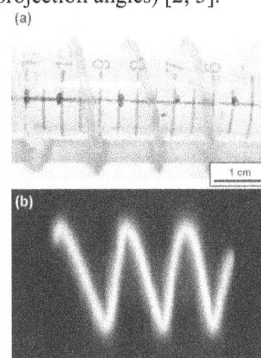

Figure 2: (a) Spiral phantom constructed using 1.1 mm ID plastic tubing containing Lodespin SPIO nanoparticles (6.2 mmol/L). (b) 3D projection reconstruction image of the spiral phantom using the 5.5 T/m FFL MPI scanner. FOV of 5 × 5 × 4 cm, 35 projections, 15 seconds per projection.

RESULTS Our new FFL MPI scanner currently achieves a 5.5 T/m continuous magnetic field gradient, and can pulse up to 6.3 T/m FFL with our existing power supplies. A Max. Intensity Projection (MIP) of the 3D MPI image is shown in Fig. 2. This gradient improves resolution 3-fold over our prior FFL scanner. We plan to increase the gradient to 7.5 T/m with additional power supplies and cooling for a native resolution of 600 μm. Preliminary results also indicate an excellent sensitivity of (< 40 ng/voxel).

CONCLUSION We designed and constructed the world's highest resolution FFL MPI scanner. Our early images demonstrate nearly 3-fold improved resolution and excellent sensitivity.

ACKNOWLEDGEMENTS We would like to acknowledge funding support from NSF GRFP, NIH R01 EB013689, CIRM RT2-01893, Keck Foundation 009323, NIH 1R24 MH106053, NIH 1R01 EB019458, and ACTG 037829.

REFERENCES
[1] Gleich, B., & Weizenecker, J. (2005). Nature, 435(7046).
[2] Goodwill, P. W., & Conolly, S. M. (2010). IEEE TMI, 29(11).
[3] Konkle, J. J., Goodwill, P. W., Carrasco-Zevallos, O. M., & Conolly,

MPI Cube – fully 3D field free line scanner

P. Vogel [a,*], M.A. Rückert [a], V.C. Behr [a]

[a] Department of Experimental Physics 5 (Biophysics), University of Würzburg, Würzburg
* Corresponding author, email: Patrick.Vogel@physik.uni-wuerzburg.de

INTRODUCTION Since the first publication of Magnetic Particle Imaging (MPI) several types of scanners have been presented [1]. Basically two different methods for scanning a sample exist: scanning with a field free point (FFP) and scanning with a field free line (FFL). MPI-scanners using a FFP scan an entire 3D region point-by-point, scanners using a FFL acquire a projection of the signal through the sample [2]. However, all FFL scanners only can scan a single slice, which means for acquiring a full 3D volume the sample has to be moved mechanically along one axis and scanned slice-by-slice. Several hardware simplifications have been published to improve the practicability. However, the basic coil setup for generating a static FFL consists of only two coil pairs in Maxwell configuration (see Fig. 1 (a)) [3]. For moving the FFL along one axis one of the coil pairs requires an additional offset current.

In this abstract a concept for a full 3D FFL scanner is presented, which enables scanning an entire 3D volume without moving the sample.

MATERIAL AND METHODS Instead of using additional coils, which move and rotate the FFL, the 3D FFL scanner is designed to generate a static FFL in three orthogonal directions. Therefore, six individual coils are required for the static FFL generation and additional six coils for the linear movement of the FFL. The resulting cube with two coils placed at each side allows to generate a FFL in three orthogonal directions and move them arbitrarily along the respective perpendicular plane. In Fig. 1 (b) the unwrapped coil overview can be seen.

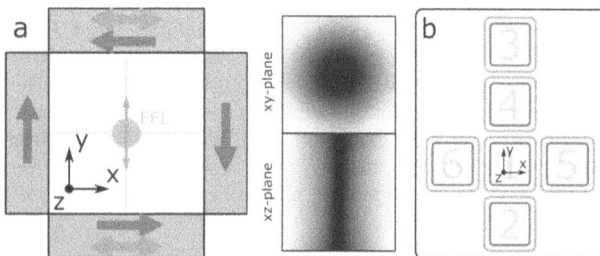

Figure 1: (a) Two perpendicular coil pairs in Maxwell configuration (blue arrows show the current flow), which are orientated in the x-y-plane, are required for the generation of a static FFL (z-direction). To move the FFL along the y-axis an additional offset current is required (red arrows). The two images show the magnetic field distribution. (b) On each side of the MPI cube, two separate coils (blue and red) are required. The unwrapped 2D image gives the overview of the coil indices.

The scanning process of an entire 3D volume is performed in three steps. Step 1: use coils 2blue, 4blue, 5blue and 6blue for the generation of a static FFL aligned in z-direction. Use coils 5red and 6red to move the FFL along the x-axis with a frequency f_1. Use coils 2red and 4red to move the FFL along the y-axis with the frequency f_2. Step 2 and step 3 are performed after the same scheme (see Fig. 2 left). In each step a FFL is generated and

moved along a 2D plane. The acquired signal represents the projection of the sample along the direction of the FFL and can be processed in the same way as a TWMPI dataset [4]. After a 2D reconstruction of the three full projections using a Wiener deconvolution the entire volume can be reconstructed using a backprojection algorithm (see Fig. 2 middle and right).

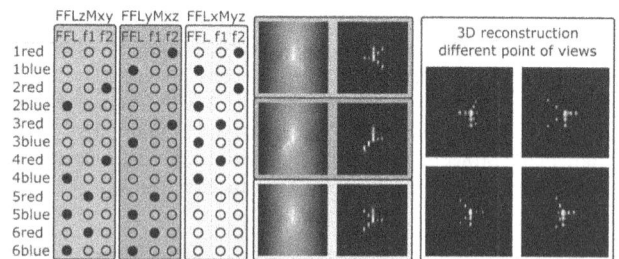

Figure 2: Left: the checklist shows the control settings of the 12 coils during the three scanning steps. FFLzMxy means generating a FFL in z direction and move them in x with f_1 and y with f_2. The simulated images in the **middle** show the raw-data of the three projections and the reconstructed data. **Right:** After the backprojection process a full 3D dataset is available.

DISCUSSION The linear simplification of the FFL movement causes a decreasing of the scanning and reconstruction time. Furthermore, it is possible to scan an entire 3D volume without moving the sample.

However, the system is massively underdetermined. During a full scan with three projections $3 \times N \times N$ datapoints are acquired, but the volume of interest has $N \times N \times N$ datapoints. This means that e.g. a symmetrical sample, which is aligned accurate along the axes, cannot be reconstructed correctly. In that case, it is possible to scan the sample using an FFP within the same scanner.

CONCLUSION In this abstract, a novel approach for a fully 3D FFL MPI scanner is presented. Because of the linear simplification the hardware is less sophisticated than other FFL devices and comes with a reduced scanning time.

ACKNOWLEDGEMENTS This work was partially funded by the DFG (BE-5293/1-1).

REFERENCES

[1] N. Panagiotopoulos, et al., *Int J Nanomedicine*, 10:3097—3114, 2015. doi: 10.2147/IJN.S70488.

[2] J. Weizenecker, et al., *J. Phys. D: Appl. Phys.*, 41:105009, 2008. doi: 10.1088/0022-3727

[3] T. Knopp, et al., *Appl. Phys. Lett.*, 97:092505, 2010. doi: 10.1063/1.3486118.

[4] P. Vogel and S. Lother, et al., *IEEE TMI*, 33(10):1954-1959, 2014. doi: 10.1109/TMI.2014.2327515.

Self-Shielded, High-Resolution, and High-Sensitivity MPI FFL Imager

Patrick Goodwill[a,*], Justin Konkle[a], Steven Suddarth[a], Anna Christensen[a]

[a] Magnetic Insight, Inc., Newark, CA, USA
* Corresponding author, email: goodwill@magneticinsight.com

INTRODUCTION Magnetic Particle Imaging (MPI) is now transitioning from the engineering laboratory and into the laboratories of translational scientists and clinicians [1,2]. Here we describe the design of a self-shielded murine MPI imager optimized for quantitative, high-resolution and high-sensitivity murine imaging applications.

The system is designed to enable biologists with quantitative, deep tissue imaging to accelerate cell tracking, disease targeting, and vascular research at sub-millimeter resolutions and nanogram/voxel sensitivities. Siting costs are minimized by not requiring a shield room and enabling installation through a standard 36" door.

To achieve the system design criteria (see Table 1), we have designed a main magnet that is compact, self-shielded, and high-resolution. The main magnet is a water-cooled electromagnet that produces a 5.5 T/m Field Free Line (FFL) selection field. The strong magnetic field gradient enables sub-millimeter resolutions when imaging optimized nanoparticles [3]. The use of a FFL magnet increases the sensitivity of the system by an order of magnitude over a comparable sensitive point imager [4,5], which is crucial for applications requiring the utmost in sensitivity such as cell tracking. Although the magnet is inherently a projection imager, we produce tomographic images by mechanically rotating the magnet on a gantry. Both projection and tomographic images are fully quantitative. For ease of siting, we have designed an integrated copper shield enclosure that provides >90 dB of electromagnetic isolation in the detection band without requiring a shield room.

Table 1: System Design Parameters.

Design parameter	Value
Magnet Configuration	Field Free Line
Magnet / Imaging Free Bore	12 cm / 4.5 cm
Animal models	Mouse & Rat
Selection Field	5.5 T/m x 5.5 T/m
Drive Field	20 mTp x 20 mTp
Drive Field Frequency	45 khz
Sensitivity	<10 ng/voxel
Field Generation	Electromagnet
Power Dissipation	30 kW
RF Shielding	Integrated
Footprint (Magnet + support)	<12 m^2

MATERIAL AND METHODS Systems were designed in-house with solid modeling and magnetics modeling software. Parts were fabricated with local and international vendors.

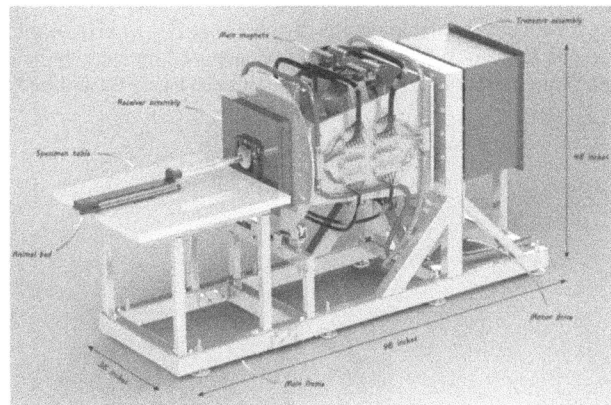

Figure 1: MPI imager rendering showing major components, including self-shielding, main magnet, and rotating gantry.

RESULTS Shown in Fig. 1 is the system overview of the MPI imager showing the main magnet and supporting equipment, water cooling, etc.

CONCLUSION The design and construction of a pre-clinical MPI imager suitable for commercialization is a substantial undertaking. Here we have presented the design overview of a self-shielded MPI imager that can be easily sited in core imaging facilities. This design will offer the enablement of MPI technology with the cost and feature set expected by the pre-clinical community.

ACKNOWLEDGEMENTS We thank the Conolly laboratory and UC Berkeley for their unwavering support of our commercialization efforts of technology developed there. Research reported in this publication was supported by NIBIB and NIDA of the National Institutes of Health under award numbers R43EB020463 and R43DA041814, respectively.

REFERENCES
[1] Zheng, et al. Scientific Reports, 5, 14055, 2015.
[2] Kaul, et al. *RoFo : Fortschritte auf dem Gebiete der Rontgenstrahlen und der Nuklearmedizin.* 187(5):347-352, 2015.
[3] Ferguson, et al. *IEEE Trans. Med. Imag.* **99**, 1 – 9, 2014. doi: 10.1109/TMI.2014.2375065
[4] Weizenecker et al. *J. Phys. D: Appl. Phys.* 41 105009. 2008. Doi: 10.1088/0022-3727/41/10/105009
[5] Konkle, et al. *IEEE Trans. Med. Imag.* vol.32, 2, 338-347, Feb. 2013. doi: 10.1109/TMI.2012.2227121

Modular mobility MPI system

Sebastian Draack [a,*], Christian Kuhlmann [a], Thilo Viereck [a], Frank Ludwig [a], Meinhard Schilling [a]

[a] Institut für Elektrische Messtechnik und Grundlagen der Elektrotechnik, TU Braunschweig, Germany
* Corresponding author, email: s.draack@tu-bs.de

INTRODUCTION Mobility Magnetic Particle Imaging (mMPI) is a new functional imaging approach which includes information about the binding state of the nanoparticles [1]. To examine the mobility of the nanoparticles in an MPI [2] imaging process one can use two different drive field frequencies f_l and f_h where ideally f_l is lower and f_h higher than the Brown-Néel-transition frequency f_t. We have designed the system to be comparable to commercially available devices by choosing $f_h = 25$ kHz as our second frequency. It was decided to construct a modular system platform in which individual components can easily be replaced.

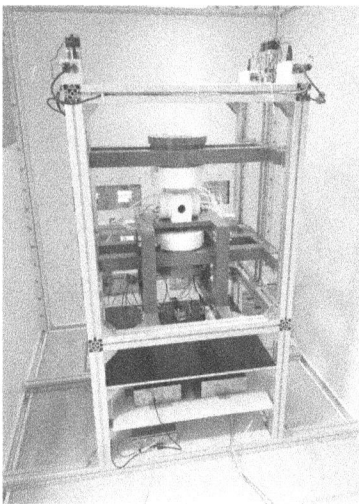

Figure 1: Modular mobility MPI system in commissioning state.

MATERIAL AND METHODS A new modular MPI system was developed operating at two frequencies $f_l = 10$ kHz and $f_h = 25$ kHz to demonstrate mMPI and study other new imaging approaches (see Fig. 1). The maintenance-free selection field was realized by two NdFeB N45H permanent magnets. The maximum gradient strength in main direction is $G_z = 6.72$ T/m and $G_x = G_y = 3.08$ T/m in the horizontal imaging plane. The magnets can be shifted independently of each other by a mechanical positioning system via geared high-torque step motors. Thus, one is able to adjust the magnetic gradient strength or shift the field free point (FFP) along the z-direction. This allows a mechanical slice selection and thus a slow 3-D imaging option by stacking several 2-D images.

The water-cooled drive field coils are installed in a modular housing manufactured from Polyamid 6. The drive field coil in x-direction was realized as a solenoid including space for a modular detection coil carrier with a bore-diameter $d_b = 35$ mm. A Helmholtz coil pair forms the drive field coils in y-direction. Water as a cooling medium increases the self-capacitance of a coil dramatically because of its dielectric constant. Care has been taken to obtain high self-resonance frequencies in order to keep thermal losses to a minimum. The nominal isotropic field of view (FOV) in the fast horizontal imaging x-y-plane covers an area of 20×10 mm^2 for a gradient strength of $G_{x,y} = 3$ T/m. The drive field coils are connected to a series resonant circuit with sinusoidal currents at $I_x = 17.69$ A and $I_y = 29.31$ A peak, which correspond to $B_x = 30$ mT and $B_y = 15$ mT. At this operating point, power losses of about $P_x \approx 110$ W and $P_y \approx 260$ W were calculated, the effective power dissipation depends on the pulse duration of the operating mode.

The water-cooled drive field coil system includes thermal reserves, so a larger FOV could be covered by increasing the drive field amplitudes.

RESULTS Both imaging axes have been put into operation using Lissajous trajectory frequencies $f_x = 25.0$ kHz and $f_y = 25.64$ kHz. Our preliminary results show that the imaging process is possible in both axes. First images in x and y directions were acquired to demonstrate the operability of our new system (see Fig. 2).

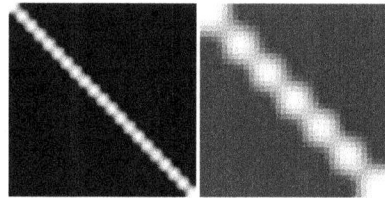

Figure 2: First 1-D images of Resovist calibration samples acquired with our new system at a gradient strength of $G_{x,y} = 1$ T/m, using a 2 mm grid.
left: x-axis, 22 mT$_{pk}$ drive field, $f_x = 25.0$ kHz, 2 mm Resovist calibration sample, image height corresponds to 40 mm.
right: y-axis, 12.5 mT$_{pk}$ drive field, $f_y = 25.64$ kHz, 3 mm Resovist calibration sample, image height corresponds to 12 mm.

CONCLUSION We developed a 2-D MPI scanner operating at $f_l = 10$ kHz and $f_h = 25$ kHz to demonstrate mMPI improving our previous MPI system design [3]. The new system has the ability to vary the position of the NdFeB permanent magnets mechanically independently of each other. Thus, our new system features an adjustable gradient and mechanical slice selection. We successfully commissioned both imaging axes at drive field frequencies around 25 kHz. Next steps are improvements regarding SNR as well as the acquisition of 2-D mMPI images.

ACKNOWLEDGEMENTS Financial support by the German Research Foundation DFG (LU 800/5-1) is acknowledged.

REFERENCES
[1] T. Wawrzik, C. Kuhlmann, F. Ludwig, and M. Schilling. International Workshop on Magnetic Particle Imaging, IWMPI 2013, Berkeley, CA art.no. 6528372, *IEEE XPlore*, 2013. doi: 10.1109/IWMPI.2013.6528372.
[2] B. Gleich and J. Weizenecker. *Nature*, 435(7046):1217—1217, 2005. doi: 10.1038/nature03808.
[3] M. Schilling, F. Ludwig, C. Kuhlmann, and T. Wawrzik, *Biomed. Tech.* 58(6):557—563, 2013. doi: 10.1515/bmt-2013-0014.

Bimodal TWMPI-MRI hybrid scanner – first MRI results

Peter Klauer[a*], Eberhard Rommel[a], Patrick Vogel[a], Martin A. Rückert[a], Volker C. Behr[a]

[a] Experimental Physics V, University of Würzburg, Würzburg, Germany
[*] Corresponding author, email: peter.klauer@physik.uni-wuerzburg.de

INTRODUCTION MPI (Magnetic Particle Imaging [1]) uses the nonlinearity of superparamagnetic nanoparticles and the fact that the magnetization saturates at sufficiently high magnetic fields. With the property of only detecting the tracer but not the background of any tissue comes the drawback of requiring a complementary imaging technique. The concept for an integrated MPI–MRI-hybrid system was presented in [2]. In [3] the first successful combination of an MPI with an MRI (Magnetic Resonance Imaging) system was demonstrated. A higher level of integration was achieved in [4] where the TWMPI scanner [5] was modified to do MRI measurements as well. This abstract presents the first MRI results of the TWMPI-MRI hybrid scanner.

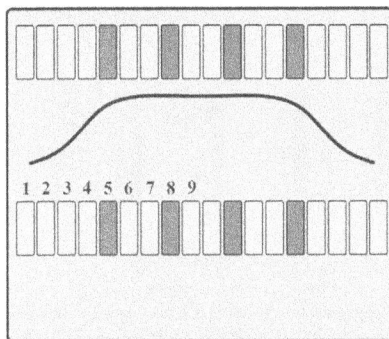

Figure 1: Schematic of the dLGA. Yellow (bright) coils are used for B_0 generation. The red (dark) coils are not necessary. They are used for generating a magnetic field gradient.

MATERIAL AND METHODS Starting point for the TWMPI-MRI-hybrid scanner was the TWMPI scanner. For generating the strong magnetic field gradient required for MPI, 20 individual copper coils were used (= dLGA (dynamic Linear Gradient Array). The idea was to change the current distribution in these coils (fig. 1) and to create a homogenous magnetic field (B_0) with a field strength of 235 mT. This field was sufficiently stable to perform NMR measurements (spin echo). Because not all of the dLGA coils are necessary for the B_0 generation, the additional coils can be used for spatial encoding in one direction. The objective in both approaches was the separation two samples.

RESULTS Four coils of the dLGA (dark coils in fig. 1) were used to create a linear field gradient along B_0. It was firstly tested whether two oil samples can be separated by frequency encoding. For this purpose a saddle coil with an inner diameter of 5 mm was used. The two 8 mm long oil samples were placed inside the saddle coil with a distance of 12 mm. In fig. 2 the spectra with and without frequency encoding are shown. It is possible to separate the two samples. The same four coils of the dLGA were also used for phase encoding. This time a solenoid coil with an inner diameter of 10 mm served as receive coil. The sample was a 10 mm NMR glass tube filled with oil and a 3 mm gap in the center to create two compartments. 64 phase steps and 4 averages

were acquired. The raw data was corrected for errors in B_0 and phase.

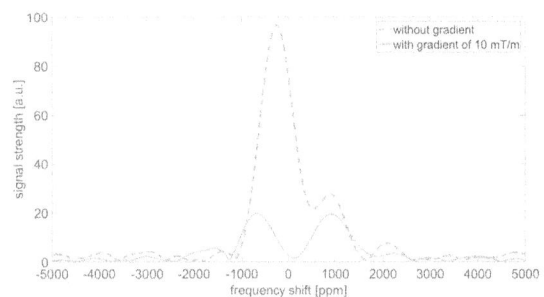

Figure 2: Spectra of two oil samples with and without frequency encoding.

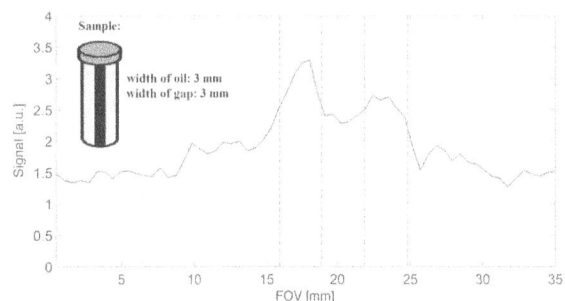

Figure 3: Profile along phase encoding direction of two oil samples with a length of 3mm and a distance of 3 mm.

CONCLUSION The introduced system is based on a regular TWMPI scanner [5]. By only changing the current distribution in the dLGA also MRI is possible with the same setup. The next step will be to apply MPI and MRI in one sample and to add another set of gradient coils for 2D encoding.

ACKNOWLEDGEMENTS This work was partially funded by the DFG (BE 5293/1-1).

REFERENCES

[1] B. Gleich and J. Weizenecker. "Tomographic imaging using the nonlinear response of magnetic particles", *Nature*, 435(7046):1217-1217, 2005. doi: 10.1038/nature03808.

[2] J. Franke et al. "First hybrid MPI-MRI imaging system as integrated design for mice and rats: description of the instrumentation setup", *Proc. IWMPI*, Berkeley, 2013 (IEEE – IWMPI 2013). doi: 10.1109/IWMPI.2013.6528363

[3] P. Vogel and S. Lother et al. "MPI meets MRI", *IEEE TMI*, 33(10):1954-1959, 2014. doi: 10.1109/TMI.2014.2327515

[4] P. Klauer et al. "Bimodal TWMPI-MRI Hybrid Scanner - Coil Setup and Electronics", *Proc. on IWMPI*, Berlin, 2014. doi: 10.1109/TMAG.2014.2324180

[5] P. Vogel, et al. "Traveling Wave Magnetic Particle Imaging", *IEEE TMI*, 33(2):400-407, 2013. doi: 10.1109/TMI.2013.2285472.

Studies on the Optimization of Efficient Selection and Focus Field Coil Configurations

Julia Mrongowius, Christian Kaethner*, Thorsten M. Buzug*

Institute of Medical Engineering, Universität zu Lübeck
* Corresponding author, email: {kaethner, buzug}@imt.uni-luebeck.de

INTRODUCTION Magnetic Particle Imaging (MPI) is an imaging method that is capable of detecting the distribution of superparamagnetic iron oxide particles due to their nonlinear magnetization behavior [2]. In order to specify the region, where a signal can be detected, a selection field can be used that features a field free point (FFP). Due to technical and medical limitations, the FOV that can be covered by such an FFP is very limited [5]. However, an enlargement of the FOV is possible by use of focus fields [2,7]. Instead of using separate electromagnetic coils to generate the selection and the focus field, a combined SeFo coil assembly that generates both fields can be used [1]. A power efficient concept for a combined SeFo coil configuration was presented for a 1D scenario in [4] and for multiple dimensions in [3]. The concept combines a classical Maxwell coil configuration and a single-sided coil setup [6] in one coil assembly. Both works focused primarily on an efficient transition from preclinical developments to human sized MPI systems [1]. Such an upscaling of an MPI scanning device and the involved electromagnetic coils leads to an immensely increased electrical power loss for the generation of a selection and focus field.

In this work, different simulation studies are performed that focus on the optimization possibilities for efficient SeFo field coil arrangements. In this context, a curved rectangular coil featuring a radial curvature (see Fig. 1) is introduced to allow for a further reduction of the electrical power loss.

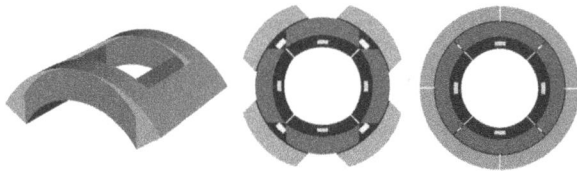

Figure 1: (*left*) Visual comparison of a conventionally (dark) and radially (light) curved rectangular coil. (*middle/right*) Front view on the used reference setup (middle) and setup 1.4 (right).

MATERIAL AND METHODS The optimizations are based on a reference setup, which was presented in [3] and can be seen in Fig. 1 from a frontal point of view. The optimization is divided into two parts, where in the first the volume of the coil assembly with respect to the reference setup is kept constant. In the second study the volume is a variabale parameter. Note, that in this contribution only the results of the first study are shown. Over all optimization steps, the bore diameter is kept constant at a value of 40cm. The calculations are done in FOV with the size of 28×28 cm², that is discretized in 101×101 points. The gradient strength is given by 1.5 $Tm^{-1}\mu_0^{-1}$ in y direction and 0.75 $Tm^{-1}\mu_0^{-1}$ in x and z direction.

RESULTS In Fig. 2, an overview over the maximal electrical power consumption of each optimization step is given. An example visualization of the optimized coil configuration of setup 1.4 is depicted in Fig. 1.

Closing the gaps between the coils on the outer ring in the first step and redistributing the volume reduction in the second step results in an electrical power loss of 117.08 kW. In the third step, the radially curved coil geometry was introduced and embedded in the reference setup. In the last step, the side length of the coils on the outer ring again is increased. The reduction of the length of all coils, in terms of keeping the same volume, results with 71.73 kW in the lowest power loss of the presented setups.

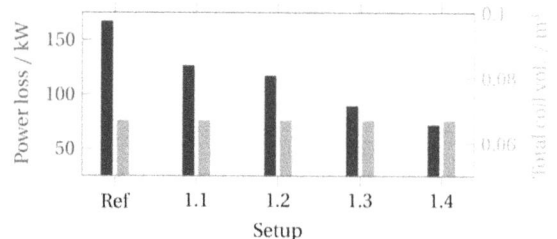

Figure 2: Setups of the optimization based on a constant coil volume, both power loss and total coil volume are depicted.

CONCLUSION In this optimization study, the electrical power loss of an efficient SeFo coil arrangement has been reduced by changing its geometrical assembly. A drastical reduction is possible, when cylindrical patient access is fully enclosed by the electromagnetic coils in combination with radially curved rectangular coils are used. As a result of the optimization study considering a constant coil volume, a reduction of the electrical power loss by 57% from 166.95 kW to 71.73 kW is possible.

REFERENCES

[1] C. Bontus, B. Gleich, B. David, O. Mende, and J. Borgert. *IEEE Trans. Magn.*, 51(2):6502004, 2015. doi: 10.1109/ TMAG.2014.2326003. doi: 10.1109/ TMAG.2014.2326003.

[2] B. Gleich and J. Weizenecker. *Nature*, 435(7046):1214– 1217, 2005. doi: 10.1038/nature03808.

[3] C. Kaethner, M. Ahlborg, T. Knopp, T. F. Sattel, and T. M. Buzug. *J. Appl. Phys.*, 115(4):044910–1–044910–5, 2014. doi: 10.1063/1.4863177.

[4] T. Knopp, T. F. Sattel, and T. M. Buzug. *IEEE Magn. Lett.*, 3:6500104, 2012. doi: 10.1109/LMAG.2011.2181341

[5] E. U. Saritas, P. W. Goodwill, G. Z. Zhang, and S. M. Conolly. *IEEE Trans. Med. Imag.*, 32(9):1600–1610, 2013. doi: 10.1109/TMI.2013.2260764

[6] T. F. Sattel, T. Knopp, S. Biederer, B. Gleich, J. Weizenecker, J. Borgert, and T. M. Buzug. *J. Phys. D: Appl. Phys.*, 42(1):1–5, 2009. doi: 10.1088/0022-3727/42/2/022001

[7] I. Schmale, J. Rahmer, B. Gleich, J. Kanzenbach, J. D. Schmidt, C. Bontus, O. Woywode, and J. Borgert. In *SPIE Medical Imaging*, 2011. doi: 10.1117/12.877339

Magnetic Particle Imaging by Using Multichannel Coil Arrays

Shu-Hsien Liao[a], Jen-Jie Chieh[a], Herng-Er Horng[a], Hong-Chang Yang[a], Saburo Tanaka[b]

[a] Institute of Electro-optical Science and Technology, National Taiwan Normal University, Taipei 116, Taiwan
[b] Department of Environmental and Life Sciences, Toyohashi University of Technology, Toyohashi, Aichi 441-8580, Japan
[*] Corresponding author, email: shliao@ntnu.edu.tw

INTRODUCTION Magnetic particle imaging was introduced to an noninvasive method for 3-dimensional imaging tool for tracing the distribution of magnetic particle in vivo. However the proposed method should not only applying a high strength AC magnetic field to induce the third harmonic magnetic signal form the nonlinear magnetized magnetic nanoparticles but also applying three strong gradient fields and three strong magnetic fields to construct a small field-free point (FFP) and scan the FFP for imaging. The strength of gradient field should be several T/m for a high spatial resolution about several mm. It limits the imaging size and the cost of high power supplies for strong scanning magnetic fields is expensive. Here, we propose a new magnetic particle imaging method by using pickup coil array and a strong ac exciting field instead of the gradient fields and scanning field. This method reaches the real time imaging and low cost and promise for the industrialization in the new future.

MATERIAL AND METHODS The schematic of 16 channel magnetic particle imaging system set up in a 3-layer Aluminum shielded box is shown in figure 1. This system consists by an ac magnetic field to exciting the third harmonic magnetization signal from magnetic particle. The pick-to-pick strength of the ac field is 100 Oe at a frequency of 3.1 kHz. Therefore the third harmonic signal would be with a frequency of 9.3 kHz. The pick-up coil array is consisted by 16-channel axial gradiometer. The outside diameter of each coil is 10 mm with a baseline of 13 cm for canceling the environmental noise and reduces the artifact from the exciting magnetic field. The 16-channel pick-up coils form a four by four array with the separation of 13 mm. The sample of magnetic fluid was put inside the excitation coil and measures the magnetizing signal by pick-up coil array to obtain the magnetic field distribution. After measurement, a minimum-norm estimation method was applied for reconstructing the magnetic particle imaging.

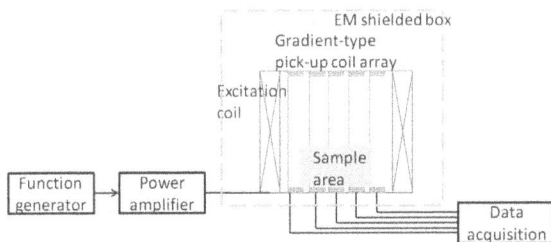

Figure 1 The schematic of 16 channel magnetic particle imaging system.

RESULTS The sample of magnetic fluid and the arrangement was shown in figure 2(a). The magnetic fluid was sealed in a L- sharp container and put inside the magnetic particle imaging system for detecting the field distribution. The reconstructed 3-D imaging was obtained by MNE analysis shown in figure 2(b). The reconstructed imaging shows a clear imaging with high consistence with the origin sharp. Beside, our MPI system also perform a real-time imaging as shown in figure 3. The moving trace of the sample and the corresponding images were shown in figure 3. The moving sample could be imaging with time. The temporal resolution of our MPI system is 160 ms.

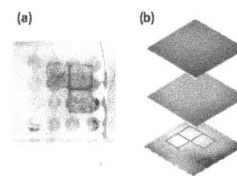

Figure 2 The 3-D magnetic particle imaging for a "L" sharp magnetic fluid sample.

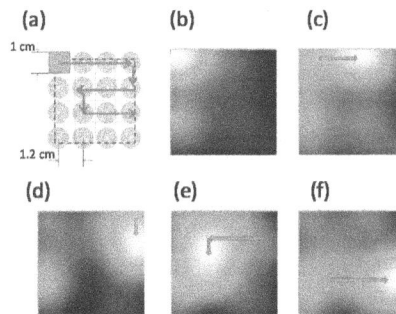

Figure 3 The real-time imaging of magnetic fluid.

CONCLUSION Here we demonstrate the 16 channel magnetic particle imaging system for imaging magnetic fluid. The minimum-norm estimation method is applied for resolving the source localization of magnetic nanoparticles. In this system we realize not only 3-D imaging but also real-time imaging. It shows the feasibility to construct a larger system to enlarge the imaging area for biomedical application in the future.

ACKNOWLEDGEMENTS This work is supported by the National Science Council of Taiwan under grant number: NSC 103-2420-H-003 -006-, NSC 103-2923-M-003 -002 -, NSC 102-2120-M-168-001- and by "Aim for the Top University Plan" of the National Taiwan Normal University and the Ministry of Education, Taiwan, R.O.C under grant number 104J1A27 and 105J1A27

REFERENCES
[1] B. Gleich, and J. Weizenecker, Nature 435, 1214-1217, Jun. 2005.
[2] J. Rahmer, J. Weizenecker, B. Gleich, and J. Borgert, IEEE Trans. Med. Imaging, vol. 31, no. 6, pp. 1289-99, Jun. 2012.
[3] J. Rahmer, J. Weizenecker, B. Gleich, and J. Borgert, BMC Med. Imaging, vol. 9, no. 1, p. 4, Jan. 2009.

Designing coils to minimize the maximal induced electrical field amplitude in a patient

Gael Bringout[a,*], Johan Löfberg[b], Patricia Ulloa[a], Martin A. Koch[a], Thorsten M. Buzug[a]

[a] Institue of Medical Engineering, Universität zu Lübeck, Germany
[b] Department of Electrical Engineering, Linköping University, Sweden
[*] Corresponding author, email: bringout@imt.uni-luebeck.de

INTRODUCTION Peripheral nerve stimulation (PNS) due to the electrical field $\vec{E}$ induced in the patient body limits the increase of the drive field amplitude and thus the imaging field of view (FOV) and speed [1-4]. In order to maximize the drive field amplitude, the minimization of the normalized maximal induced electrical field amplitude (nMIEFA) [5] is done during the design of x-, y- and z-drive coils.

MATERIAL AND METHODS The coil design can be formulated as a problem minimizing the dissipated power P_{QCQP} and constraining the field quality [6], typically called quadratically constrained quadratic program (QCQP). To minimize the nMIEFA, a Second-Order Conic Program (SOCP) problem is introduced and formulated as

$$\min_{\vec{s}} \max_{k}(\|\vec{E_k}(\vec{s})\|)$$
$$\text{subject to: } P_{SOCP} \le n\, P_{QCQP}$$

with $\vec{s}$ the surface current density, $k = 1, ..., K$ with K the number of considered surface elements forming the patient model, P_{SOCP} the dissipated power for the solution and n an integer equal to 2, 4, 6 or 8. Additionally, the same constraint on the field quality is applied as the one used for the QCQP problem. All problems are solved using the YALMIP toolbox [7] and the MOSEK (7.0, Mosek ApS, Denmark) solver.

The induced electrical field is evaluated according to [8].

A cylindrical surface with a diameter of 0.640 m and a length of 0.4 m is discretised in 400 triangles and is used as support for the coil.

The surfaces of two patients are modelled either from photos or via the segmentation of the body/air interface from a whole body MRI acquired with a voxel size of 0.8x0.8 mm², a slice thickness of 5 mm and a distance between slices of 5.5 mm. The mesh coming from the MRI acquisition is segmented using 3DSlicer [9]. Both meshes are then post-processed using Meshmixer (v10.7.84, Autodesk Inc, USA) to obtain fully connected meshes.

RESULTS The nMIEFA on 2 human models are calculated at 22 different positions for 3 magnetic field topologies und 5 different optimisation problems resulting in 660 optimisations. The results for the optimisation of y-drive coils in relation with the MRI-based model is presented in Fig. 1. Four surface current densities are shown in Fig. 2, which illustrates the variation of the coils.

CONCLUSION An nMIEFA reduction of up to 50 % for a given position can be expected for a coil optimised to minimize it. This result is patient and position dependent. The technical difficulties associated with such coils may outweigh the reduction of PNS and the increase of FOV.

ACKNOWLEDGEMENTS The authors gratefully acknowledge the financial support of the German Federal Ministry of Education and Research (BMBF, grant number 13N11090) and of the European Union and the State Schleswig-Holstein (Programme for the Future – Economy, grant number 122-10-004).

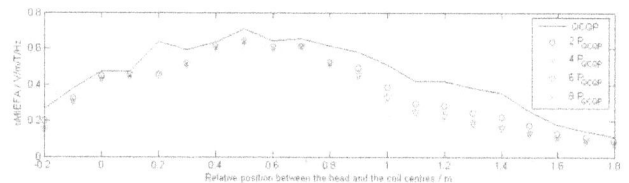

Figure 1: nMIEFA for 5 y-drive coil optimisations on the mesh based on an MRI acquisition of a human subject.

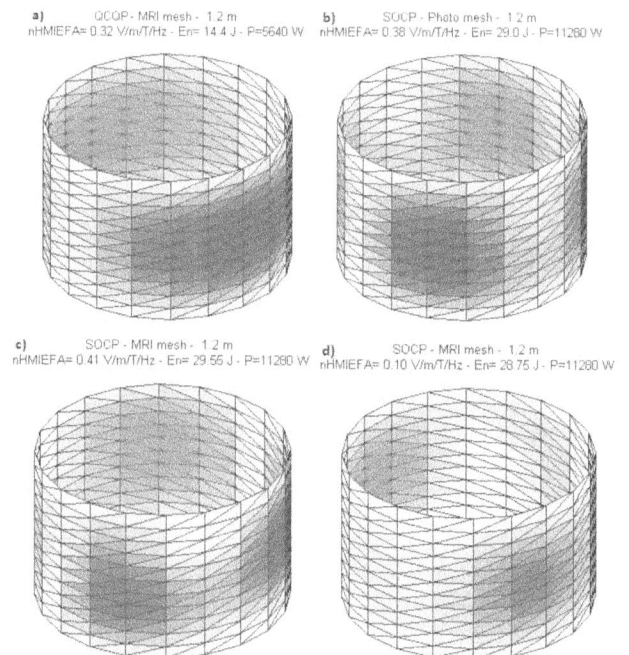

Figure 2: Surface current density for y-drive coils obtained from QCQP (a) and SOCP (b, c, d) problems at position 1.2 m (a, b, c) or 0.5 m (d) from photo (b) and MRI (a, c, d) based meshes.

REFERENCES

[1] J. Bohnert. KIT Scientific Publishing, Karlsruhe, 2011.
[2] G. Bringout et al.. *IWMPI*, 2012. doi: 10.1007/978-3-642-24133-8_57.
[3] I. Schmale et al.. *IWMPI*, 2013. doi: 10.1109/IWMPI.2013.6528346.
[4] E. U. Saritas et al.. *IEEE Transactions on Medical Imaging*, 32(9):1600-1610, 2013. doi: 10.1109/TMI.2013.2260764.
[5] G. Bringout and T. M. Buzug. 2015 *5th International Workshop on Magnetic Particle Imaging (IWMPI)*. doi: 10.1109/IWMPI.2015.7107077.
[6] G. Bringout and T. M. Buzug. *IEEE Transactions on Magnetics*, 51(2), 2014. doi: 10.1109/TMAG.2014.2344917.
[7] J. Löfberg. *2004 IEEE International Symposium on Computer Aided Control Systems Design*, 2004. doi:10.1109/CACSD.2004.1393890.
[8] C. C. Sanchez et al.. *Physics in Medicine and Biology*, 55:3087-3100, 2010. doi: 10.1088/0031-9155/55/11/007.
[9] A. Fedorov et al.. *Magnetic Resonance Imaging*, 30(9):1323-1341, 2012. doi: 10.1016/j.mri.2012.05.001.

Novel Selection Coils Design for 3D FFL-based MPI

Alexey Tonyushkin

Department of Physics, University of Massachussettss Boston, Boston, Massachusetts 02125, USA
email: alexey.tonyushkin@umb.edu

INTRODUCTION Magnetic Particle Imaging (MPI) is a new noninvasive medical imaging modality. The two different types of the magnetic gradient geometries are: field-free-point (FFP) [1] and field-free-line (FFL) [2]. The FFL-based device could potentially produce better image quality than FFP-based one at the same nanoparticles concentration [2]. However, from a technical point of view, creation of a required high strength magnetic gradient with FFL, which is capable to encode 3D volume is challenging thus limiting the expected resolution of such devices. Here we offer a novel robust design of the selection coils with FFL that could overcome the major challenges of FFL MPI scanners.

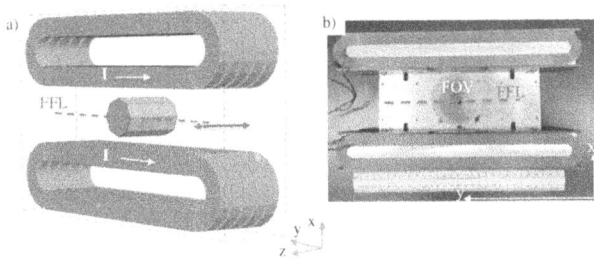

Figure 1: Selection coil array with FFL and cylindrical FOV in the middle: a) five elements array concept, b) two elements prototype. FFL (shown by dashed line) is rapidly translated along z-axis; the current I is shown by arrows on the coils.

MATERIAL AND METHODS In our design the selection coils are based on the electromagnets that are arranged in symmetrical pairs (see Fig. 1). The coils are separated by the distance d, which defines in-plane FOV. The FFL is created at the symmetry axis of the pair of coils along y-axis with the magnetic gradient scaled as $\sim 1/d^2$. The coil pairs are arranged in linear array to enable lateral (z-axis) rapid translation of the FFL with oscillating pattern. The current I in each pair of coils in the array is independently controlled according to the desired FFL translation pattern as a function of time. For example, if coil pair one has a maximum current $I_1 = I_{max}$ and all the other pairs have zero current the FFL will be located in the symmetry axis between the first pair of coils. If the current switched so that $I_1 = 0$ and $I_2 = I_{max}$ then the FFL advances along z-axis to the symmetry axis of the second pair of coils. The number of coil pairs in the array, the gap between them, and the thickness of the conductor define the lateral FOV that has no fundamental constraints. The in-plane image encoding is done by projection imaging. Specifically, the coil array is mechanically rotated around z-axis up to 180°. If we simultaneously oscillate the current in the array and rotate the array we can, for example, encode 3D volume.

RESULTS The design and simulations are based on mathematical modeling using Mathematica (Wolfram) software and Radia package (ESRF, France) and experimental analysis of a scaled prototype coil structures [3].

The simulations of proof-of-principle 30cm-long selection coils that consist of five elements array each with N=40 turns of a copper tape are shown in Fig.2. With small animal FOV d=4cm and I_{max}=100A the magnetic gradient reaches 2.5 T/m and very flat FFL within FOV. Such five-element array provides Δz=3cm FFL translation corresponding to FOV=3cm along z-axis (see Fig.3).

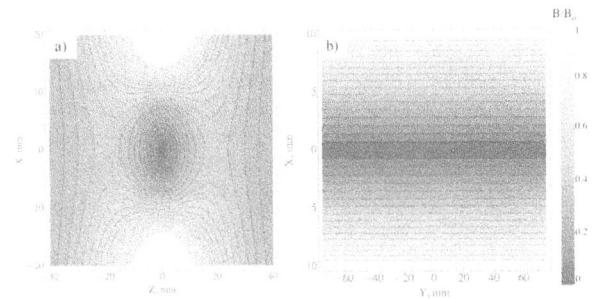

Figure 2: Normalized magnetic field contour plots for a) XZ and b) XY planes. Here, the current is on for one set of coils.

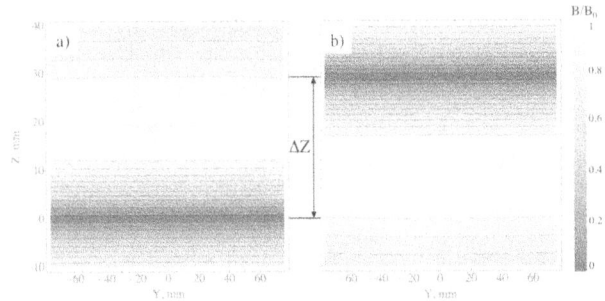

Figure 3: Normalized magnetic field contour plots showing $\Box$z=3cm translation of FFL along z-axis by oscillating the current in the array between a) the first and b) the last sets of coils.

CONCLUSION We presented a novel design of the selection coils with FFL that allow 3D spatial encoding without additional drive coils and provides relatively large FOV and field gradient with very flat FFL.

ACKNOWLEDGEMENTS We would like to thank Prof. M. Prentiss from Harvard University for lending the equipment.

REFERENCES
[1] B. Gleich and J. Weizenecker. *Nature*, 435(7046):1217—1217, 2005.
[2] J. Weizenecker, B. Gleich and J. Borgert, *J. Phys. D: Appl. Phys.* 41 105009, 2008.
[3] A. Tonyushkin and M. Prentiss. *J. Appl. Phys.* 108, 094904, 2010.

Evaluation of the spatial confidence and dual modal FOV-center conformity of a highly integrated MPI-MRI hybrid system

Jochen Franke [a,b,*], Ulrich Heinen [a], Alexander Weber [a,c], Heinrich Lehr [a], Michael Heidenreich [a], Wolfgang Ruhm [a], Volkmar Schulz [b]

[a] Bruker BioSpin MRI GmbH, Preclinical Imaging Devison, Ettlingen, Germany
[b] Physics of Molecular Imaging Systems, University RWTH Aachen, Aachen, Germany
[c] Institute of Medical Engineering, University of Lübeck, Germany
[*] Corresponding author, email: jochen.franke@bruker.com

INTRODUCTION MPI is a tracer-based imaging modality [1] which is highly sensitive to the non-linear SPIO response. The acquired data exhibit a high signal-to-noise-ratio (SNR) with no background signal generated by e.g. biological tissue. This, however, requests for additional morphological information. In [2] this was achieved by two stand-alone MPI and Magnetic Resonance Imaging (MRI) systems. Reducing co-registration errors induced by object transportation and/or repositioning, hybrid MPI-MRI scanner designs were presented that allow for sequential data acquisition [3-6]. In this study, we assess the spatial confidence and the FoV-center conformity of the hybrid system [3,6] by a dual-modal phantom experiment.

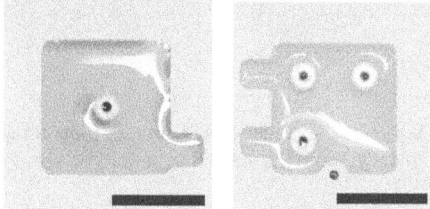

Figure 1: Photograph of the two layer FOV center evaluation phantom. The black scale bar in each of the right lower corner corresponds to a length of 10 mm.

MATERIAL AND METHODS: A generic multi-purpose multi-modality phantom [7] with two non-symmetric inlays (c.f. Fig. 1) was prepared. The two compartments of each layer were filled either with T1 standard fluid ($CuSO_4$ $5 \cdot H_2O$) containing 1% agarose or diluted Resovist (Bayer Schering Pharma AG, Germany) with an iron concentration of 0.083 mol(Fe)/ l. In the assembled phantom, the fluid layers are separated by a 3.5 mm gap. Using a highly integrated MPI-MRI hybrid system [3,6], the phantom was scanned sequentially within the same study using ParaVision6 (Bruker BioSpin MRI, Germany) in the MRI- and the MPI-mode without object movement. **1) MRI mode:** Prior to the high resolution MRI data acquisition, an automatic FID based B_0 field shimming algorithm which adjusts the 1st and 2nd order shim coil sets was performed including a re-adjustment of the basic frequency. A high resolution 3D spin echo sequence (MSME, TE/TR=32.01/500 ms, echo spacing=5.82 ms, echo averaging=10, averaging=10, scan time=14.5 min, matrix=48×48×12, FOV=24×24×12 mm³, BW=50 kHz) was performed. **2) MPI mode:** Prior to MPI data acquisition ($A_{DFx,y,z}$=12 mT, A_{SFmax}=2.2 T/m, BW=625 kHz, repetitions=40, scan time=862 ms), the modality transition was performed. For the system function based image reconstruction using ParaVision6, the 40 acquired repetitions were used for averaging while a Kaczmarz iterative algorithm [8] with 10 iterations, a relative regularization of $\lambda=10^{-7}$ resulted in one 3D dataset (matrix=28×28×16, FOV=28×28×16 mm³). **3) Data fusion:** Using the fusion toolbox of PMOD 3.6 (PMOD Technologies Ltd, Switzerland), both MRI and MPI 3D datasets were manually rigid co-registered to each other.

RESULTS Neither scaling nor rotation was necessary for hybrid MPI-MRI data fusion. A global translations (left-right (x): -2.5 mm, cranial-caudal (z): -1 mm, dorsal-ventral (y): -1.5 mm) led to a perfect match of both 3D datasets. The fused and translated 3D MPI (light grayscale) and 3D MRI (dark grayscale) phantom data can be seen in Fig. 2.

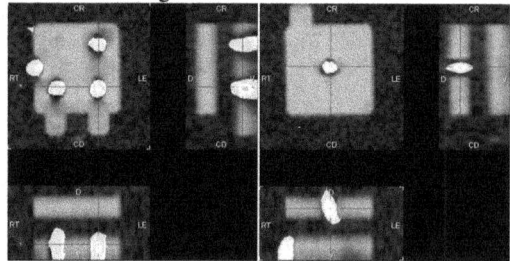

Figure 2: Two different orthogonal cross-sections of the fused and manually co-registered 3D MPI (light grayscale) and 3D MRI (dark grayscale) data of the FOV center evaluation phantom.

CONCLUSION Neither scaling nor rotation was used to co-register both complementary datasets. To fully align both datasets, a global rigid translation in all three directions was determined manually. In the MPI mode the spatial fidelity is dependent on a precise robot movement during the system function acquisition, while in the MRI mode the MRI gradient strength calibration fidelity is the crucial factor. The spatial match of the imaging centers is dependent on the alignment accuracy of the robot coordinate system origin with the MRI gradient iso-center. The latter alignment turned out to have a significant impact onto the residual mismatch of both datasets. Thus, these global rigid translation coefficients can be expected to be stable in time as long as the MRI datasets are acquired with a recently re-calibrated basic frequency. For future experiments, this offset can be omitted by aligning the robot coordinate system with the MRI gradient iso-center prior to a new system function acquisition.

ACKNOWLEDGEMENTS The authors thankfully acknowledge the financial support by the German Federal Ministry of Education and Research, FKZ 13N11088.

REFERENCES
[1] B. Gleich and J. Weizenecker. *Nature*; 10.1038/nature03808 [2] J. Weizenecker et al. Physics in Medicine and Biology, 10.1088/0031-9155/54/5/L01. [3] J. Franke et al. Proc. IWMPI 2013; 10.1109/IWMPI.2013.6528363 [4] J. Franke et al. Proc. IWMPI 2013; 10.1109/IWMPI.2013.6528367 [5] P. Vogel et al. IEEE TMI, 10.1109/TMI.2014.2327515.[6] J. Franke et al. Proc. IWMPI 2015; 10.1109/IWMPI.2015.7106990 .[7] U. Heinen et al. Proc. IWMPI 2015; 10.1109/IWMPI.2015.7107033 [8] S. Kaczmarz. Bull. Internat. Acad. Polon. Sci. Lett., 35:355357, 1937.

Metallic artefact suppression in intraoperative magnetometers

Sebastiaan Waanders[a,*], Rogier Wildeboer[a], Erik Krooshoop[a], Bennie ten Haken[a]

[a] MIRA Institute for Biomedical Technology and Technical Medicine, University of Twente, The Netherlands
* Corresponding author, email: s.waanders@utwente.nl

INTRODUCTION In addition to their use as contrast agents or tracers in imaging modalities, superparamagnetic nanoparticles are increasingly used in novel biomedical applications. One of these emerging applications is the sentinel lymph node procedure, which is traditionally applied using a radiocolloid and a blue dye. Because of the logistical issues surrounding the practical application of radioisotopes in the clinic, a magnetic alternative is being developed by different groups[1-2], using different approaches for the intra-operative detection of the injected nanoparticles. Conventional methods relying on SQUID magnetometry or AC magnetometry are limited in their applicability because of their sensitivity for external noise sources, for example the diamagnetic nature of the tissue surrounding the tracer particles, which limits their maximum attainable resolution. Differential Magnetometry or DiffMag is a variation on the AC magnetometry process that exploits the strong nonlinear magnetization characteristics of the SPIONs, resulting in a signal that is independent of the background and is selective. In addition to the problem of diamagnetism, conventional magnetometers are severely limited in their applicability in a standard surgeon's operating theatre because of the ubiquity of metallic instruments employed in the procedure. Here, we describe how DiffMag suppresses not just the diamagnetic signal from the surrounding tissue, but is also surprisingly effective at suppressing the signal artefacts originating from the use of metallic surgical equipment.

MATERIALS AND METHODS A handheld magnetometer was developed, suitable for intra-operative use. This sensor can be operated in both conventional AC magnetometry mode and DiffMag mode, allowing for an accurate comparison of the two detection schemes. The magnetometer consists of a solenoidal excitation coil with a coaxial gradiometer detection coil pair, wound on a body machined out of a ceramic composite material. The excitation coil is driven by a purpose-built power amplifier (ServoWatt 24V 2A continuous, 4A peak output), connected by a shielded cable. To accommodate for the strict hygiene requirements posed by the surgical environment in which the probe is to operate, the entire probe body is enclosed in a Delrin shell. The probe is connected to a base unit for signal conditioning and data acquisition. Signals are generated and subsequently processed in MATLAB®. For signal processing, phase sensitive detection is employed, using the excitation current as a reference, after which DiffMag processing (described in [3]) is performed. The probe can operate in simple AC magnetometry mode at 2.5 kHz, or in DiffMag mode, where we apply a quasi-periodic offset field to the alternating field to modulate the SPION magnetization. Because of the differential nature of the DiffMag protocol, any quasi-constant baseline may be subtracted from the signal, without interfering with the measurement result. This allows for dynamic rebalancing of the probe to improve the dynamic range of the setup. This is achieved by means of a compensation coil wound around the detection gradiometer, allowing for extra magnetic flux coupling without interfering with the DiffMag measurement. Signal intensities were obtained for static metallic artefacts (surgical steel retractor placed at 4cm from the probe) and small lymph node phantoms containing Resovist®, and compared for AC and DiffMag modes.

RESULTS Application of AC magnetometers close to metallic objects results in a huge inductive signal on the magnetometer, originating from eddy currents induced in the metallic object due to the probe's excitation field. Figure 1 shows obtained signal intensities from the probe when measuring a sample of SPIONs, a human hand held under the probe, and a metallic surgical retractor placed under the probe. In AC mode, the signal from the metallic object is significantly larger than that of the SPION sample, and we clearly see a negative signal for the human hand. When the probe is operated in DiffMag mode, a strong attenuation of both the diamagnetic tissue signal and the metallic object is observed, which can be attributed to the low nonlinearity of the retractor material's magnetization curve in the field region we're measuring in, and the linear nature of the diamagnetic tissue, which contrasts with the strong nonlinearity of the SPIONs.

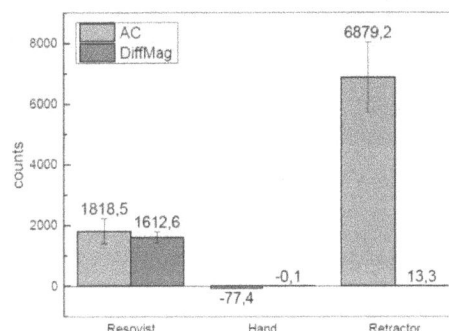

Figure 1: Signal intensities obtained for both AC and DiffMag operation modes, measuring a 250µg Resovist® sample (a), a human hand (b) and a surgical steel retractor (c)

CONCLUSION In addition to the lack of selectivity of conventional magnetometry schemes used to detect SPIONs *in vivo*, operation of these devices in clinical situations where metallic objects are abundant is very limited. We show that signals from mockup lymph nodes containing SPIONs are easily obfuscated by the induced noise generated by metallic objects present around the surgical area. Differential Magnetometry improves both SPION detection limits and the attenuation of metallic artefacts, which clears the way for succesful implementation of these techniques in clinical practice.

REFERENCES
[1] M. Douek, J. Klaase, I. Monypenny, A. Kothari, K. Zechmeister, D. Brown, et al., Ann. Surg. Oncol. 21 (2013) 1237–45. doi:10.1245/s10434-013-3379-6.
[2] K. Imai, Y. Kawaharada, J. -i. Ogawa, H. Saito, S. Kudo, S. Takashima, et al., Surg. Innov. 22 (2015) 401–405. doi:10.1177/1553350615585421.
[3] M. Visscher, S. Waanders, H.J.G. Krooshoop, B. ten Haken, J. Magn. Magn. Mater. 365 (2014) 31–39. doi:10.1016/j.jmmm.2014.04.044.
[4] S. Waanders, M. Visscher, T. Oderkerk, B. Haken, 2013 Int. Work. Magn. Part. Imaging.

Systematic Background Estimation

Marcel Straub[a,*], Bernhard Gleich[b], Jürgen Rahmer[b], Volkmar Schulz[a,b]

[a] Department of Physics of Molecular Imaging, Institute of Experimental Molecular Imaging, RWTH Aachen University, Aachen, Germany
[b] Philips Research Europe, Germany
* Corresponding author, email: marcel.straub@pmi.rwth-aachen.de

INTRODUCTION Magnetic Particle Imaging (MPI) as well as other imaging modalities such as Fluorescent Molecular Tomography (FMT) suffer from systematic background [1]. Usually one measures the background signal just before/after the measurement of the actual sample. However, for a drifting background, as it is given for our MPI scanner [2], this is not sufficient. Hence, it is crucial to measure the background about once every minute to maintain a high image quality. As this might be feasible for examining non-living probes this is not feasible for examining humans, as it requires the full removal of the subject from the field of view (FOV).

In this paper, we are going to present a method to estimate the systematic background from two consecutive volumes with displaced image information.

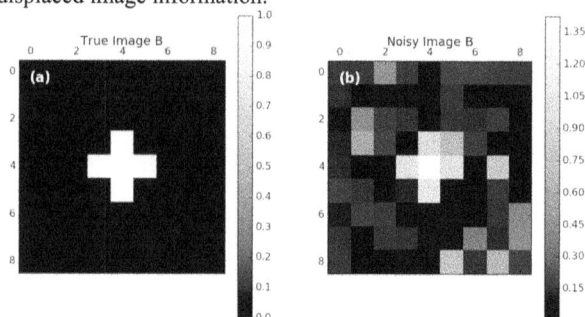

Figure 1: (a) True noise-free image used for the simulation study. (b) Image (a) with added systematic background b and Gaussian noise s_i^t.

MATERIAL AND METHODS The main idea of our method is, that it is possible to estimate the systematic background from a set of two given images, which are recorded consecutively and are shifted by a well-defined offset. Furthermore, it is required, that the shift between the two images does not shift the background itself. One can shift the object between two images by physically shifting it, e.g. via a robot, or by superimposing a homogeneous field slightly shifting the FOV [3].

We use a least-square fit to estimate the common systematic background of both images. T. Chen et.al. describe a similar method for FMT in 2006 [1]. The cost function for the two images $x_0 \in \mathbb{R}^{n \times m}$ (not shifted image) and $x_1 \in \mathbb{R}^{n \times m}$ (shifted image), an image shifting operator Δ, and the systematic background $b \in \mathbb{R}^{n \times m}$ is given by:

$$\| \Delta(x_0 - b) - (x_1 - b) \|^2 \to \min$$

To show the feasibility of this method we test the algorithm with synthetic data. We generate two noise free images x_0^t and x_1^t with superimposed white noise s_0^t and s_1^t ($s_0^t \neq s_1^t$) to which we add the same systematic background b^t. Hence, both images ($i = 0, 1$) constructed of three components:

$$x_i = x_i^t + s_i^t + b^t$$

All images have a size of 9×9 pixels. The pixels of the images x_i^t that belong to the cross are set to 1.0 (cf. Fig. 1a) all other pixels are set to 0.0. The systematic noise b^t (cf. Fig. 2a) is generated by a uniform random number generator with a pixel value in the range of $[0.0, 0.8)$. The additional statistic noise s_i^t is uniformly distributed in the range of $[0.0, 0.2)$. Fig. 1b exemplarily shows image x_1. The objective function is minimized with the BFGS algorithm without preinitialization.

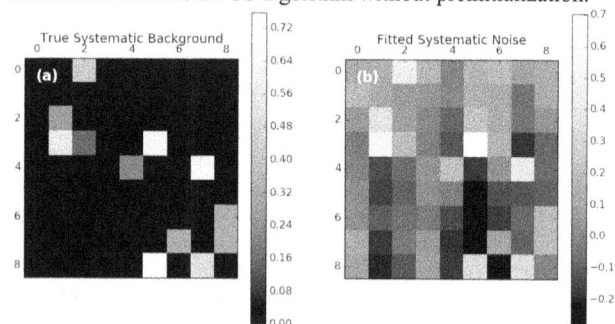

Figure 2: (a) Generated true systematic noise b^t. (b) Systematic noise b determined by minimizing the objective function.

RESULTS Fig. 2b shows the fitted systematic background b of the generated noisy images x_0 and x_1 (cf. Fig. 1b). The systematic background corrected image of x_1 is shown in Fig. 3.

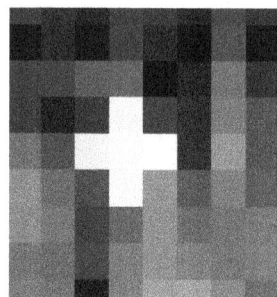

Figure 3: Noisy image x_1 (cf. Fig. 1b) corrected for the fitted background b (cf. Fig. 2b).

CONCLUSION Fig. 2a and b show that our approach is capable of identifying the main features of the systematic background. However, due to the fitting algorithm as well as the additional statistical noise s_i^t, some blurring might become visible. On the other hand, the background corrected image (cf. Fig. 3) is visibly improved over the uncorrected one (cf. Fig. 1b)

ACKNOWLEDGEMENTS The authors would like to thank Philips for their financial support of the Ph.D. position of Marcel Straub and for donating us their first MPI scanner.

REFERENCES

[1] T. Chen, B. Lin, E. Brunner and D. Schild. *Biophysical Journal*, 90: 2534–2547, 2006. doi: 10.1529/biophysj.105.070854
[2] B. Gleich and J. Weizenecker. *Nature*, 435(7046):1217—1217, 2005. doi: 10.1038/nature03808.
[3] J. Weizenecker, B. Gleich, J. Rahmer, H. Dahnke and J. Borgert. *Phys. Med. Biol.*, 54: L1-L10, 2009. doi: 10.1088/0031-9155/54/5/L01

Controlling the Position of the Field-Free-Point in Magnetic Particle Imaging

A. Weber[a,b*], J. Weizenecker[c], R. Pietig[d], U. Heinen[a], T.M. Buzug[b]

[a] Bruker Biospin MRI GmbH, Rudolf-Plank-Str. 23, 76275 Ettlingen, Germany
[b] Institute of Medical Engineering, University of Lübeck, Ratzeburger Allee 160, 23562 Lübeck, Germany
[c] University of Applied Sciences, Moltkestr. 30, 76133 Karlsruhe, Germany
[d] Bruker Biospin GmbH, Wikingerstr. 13, 76189 Karlsruhe, Germany
* Corresponding author, email: alexander.weber@bruker.com

INTRODUCTION In Magnetic Particle Imaging, the center of the field-of-view (FOV) is determined by the field-free-point (FFP) of the gradient field. Since the size of the FOV is limited, additional focus fields are necessary to shift the FOV to the desired region-of-interest or to cover a large FOV in a multipatch approach [1]. Due to intrinsic inhomogeneity of the magnetic fields, the FFP is not shifted linearly by a linear increase of the focus field currents. Furthermore, due to the presence of soft-magnetic shielding in the Bruker Preclinical MPI System, the magnetic field strength of the Selection and Focus fields deviate from a strictly linear current dependency and exhibit cross correlations. Thus, to control the actual position of the FFP, an exact model of the Selection and Focus fields and their current dependencies has to be established.

MATERIAL AND METHODS Every component of a static magnetic field B can be described by

$$B_i(r,\theta,\varphi) = \sum_{l=0}^{\infty} \sum_{m=-l}^{l} c_{lm}^i \, r^l Y_{lm}(\theta,\varphi), \quad i = x,y,z \quad (1)$$

a series of spherical harmonics Y_{lm} [2]. Hereby Y_{00} describes the homogeneous part and Y_{1m}, $m = -1,0,1$, the gradient parts. The spherical harmonics with higher order correspond to the inhomogeneity of the magnetic field component. For typical fields, the coefficients c_{lm}^i quickly decrease with increasing order and a series truncation is possible. To model the current dependencies of the magnetic field, the coefficients c_{lm}^i are considered current dependent and Eq. 1 is extended to

$$B_i(\boldsymbol{I},r,\theta,\varphi) = \sum_{l=0}^{\infty} \sum_{m=-l}^{l} c_{lm}^i(\boldsymbol{I}) \, r^l Y_{lm}(\theta,\varphi), \quad i = x,y,z \quad (2)$$

where the current vector $\boldsymbol{I} = (I_{SF}, I_{FFX}, I_{FFY}, I_{FFZ})^T$ includes the currents of the Selection and the three Focus fields. By performing the spherical harmonic decomposition for several different current settings, a suitable model for the current dependency of the coefficients $c_{lm}^i(\boldsymbol{I})$ can be calibrated. Using this model, the required currents for shifting the FFP to $(r_{FFP}, \theta_{FFP}, \varphi_{FFP})$ with a target gradient of g_{FFP} at the FFP can be determined from a system of nonlinear equations that can be solved by the Newton method.

The presented measurements have been performed at the Bruker/Philips PreClinical MPI Scanner.

RESULTS Fig.1 shows the coefficients of a spherical harmonics series of B_x for $I_{SF} = 300A$ and $I_{FFX} = 150A$. Notable levels of the 2nd order coefficients indicate non-negligible magnetic field inhomogeneity. However, with increasing order, the coefficients are quickly decreasing, allowing proper modeling by a truncated series.

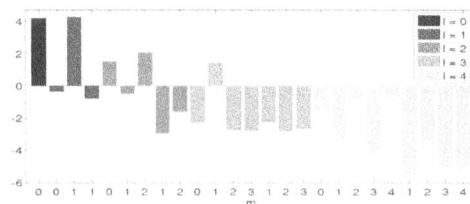

Figure 1: Spherical harmonics coefficients in a logarithmic scale of B_x for $I_{SF} = 300A$ and $I_{FFX} = 150A$.

As an example, the current dependency is presented by the I_{SF} dependency of the x-gradient of B_x. A linear model would lead to a gradient overestimation. With a 2nd order polynomial model a good approximation can be found.

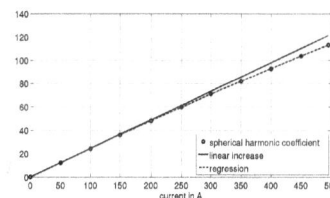

Figure 2: Dependency of I_{SF} of the x-gradient of B_x.

With the help of this model, the current setting could be predicted to shift the FFP to the desired position with high accuracy.

CONCLUSION An appropriate model of the Selection and Focus fields has been established. This model enables a precise positioning of the FOV, which is a requirement of an accurate coverage of large FOVs.

ACKNOWLEDGEMENTS The authors thankfully acknowledge the financial support by the German Federal Ministry of Education and Research, FKZ 13N11088.

REFERENCES
[1] T. Knopp et al, *Phys. Med. Biol.*, 60:L15, 2015.
[2] G. B. Arfken, Academic Press, San Diego, 2001.

Force analysis device for magnetic manipulation

David Weller, Thorsten M. Buzug and Thomas Friedrich*

Institute of Medical Engineering, University of Lübeck, Germany
* Corresponding author, email: {friedrich,buzug}@imt.uni-luebeck.de

INTRODUCTION As various other fields, medical engineering often has to deal with navigation and placement problems. Considering the treatment of vascular diseases by means of a catheter for example, the surgeon has to take the right turns inside the vasculature. In order to track the instrument during treatment, additional imaging techniques – like X-ray fluoroscopy are required. For MPI [1,2], tracer material needs to be delivered to specific locations inside the patient's body. For both scenarios, a method that allows for the tracking and the navigation of an instrument simultaneously would be beneficial. As previous work [3] shows, the idea of using an MPI scanner to magnetically navigate an object is not new. However, the choice of the right materials and an optimized design has the potential to improve the usability significantly. This means, a quantitative method for analyzing the magnetic force acting on a specific object while subjected to the magnetic fields generated by an MPI scanner, is desirable. In the presented work, we describe a magneto mechanical setup for this purpose.

SETUP The inhomogeneous field in between two permanent magnets oriented antiparallel is used to generate a field-free point. By adjusting the distance of the magnets, the gradient strength can be controlled. On the one side of a lever of 60 cm in length, the sample to analyze is mounted. The other side of this lever is connected to an electronic balance, which is capable of resolving 0.1 mg. The position of the sample is controlled by moving the supporting point of the lever up and down by means of a translation stage. *Figure 1* depicts the setup.

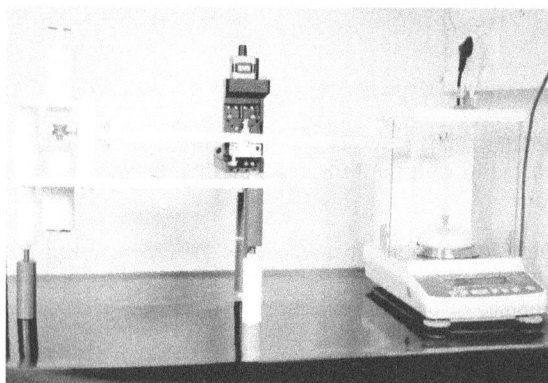

Figure 1: Photographic picture of the setup, with permanent magnets on the left side, translation stage in the center, and balance on the right side.

A holder, made of nonmagnetic plastic (3D-printed PLA), holds the sample in position. A magnetic force, acting on the sample, changes the reading of the balance. As the length ratio of both sides of the lever can be measured easily, this change of the weight reading is directly connected to the magnetic force acting on the sample. To account for the small influence of the lever-rotation, an empty measurement in absence of any sample is subtracted.

In order to determine the magnetic force as a function of the sample position, a computer program controls the stepper motor of the translation stage to change the sample position in equidistant steps. For each position, the reading of the balance is recorded via an RS232 connection.

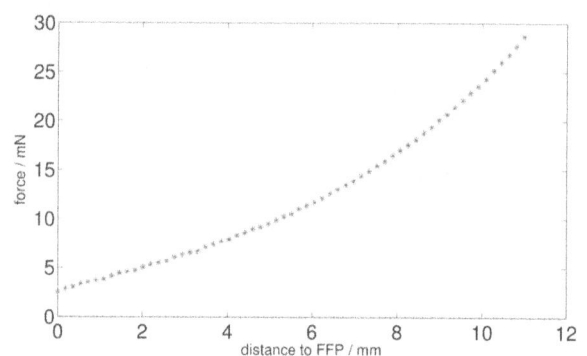

Figure 2: Magnetic force acting on a sphere of stainless steel with a diameter of 6 mm versus the distance to the field-free point.

RESULTS Objects of different shape, made of various ferro- and superparamagnetic materials are compared with respect to the magnetic force they experience for different magnitudes of the field gradient. *Figure 2* shows an example for the force acting on a sphere of stainless steel with a diameter of 6 mm.

CONCLUSION The requirements for the sample, in order to achieve good actuation differ from those for MPI imaging. This means, that careful studies have to be carried out to find appropriate objects for both, actuation and imaging. For this task, the described device is a convenient addition to an MPI scanner for the force analysis.

ACKNOWLEDGEMENTS This work is supported by the German Federal Ministry of Education and Research under grant number BMBF 13GW0069A (SAMBA PATI).

REFERENCES
[1] B. Gleich and J. Weizenecker. *Nature*, 435(7046):1217—1217, 2005. doi: 10.1038/nature03808.
[2] T. Knopp and T. M. Buzug. Springer, Berlin/Heidelberg, 2012. doi: 10.1007/978-3-642-04199-0.
[3] N. Nothnagel et. al., International Workshop on Magnetic Particle Imaging (IWMPI) 2013. doi:10.1109/IWMPI.2013.6528358.

Study of temperature measurement on pn-junction of light-emitting diodes using magnetic nanothermometer

Zhongzhou Du[a], Kai Wei[b], Rijian Su[a], Yong Gan[a], and Wenzhong Liu[b,*]

[a] School of Computer and Communication Engineering, Zhengzhou University of Light Industry, Zhengzhou 450002, China
[b] School of Automation, Huazhong University of Science and Technology, Wuhan, 430074, China
* Corresponding author, email: lwz7410@hust.edu.cn

INTRODUCTION Magnetic nanothermometer (MNT) has attracted more and more attention for the advantages of noninvasive temperature probing in industrial research field, e.g. temperature probing inside LEDs. In 2005, B.Gleich and J.Weizenecker proposed a new method that tomographic imaging can be obtained by using the nonlinear response of magnetic particles [1], which was a pioneering research. Weaver, J.B [2] and Wenzhong, L. [3, 4] reported a method of temperature measurement using the thermal-sensitive characteristic and nonlinear response of magnetic nanoparticles. In 2014, the scientific group led by Wenzhong L. has developed the thermometer prototypes [5]. MNT is of great significance to industrial and biomedical applications.

Light-emitting diodes (LEDs) are well known as the next generation of green light sources for the advantages of energy saving, environmental protection, long life and low lost, over incandescent and fluorescent lamp [6]. However, the high pn-junction temperature lead to performance degradation of the LEDs, e. g. high pn-junction temperature can shorten the device lifetime, reduce the device reliability, and accelerate the deviation of light wavelength. The high pn-junction temperature is important factor to hinder development of the LEDs. The accurate and true data on the pn-junction temperature of the LEDs was basis and prerequisite for decreasing the pn-junction temperature of the LEDs.

In this study, we report pn-junction temperature of LEDs can be obtained by using MNT. MNT may provide an efficient tool for engineers to improve the design of LEDs to decrease the influence of temperature to LEDs performance.

METHODS The first-order Langevin function describing the superparamagnetism of the magnetic nanoparticles (MNPs) was described as [1-5, 7]

$$M = NM_s \left(\coth\left(\frac{M_s V H}{kT}\right) - \frac{kT}{M_s V H} \right)$$

Where N is the volume fraction of MNPs, Ms is the saturation magnetization, V is the particle's volume, k is the Boltzmann constant, T is the absolute temperature of MNPs, H is the external excitation magnetic field. The nonlinear response of the MNPs in an AC time-varying magnetic field contains the 1st, 3rd, 5th and other odd harmonics. The 1st and 3rd harmonic amplitudes are substituted for the temperature by inversion algorithm [3-5, 7].

EXPERIMENTS AND RESULTS The new LEDs should be fabricated before measuring the junction temperature. A single layer of MNPs (EMG1300, Ferrortec, Inc.) was placed between the LED-chip layer and chip-fixing layer. The fabricated LEDs was putted in the test zone, and the DC voltage to light up LEDs was 5.0V, 5.1V, and 5.2V, respectively. When the pn-junction temperature of LEDs became stable, we turned off the LEDs. The experimental results were shown as Figure 1. When the LEDs was lighted up, the pn-junction temperature of LEDs increased gradually from room temperature (298 K) and became constant eventually. The highest pn-junction temperatures of LEDs were 304 K, 314 K, and 327 K at 5.0 V, 5.1 V, and 5.2 V, respectively. Then the LEDs was turned off, the pn-junction temperature of LEDs decreased greatly at first and then decreased to room temperature eventually.

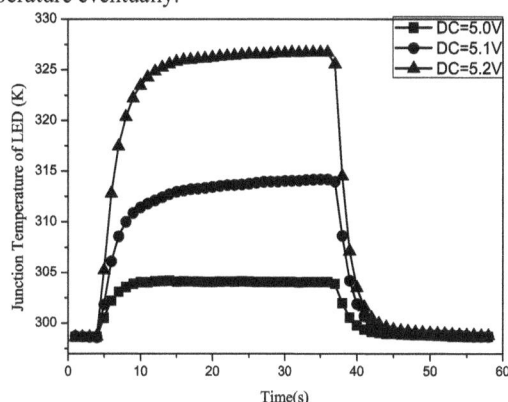

Figure 1: Experiment results on pn-junction temperature of LEDs measured by MNT under different DC voltages.

CONCLUSION The pn-junction temperature of LEDs was measured by using MNT, and the MNT may provide an accurate and true data for engineers to decrease the pn-junction temperature of the LEDs. Furthermore, the MNT not only can be used to measure the pn-junction temperature of LEDs but also can be used to others, including phosphors temperature of LEDs, IC (circuit chip) packages, and IGBT (insulated gate bipolar transistor) packages.

ACKNOWLEDGEMENTS This work was supported by the National Natural Science Foundation of China (61374014, 61571199, 81501547) and the Science and Technology Program of Henan, China (132102210056).

REFERENCES
[1] B. Gleich and J. Weizenecker. Nature, 435(7046):1217—1217, 2005. doi: 10.1038/nature03808.
[2] Rauwerdink A M, Hansen E W, Weaver J B. Physics in medicine and biology,54(19): L51,2009. doi:10.1088/0031-9155/54/19/L01.
[3] Zhong J, Liu W, Du Z, et al. Nanotechnology, 23(7): 075703, 2012. doi: 10.1088/0957-4484/23/7/075703.
[4] Zhong J, Liu W, Kong L, et al. Scientific reports, 4: 6338, 2014. doi:10.1038/srep06338.
[5] Du Z, Su R, Liu W, et al. Sensors, 15(4): 8624-8641, 2015. doi: 10.3390/s150408624.
[6] Narukawa Y, Narita J, Sakamoto T, et al. Japanese Journal of Applied Physics, 45(10L): L1084, 2006. doi: 10.1143/JJAP.45.L1084.
[7] T. Knopp and T. M. Buzug. Springer, Berlin/Heidelberg, 2012. doi: 10.1007/978-3-642-04199-0.

Elevator speeches 1:

Applications

In vitro MPI iron quantification of labeled cells for a metastasis-tracking study

Vera Paefgen [a,*], Marcel Straub [b], Fabian Kießling [a], Volkmar Schulz [b,c]

[a] Institute of Experimental Molecular Imaging, RWTH Aachen University, Aachen, Germany
[b] Department of Physics of Molecular Imaging, Institute of Experimental Molecular Imaging, RWTH Aachen University, Aachen, Germany
[c] Philips Research Europe, Germany
[*] Corresponding author email: vpaefgen@ukaachen.de

INTRODUCTION Magnetic Particle Imaging (MPI) is a non-invasive and highly sensitive tracer-based imaging modality [1]. Since the MPI signal is not attenuated by the surrounding tissue, those tracers can be used to label cells *ex vivo* and then administered to the target. The distribution, proliferation and accumulation of the labeled cells can be measured *in vivo* after injection. We develop a labeling protocol for 4T1 cells with Resovist® and a reliable iron concentration quantification *in vitro* with MPI as a comparison to photometric measurements. Later on, we follow up tumor and metastasis development *in vivo* with MPI quantification.

MATERIAL AND METHODS Murine 4T1 cells were grown in RPMI medium with 10% FCS and 1% Pen/Strep at 37°C and 5% CO_2. For Resovist® (Bayer Schering) labeling, cells were seeded in 6- or 24-well plates. A growth medium with 1/5/10 µl Resovist® ± 1.5 µg Poly(L-lysine) (PLL) per ml was incubated for 1h under constant rolling conditions. The cells were then incubated with the different media for 48 h and washed 3 times. For MPI measurements, the cells were trypsinized and centrifuged. The supernatant was removed and the samples placed in the MPI scanner. The MPI scanner was operated with a high gradient of 5.5 T/m and drive field amplitudes of 20 mT. For image reconstruction a system matrix is recorded for Resovist® (27.9 mg_{Fe}/ml) with an overall volume of 22×20×5.6 mm. Signal to noise ratio (SNR) calculations have been done similar to J. Rahmer *et al.* [2]. To check iron uptake histologically, cells were stained with Prussian Blue (freshly prepared 1:1 mixture of 10% potassium ferrocyanide and 20% HCl) and microscopy images were taken. Photometric analysis of the cellular iron content was done with a TECAN Microplate Reader as described previously after preparation of a standard curve [3].

RESULTS In Fig. 1, Resovist® labeled 4T1 cells after Prussian Blue staining are shown. Except for Fig. 1A, intracellular (blue) particle accumulation can be seen, seemingly more with PLL added and increasing iron concentration. PLL also increased the amount of particles stuck to the outer cell membrane.

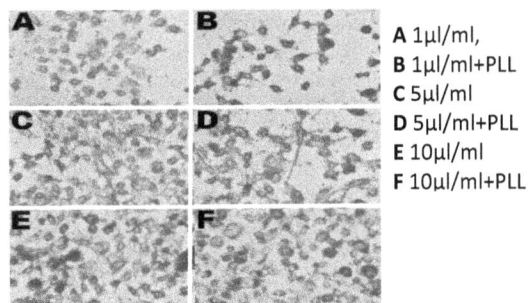

A 1µl/ml,
B 1µl/ml+PLL
C 5µl/ml
D 5µl/ml+PLL
E 10µl/ml
F 10µl/ml+PLL

Figure 1: Iron-labeled 4T1 cells after Prussian Blue staining

Fig. 2 shows the absolute amount of iron in cell pellets (white bars) as determined by MPI, as well as the iron concentration in SDS-lysed cells after photometric quantification (black bars). Both quantification methods highly correlated (r^2=0.9926).

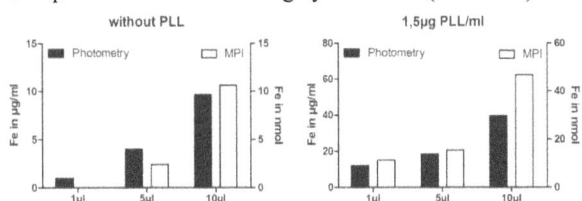

Figure 2: Iron quantification by MPI (white, Fe in nmol) and photometry (black, Fe in µg/ml) of cells +/- PLL

The high sensitivity of our MPI system can be seen in Fig. 3, where scans of cell pellets (1 and 5µl/ml+PLL) have been reconstructed. The quantification in Fig. 2 shows only little difference in the iron content, but the signal intensities after reconstruction differ notably.

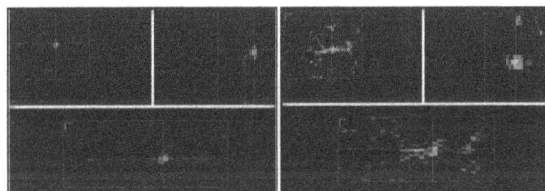

Figure 3: MPI reconstructions of cell pellets with 1µl/ml+PLL (left) and 5µl/ml+PLL (right).

CONCLUSION We successfully labeled murine 4T1 cells with Resovist®. Addition of PLL as a transfection agent increased the measured iron content in samples for photometric and MPI quantification. Strong differences in signal intensities can also be detected in MPI images after reconstruction.

ACKNOWLEDGEMENTS The authors would like to thank Philips for their financial support of the Ph.D. position of Marcel Straub and for donating us their first MPI scanner

REFERENCES
[1] B. Gleich and J. Weizenecker, "Tomographic imaging using the nonlinear response of magnetic particles.," *Nature*, vol. 435, no. 7046, pp. 1214–7, Jun. 2005.
[2] J. Rahmer, J. Weizenecker, B. Gleich, and J. Borgert, "Analysis of a 3-D System Function Measured for Magnetic Particle Imaging," *IEEE Trans. Med. Imaging*, vol. 31, no. 6, pp. 1289–1299, Jun. 2012.
[3] E. R. Dadashzadeh, M. Hobson, L. Henry Bryant, D. D. Dean, and J. A. Frank, "Rapid spectrophotometric technique for quantifying iron in cells labeled with superparamagnetic iron oxide nanoparticles: potential translation to the clinic," *Contrast Media Mol. Imaging*, vol. 8, no. 1, pp. 50–56, Jan. 2013.

MPI-Detection of Multicore Iron Oxide Nanoparticles dedicated for Magnetic Drug Targeting

Stefan Iyer[a,*], Tobias Knopp[b], Franziska Werner[b], Lutz Trahms[c], Frank Wiekhorst[c], Tobias Struffert[d], Tobias Engelhorn[d], Arndt Dörfler[d], Tobias Bäuerle[d], Michael Uder[d], Christoph Alexiou[a]

[a] Universitätsklinikum Erlangen, ENT-Department, Section of Experimental Oncology and Nanomedicine (SEON), Else-Kröner-Fresenius Stiftung-Professorship
[b] Universitätsklinikum Hamburg-Eppendorf, Diagnostic and Interventional Radiology Department and Clinic
[c] Physikalisch Technische Bundesanstalt Berlin
[d] Universitätsklinikum Erlangen, Department of Radiology/Neuroradiology
* Corresponding author, email: stefan.iyer@uk-erlangen.de

INTRODUCTION Superparamagnetic Iron Oxide Particles (SPIONs) offer a great potential for a variety of biomedical applications. Among those are in vitro diagnostics, drug delivery in different disease patterns like infections, arteriosclerosis and cancer and diagnostics. Due to their big surface SPIONS can bind and deliver drugs in high amounts. In addition to that, targeting of the diseased area can be achieved by either secondary surface modifications capable of recognizing molecular target structures, e.g. on cancer cells, or by magnetic fields, because of their magnetic properties. But SPIONs can also be used for imaging in magnetic resonance imaging (MRI) (Fig. 1) and are the tracers for the new imaging technology of magnetic particle imaging (MPI). Taking all these properties together, SPIONs can be seen as a unique platform for theranostic applications. Here, the great advantage of magnetic particle imaging is that SPIONs can be imaged and quantified. Therefore, with MPI it could be possible to estimate the drug load in the diseased area after the application by measuring the content of the SPIONs used for delivering the drug.

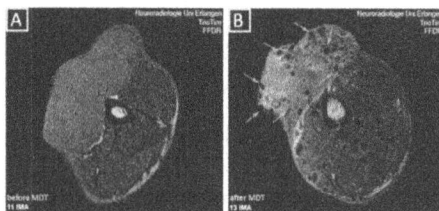

Figure 1: MRI imaging of SPIONs in a VX2-rabbit tumor. MRI-imaging before (A) and after (B) the accumulation of SPIONs in a tumor by Magnetic Drug Targeting. The signal extinction due to SPIONs is marked by green arrows.

The aim of the Section of Experimental Oncology and Nanomedicine (SEON) is to utilize SPIONs for the treatment of cancer and arteriosclerosis by MDT. Therefore, SEON over a period of several years developed SPIONs optimized for the purpose of magnetic drug delivery. These particles are very stable in human and animal blood, can carry a more than sufficient drug load [1], be accumulated in a target area by magnetic fields and first results show, that these particles even can be heated after magnetic accumulation in tumors. These SPIONs were also suitable for MRI-imaging but from the theoretical point of view not optimal for MPI, because they are clusters of a size between 60 nm and 70 nm with a single core diameter of approximately 7,6 nm.

Therefore, the aim of this preliminary study was to investigate the potential of these multicore nanoparticles for magnetic particle imaging.

MATERIAL AND METHODS SEON^{LA-BSA}-nanoparticles were synthesized according to [1]. In short, the nanoparticles were synthesized by coprecipitation of iron salts and stabilized by a dual coting with lauric acid and bovine serum albumin.

Subsequently, a dilution series was measured with Magnetic Particle Spectroscopy (MPS) at Universitätsklinikum Hamburg-Eppendorf and after that a sample of 20 µl (2 mm * 2mm * 1mm) was measured with MPI.

RESULTS The MPS-spectrum showed that in comparison to the MPS-signal of Resovist® the signal of the SEON^{LA-BSA}-nanoparticles was weaker at higher frequencies but comparable at frequencies below 100kHz (Fig. 2).

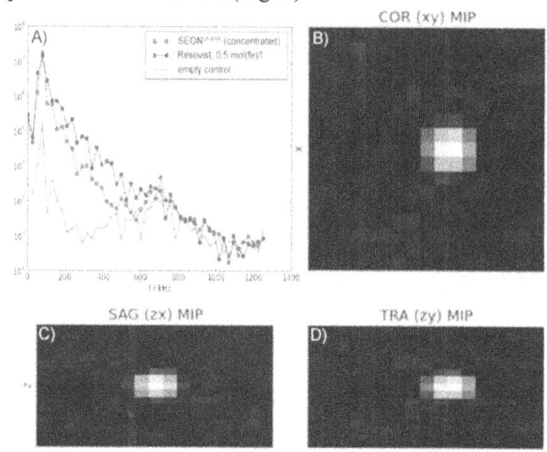

Figure 2: A) Comparison of the MPS signal of SEON^{LA-BSA}-nanoparticles and Resovist® at different frequencies. B) – D) MPI-signal of a point sample of SEON^{LA-BSA}-nanoparticles.

CONCLUSION The experiments showed that the MPS signal of SEON^{LA-BSA}-nanoparticles is comparable to that of Resovist® at frequencies below 100 kHz, if a concentrated sample is taken for the tests. The first MPI-measurements also showed that with this concentration a MPI-signal can be generated.

Further experiments now have to elucidate, if this signal is enough to measure the distribution of these nanoparticles in a tumor after their accumulation by MDT.

ACKNOWLEDGEMENTS The authors would like to thank the German Research Foundation (SPP1681) for financial support.

REFERENCES
[1] Zaloga J, Lee G, Alexiou C. Int J Nanomedicine. 2014 Oct 20;9:4847-66. doi: 10.2147/IJN.S68539

Different Behavior of MPI Signals from Magnetic Nanoparticles Internalized by Macrophages and Colon Cancer Cells

Hisaaki Suzuka, Atsushi Mimura, Yoshimi Inaoka, Kohei Nishimoto, Natsuo Banura, Kenya Murase*

Department of Medical Physics and Engineering, Division of Medical Technology and Science, Faculty of Health Science, Graduate School of Medicine, Osaka University, Osaka, Japan
* Corresponding author, email: murase@sahs.med.osaka-u.ac.jp

INTRODUCTION Tracking cells labeled with magnetic nanoparticles (MNPs) is one of the most promising applications of magnetic particle imaging (MPI) [1]. MNP-cell labeling combined with MPI will enable quantitative cell mapping in positive contrast even at deep organs, which is applicable to the evaluation of cell transplantation therapy. Besides, the original advantages of MNP-cell labeling such as long-term stability and cell guidance with strong magnets will also make cell tracking using MPI attractive. However, it has been known that the MPI signal is vulnerable to the factors in the MNPs' environment such as viscosity [2], chemical conjugation to fixated molecules, and cell uptake [3]. Although the preceding signal acquisition might normalize the affected signal before administration, the breaking of quantitativity can matter when the intracellular MNPs transfer to another environment. In addition, further investigations on the behavior of the MPI signals from the MNPs in cells are crucial for quantitative cell-tracking studies with *in situ* labeling. In this study, we investigated the difference in the behavior of the MPI signals from the MNPs internalized by two types of cells and its dependency on the labeling methods.

MATERIAL AND METHODS First, the samples containing carbodextran-coated MNPs (Resovist®) were prepared in the presence of heparin and/or protamine, both of which are protein vectors utilized to label cells with MNPs by different mechanisms. The MNPs were initially complexed with the various amounts of protamine (positive-charged protein) to get MNPs close to negative-charged cell surfaces by electrostatic interaction. The various amounts of heparin were then mixed with the MNPs to enhance the engulfment by cells. The third harmonic signals from the above samples, filled in 1.5 mL microtubes, were measured using magnetic particle spectroscopy (MPS). The hydrodynamic diameters and zeta potentials of the MNPs in Roswell Park Memorial Institute 1640 medium were also measured using the dynamic light scattering method.

Second, the quantitative performance of our MPI scanner [2, 4] was evaluated. The cylindrical phantoms (6 mm in diameter and 5 mm in length) filled with various concentrations of MNP solution were imaged using our MPI scanner [2, 4], and the relationship between the concentration of the MNPs and the pixel value of the MPI image was investigated.

In the cell-labeling studies, Raw264 macrophage cells were labeled with the MNPs by overnight co-culture at various MNP concentrations. Colon26 carcinoma cells were also incubated for 24 hours with the MNPs or MNP-protamine complexes in the presence or absence of heparin. The macrophages and cancer cells were washed three times with 10 mL phosphate buffered saline and collected with use of cell scrapers and trypsin-EDTE, respectively. The collected cells were suspended in the cylindrical phantom (100 µL) described above and were imaged using our MPI scanner. The number of cells within the sample was adjusted to 10^7. A region of interest (ROI) was drawn on the area corresponding to the section of the phantom on the MPI image, and the mean pixel value within the ROI was calculated. After the MPI studies, the absolute amount of iron in cells were determined by thiocyanate colorimetry. The behavior of the MPI signal from the MNPs internalized by cells was evaluated by normalizing the mean pixel value of the MPI image by the iron content in 10^7 cells. Finally, the labeled cells were lysed by 1% Triton X-100 and the signals from them were measured by MPS.

RESULTS In the MPS studies, the signal intensities from the MNPs mixed with protamine decreased with increasing concentration of protamine, whereas their hydrodynamic diameters increased gradually. In contrast, there was no significant change in both the MPS signal and hydrodynamic diameter in the samples supplied with only heparin. In spite of the drastic growth of the diameter of the MNPs mixed with protamine, their zeta potentials did not significantly change.

There was an excellent correlation (r=0.994) between the concentration of MNPs and mean pixel value of the MPI image, indicating that the quantitative performance of our MPI scanner is sufficient for evaluating the MNPs internalized by cells.

The macrophage cells were successfully labeled with the MNPs and the concentration of the MNPs in the cell culture medium significantly correlated with the mean pixel value of the MPI image (r=0.898). Both heparin and protamine enhanced the cellular uptake of the MNPs, resulting in the increase of the pixel value of the MPI image. When the MNPs were internalized by the cells, the mean pixel value of the MPI image normalized by the iron content significantly decreased compared to that of the MNPs in the cell culture medium. When the macrophage cells were used, this normalized pixel value was almost two times higher than the case when the colon carcinoma cells were used. These results were consistent regardless of the cell-labeling methods and culture conditions. The inhibited signals were partly recovered by the cell lysis using Triton X-100.

CONCLUSION Our results suggest that the behavior of the MPI signals from the intracellular MNPs would differ depending on the type of cells and might be characterized by the difference in the MPI pixel value.

ACKNOWLEDGEMENTS This work was supported by a Grant-in-Aid for Scientific Research from the Japan Society for the Promotion of Science.

REFERENCES
[1] B. Zheng, T. Vazin, P. W. Goodwill, A. Conway, A. Verma, E. U. Saritas, D. Schaffer, and S. M. Conolly. *Sci. Rep.*, 11(5):14055, 2015. doi: 10.1038/srep14055.
[2] K. Murase, R. Song, and S. Hiratsuka. *Appl. Phys. Lett.*, 104(25): 252409, 2014. doi: 10.1063/1.4885146.
[3] F. Wiekhorst, N. Löwa, W. Poller, S. Metzkow, A. Ludwig, and L. Trahms. *IWMPI Proc.*, 24, 2015. Doi: 10.1109/IWMPI.2015.7107009.
[4] K. Murase, S. Hiratsuka, R. Song, and Y. Takeuchi. *Jpn. J. Appl. Phys.*, 53(6):067001, 2014. doi: 10.7567/JJAP.53.067001.

Processing of SPIO in macrophages and tumor tissue for MPI lymph node imaging in breast cancer

Dominique Finas[a,*], Janine Stegmann-Frehse J[b], Benjamin Sauer[c], Gereon Hüttmann[c], Acim Rody[b], Thorsten Buzug[d], Kerstin Lüdtke-Buzug[d]

[a] Department of Obstetrics and Gynecology, Evangelical Hospital Bielefeld and University of Lübeck, Bielefeld and Lübeck Germany

[b] Department of Obstetrics and Gynecology, University of Lübeck, Lübeck, Germany

[c] Institute of Biomedical Optics, University of Lübeck, Lübeck, Germany

d Institute of Medical Engineering, University of Lübeck, Lübeck, Germany

[*] Corresponding author, email: finas.d@arcor.de

INTRODUCTION The most common cancer in women worldwide is breast cancer (BC). To avoid tissue damaging while axillary surgery in BC we aim to develop a new sentinel lymph node biopsy (SLNB) methode using superparamagnetic iron oxide nanoparticles (SPIOs) and magnetic particle imaging (MPI). It is well known from i.v. SPIO application in magnetic resonance imaging (MRI) that macrophages (MP) are key role player in processing of SPIOs (e.g. in liver) causing a drop of signal intensity. Nevertheless, knowledge lacks concerning enrichment processes of SPIOs after injection in breast tissue, the adjacent lymphatic tissues and associated cells, especially in BC and metastatic lymph nodes. Previously we evaluated the distribution of SPIOs in an in vivo healthy and tumor mouse model. Based on these studies we investigate the processing of the SPIOs in MP.

MATERIAL AND METHODS To evaluate SPIO processing we established a tumor bearing mouse model as we previously published[1]. Tumor tissue was than analyzed by conventionally hematocylin histology combined with Berlin blue staining and autofluorescence multiphoton microscopy. Additionally, the MP cell line J774A.1 was incubated either by Resovist in culture medium (RPMI, FBS), or culture medium only as control. This process was observed in vivo by multiphoton microscopy[2]. Detection of SPIOs was realized by excitation at 1200 nm.

RESULTS Resovist showed no toxic effects on macrophages after incubation. MP showed activity in phagocytosis of Resovist after incubation in multiphoton microscopy. SPIOs were detectable within tumor tissue by conventional hematoxylin microscopy as well as particle processing by 3-photon microscopy (see Fig. 1).

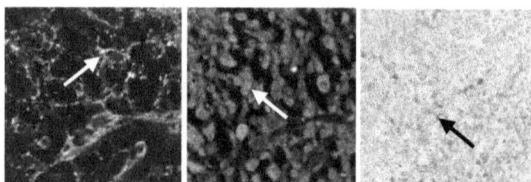

Figure 1: Tumor tissue with Resovist (arrows);
left: Resovist (3-photon microscopy, 1200nm Insight),
middle: Autofluoreszenz (MaiTai 740nm),
right: Berlin blue staining

The cell associated SPIO processing signal was detected by in vivo imaging (see Fig 2). Nanoparticle amount within cells increases over time which shows phagocytotic processes.

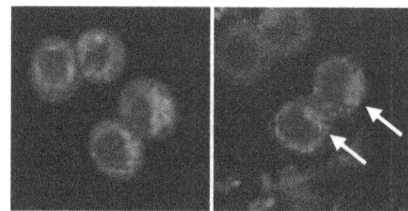

Figure 2: In vivo imaging of macrophages incubated with Resovist (3-photon microscopy: 0,55 s and 31,10 s: SPIOs within the cells (arrows)).

CONCLUSION After injection of Resovist into tumor tissue the SPIOS can be detected well distributed within the tissue by conventional histology and by a 3-photon device in a bio- medical context. System wide scanning is known (MRI, MPI), but now we are also able to identify the link to subcellular processing and localization of SPIOs. Further processing of SPIOs in MP is under development.

ACKNOWLEDGEMENTS This work was supported in by the German Federal Ministry of Education and Research (BMBF) under Grant 01EZ0912 and by the University Research Program "Imaging of Disease Processes", University of Lübeck.

REFERENCES

[1] D. Finas, J. Stegmann-Frehse, B. Sauer, G. Hüttmann, A. Rody, T. Buzug, K. Lüdtke-Buzug: Role of macrophages in SPIO processing in lymphatic tissue - further development of the breast cancer SNLB-concept using MPI. In: IEEE, 2015, pp. 110, Istanbul, Turkey

[2] D. Finas, J. Stegmann-Frehse, B. Sauer, G. Hüttmann, A. Rody, T. Buzug, K. Lüdtke-Buzug: SPIO processing in macrophages for MPI - The breast cancer MPI-SNLB-concept. Current Directions in Biomedical Engineering, 2015 (in press)

Magnetic Particle Spectrometer for the Analysis of Magnetic Particle Heating Applications

André Behrends[a,*], Thorsten M. Buzug[a], Alexander Neumann[a]

[a] Institute of Medical Engineering, University of Luebeck, Germany
* Corresponding author, email: {behrends, buzug}@imt.uni-luebeck.de

INTRODUCTION Magnetic Particle Imaging (MPI) is a fast evolving imaging technique that has been introduced about 10 years ago [1]. MPI has the potential of high-sensitivity real-time image acquisition and therefore suits multiple medical applications. One future application might be the detection of tumors in the human body [2]. However, the use of magnetic nanoparticles is not limited to the process of image acquisition only. Another well-known application is Magnetic Particle Hyperthermia, where magnetic fields are used to heat magnetic nanoparticles and perform heat-induced therapy [3]. Naturally, it arises the desire to combine both methods and create a theranostic device that is capable of Magnetic Particle Imaging as well as Magnetic Particle Hyperthermia. The realization of a device for combined diagnostic and therapy strongly depends on a deep understanding of the behavior of the magnetic nanoparticles. Theory as well as simulations of the dynamical behavior of the particles indicate a dependence on the temperature [4, 5]. To validate the temperature-dependent dynamical behavior of the magnetic nanoparticles a new type of Magnetic Particle Spectrometer (MPS) is necessary. The spectrometer has to be able to carry out both: The heating of the magnetic nanoparticles and a spectroscopic analysis. It has been suggested in literature that magnetic particle hyperthermia is best performed in a frequency-range between 100 kHz – 1 MHz, with field strengths strongly depending on the type of particles [6, 7]. This contribution seeks to present the concept of a Magnetic Particle Spectrometer for the analysis of Magnetic Particle Heating Applications as well as the current progress of the hardware realization. The spectrometer should be able to perform heating in the frequency-range of 100 kHz – 1 MHz and magnetic field-strengths of 25 kA/m.

MATERIAL AND METHODS The starting point for the setup is the work of Garaio et al. [7]. Their work has been analyzed and adapted to match the requirements of Magnetic Particle Spectrometry. A strong focus has been set on optimizing the main coil, to be able to apply the desired magnetic fields while maintaining manageable power levels. Since high currents are needed to produce the magnetic field careful selection of the materials used for connecting the system components has been conducted to minimize power losses. In contrast to [7] it is crucial to avoid harmonics produced by the system, which arises new challenges and requirements for the design and the material selection. A thermal isolation between the main coil and the sample has been added to avoid heating of the sample via conduction losses in the main coil. A gradiometer coil approach has been chosen to detect the particle signal while eliminating the excitation field produced by the main coil.

RESULTS The main coil has been optimized to have a power loss of 1.9 kW at a frequency of 1 MHz and a magnetic field strength of 25 kA/m. Low inductance connections are realized by using copper sheets instead of wires and the material used is strictly CU-OFE due to its outstanding properties. A thermal isolation is made of polyurethane and the receive coil is implemented as a gradiometer of second order. Finally, the impedance of the main coil was matched to 50 Ω for optimal power transfer. A CAD model of the proposed spectrometer is shown in Figure 1.

Figure 1: CAD model of the proposed spectrometer.

CONCLUSION A setup of a Magnetic Particle Spectrometer for the Analysis of Magnetic Particle Heating applications has been presented and is currently in the process of hardware realization.

ACKNOWLEDGEMENTS We hereby gratefully acknowledge the support of the Federal Ministry of Education and Research, Germany (BMBF) under grant number 13GW0069A.

REFERENCES
[1] B. Gleich and J. Weizenecker. *Nature*, 435(7046):1217—1217, 2005. doi: 10.1038/nature03808.
[2] L. Johnson, S. E. Pinder and M. Douek. *Histopathology*, 62(3):481—6, 2013.
[3] R. Hergt, S. Dutz, R. Müller and M. Zeisberger. *Journal of Physics: Condensed Matter*, 18:2919—2934, 2006. doi: 10.1088/0953-8984/18/38/S26.
[4] W. T. Coffey and Y. U. P. Kalmykov. *Journal of Applied Physics*, 112(12), 2012. doi: 10.1063/1.4754272
[5] J. Weizenecker, B. Gleich, J. Rahmer, J. Borgert. *Physics in Medicine and Biology*, 57(22):7317—7327, 2012. doi: 10.1088/0031-9155/57/22/7317.
[6] N. A. Usov and B. Y. Liubimov. *Journal of Applied Physics*, 112(2), 2012. doi: 10.1063/1.4737126
[7] E. Garaio et al. *Measurement Science and Technology*, 25(11), 2014. doi: 10.1088/0957-0233/25/11/115702.

Visualization and Quantification of the Intratumoral Distribution and Time-Dependent Change of Magnetic Nanoparticles in Magnetic Hyperthermia Using Magnetic Particle Imaging

Tomomi Kuboyabu, Isamu Yabata, Marina Aoki, Akiko Ohki, Mikiko Yamawaki, Yoshimi Inaoka, Kazuki Shimada, Kenya Murase*

Department of Medical Physics and Engineering, Graduate School of Medicine, Osaka University, Osaka, Japan
* Corresponding author, email: murase@sahs.med.osaka-u.ac.jp

INTRODUCTION Magnetic hyperthermia (MHT) is a strategy for cancer treatment using the temperature rise of magnetic nanoparticles (MNPs) under an alternating magnetic field (AMF). MHT is expected to be able to specifically heat tumor cells without damaging normal tissues. The temperature rise of the tumor and the therapeutic effect of MHT depend on the amount and distribution of MNPs in the tumor. Recently, a new imaging method called magnetic particle imaging (MPI) has been introduced [1], which allows imaging of the spatial distribution of MNPs. This study was undertaken to evaluate the feasibility of visualizing the intratumoral distribution of MNPs and quantifying their time-dependent change in MHT using MPI and of predicting the therapeutic effect of MHT using MPI.

MATERIAL AND METHODS First, colon-26 cells (1×10^6 cells) were injected into the back of eight-week-old male BALB/c mice. When the tumor volume reached 100-150 mm^3, 0.2 mL of MNPs (Resovist®) was injected directly into the tumor and MHT was performed using an AMF with a frequency of 600 kHz and a peak amplitude of 3 to 3.5 kA/m [2]. The treated mice were divided into three groups (A, B, and C). For comparison, we used the data of the mice without MHT (n=5) as a control. The tumors in the mice of Groups A (n=6), B (n=5), and C (n=5) were injected with Resovist® with iron concentrations of 500 mM, 400 mM, and 250 mM, respectively. We imaged the tumor-bearing mice using our MPI scanner [3, 4] immediately before, immediately after, 7 days after, and 14 days after MHT. We drew the region of interest (ROI) on the tumor in the MPI image and calculated the average MPI value by taking the threshold value for extracting the contour of the tumor as 40% of the maximum MPI value (pixel value) within the ROI. The average MPI values of the control group were taken as zero.

RESULTS Figure 1 shows the relative tumor volume growth (RTVG) in the treated and control groups, where RTVG is defined as $(V-V_0)/V_0$ with V_0 and V being the tumor volumes immediately before and after MHT, respectively. As shown in Fig. 1, the RTVG value in Group A was significantly lower than that of the control group at 5 days, 7 days, and 14 days after MHT. The RTVG value in Group B was significantly lower than that of the control group at 5 days, 7 days, and 14 days after MHT. The RTVG value in Group C was significantly lower than that of the control group at 5 days after MHT. These results indicate that the therapeutic effect of MHT largely depends on the injected dose of MNPs.

Figure 2 shows the average MPI values in the treated groups (A, B, and C) immediately before, immediately after, 7 days after, and 14 days after MHT. The average MPI values in all groups trended to decrease with time, and Resovist® remained in the

tumor even at 7 days and 14 days after MHT. These findings will be useful for the treatment planning of MHT, especially when considering the repeated application of MHT.

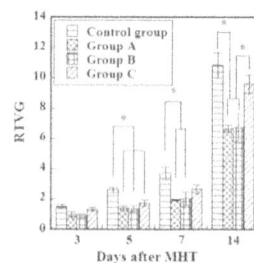

Figure 1: Relationship between RTVG and days after MHT in various groups. * $P<0.05$.

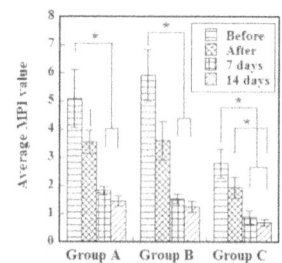

Figure 2: MPI values before, after, 7 days after, and 14 days after MHT in various groups. * $P<0.05$.

Figure 3 shows the relationship between the RTVG value at 14 days after MHT and the average MPI value immediately before MHT. There was a significant negative correlation between them (r=−0.713), suggesting that the average MPI values immediately before MHT can be used for predicting the therapeutic effect of MHT.

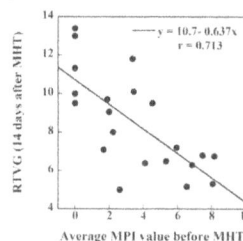

Figure 3: Correlation between the RTVG value 14 days after MHT and the average MPI value immediately before MHT.

CONCLUSION MPI can visualize the intratumoral distribution of MNPs and quantify the time-dependent change of MNPs in tumors before and after MHT. In addition, MPI will be useful for predicting the therapeutic effect of MHT.

ACKNOWLEDGEMENTS This work was supported by a Grant-in-Aid for Scientific Research (Grant No. 25282131) from the Japan Society for the Promotion of Science (JSPS).

REFERENCES
[1] B. Gleich and J. Weizenecker. *Nature*, 435(7046):1217–1217, 2005. doi: 10.1038/nature03808.
[2] K. Murase, M. Aoki, N. Banura, K. Nishimoto, A. Mimura, et al. *Open J. Med. Imaging*, 5(2):85–99, 2015. doi: 10.4236/ojmi.2015.52013.
[3] K. Murase, S. Hiratsuka, R. Song, and T. Takeuchi. *Jpn. J. Appl. Phys.*, 53(6):067001, 2014. doi: 10.7567/jjap.53.067001.
[4] K. Murase, R. Song, and S. Hiratsuka. *Appl. Phys. Lett.*, 104(25):252409, 2014. doi: 10.1063/1.4885146.

Towards Simultaneous MFH and Temperature Monitoring with MPI

Cagla Deniz Bahadir[a], Mustafa Ütkür[a,b], Emine Ulku Saritas[a,b,*]

[a] Department of Electrical and Electronics Engineering, Bilkent University, Ankara, Turkey
[b] National Magnetic Resonance Research Center (UMRAM), Bilkent University, Ankara, Turkey
* saritas@ee.bilkent.edu.tr

INTRODUCTION Magnetic Fluid Hyperthermia (MFH), one of the minimally invasive thermal therapies, can only be effective in a specific range of temperatures [1]. The physiological cooling, mainly caused by the blood flow complicates the process of maintaining a constant temperature [1] and thermometry methods are currently inadequate to yield precise temperature measurements [2]. Currently in-use non-invasive *in vivo* method of temperature monitoring is Magnetic Resonance Imaging (MRI) [1]. However, its sensitivity proves insufficient for monitoring very localized heating in the body [1]. Magnetic Particle Imaging (MPI) also has the potential to be used as a minimally invasive temperature monitoring method. Previously, the fifth-to-third harmonic ratio of the MPI signal has been tested on nanoparticles while being heated in a water bath [3]. Here, we investigate the change in MPI signal harmonics for *water-bath* heated and *magnetically* heated nanoparticles, with the goal of simultaneous MFH and temperature monitoring.

MATERIAL AND METHODS Here, we used a custom magnetic particle spectrometer (MPS) that can generate large enough fields at high frequencies. Effectively, this setup is capable of both MFH and MPS. The drive coil of 1.5 cm diameter with 50 turns of Litz wire wrapped in 5 layers and a receive coil with 13 turns in two sides and 25 turns in the middle.

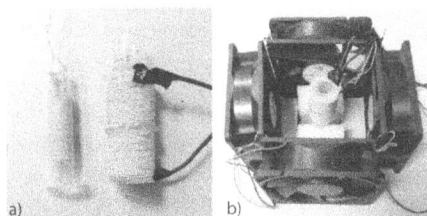

Figure 1: Custom MFH/MPS setup. a) Receive coil (left) and drive coil (right), and b) the overall experimental setup with four fans to air-cool the drive coil.

We experimented on two different-sized FeraSpin particles with 20 nm (L) and 25 nm (XL) in diameter having both 10 mM Fe/L concentration, and one undiluted nanomag-MIP (Micromod GmbH, Germany) solution with 89 mM Fe/L concentration. First, we heated the particles up to 60 °C in a water bath and measured their fifth-to-third harmonic ratio at 10.8 kHz with 8mT-peak field while they were cooling down. In the second part of the experiment, the nanomag-MIP particles were magnetically heated with 16mT-peak drive field at 150 kHz for five minutes while the setup was continuously cooled down with four fans (see Fig. 1). Average temperature rise was from 26.3°C to 44.3°C. The matching circuit for the drive coil was then switched to 10.8 kHz and 10mT-peak drive field amplitude for MPI signal acquisition, with 100 ms signal acquisition window. The particle response was measured at 1-minute intervals for 15 minutes during cool down. The fifth-to-third harmonic ratio of the signal was investigated in relation to decreasing temperature. This procedure was repeated three times.

RESULTS Three experiments with different particles were conducted. They were heated with water bath and their fifth-to-third harmonic ratio was measured while they were cooling down. FeraSpin particles, both (L) and (XL) showed a decrease in harmonics with increasing temperature as previously reported in [1] (See Fig.2.a). Additionally, three experiments were conducted with the nanomag-MIP particles magnetically heated from approximately 26.3 °C to 44.3°C. After magnetic-field-induced heating, the nanoparticle signal was measured during cool down. In Fig.2.b, the measured fifth-to-third harmonic ratio of the nanoparticle signal is plotted as a function of temperature during coold own after magnetically induced heating. Interestingly in both experiments, nanomag-MIP particles showed a monotonic increase in harmonics ratio with the increasing temperature. The reason for this different behavior, while not yet known, may be related to the nanoparticle structure or the high concentration of the nanomap-MIP particles. We plan to investigate this further.

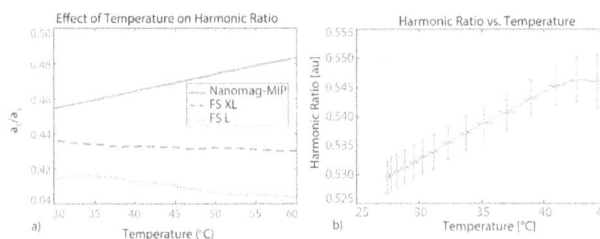

Figure 2: Experimental results for (a) water-bath heated nanoparticles, and (b) magnetically heated nanomag-MIP.

CONCLUSION The overall trend of the fifth-to-third harmonic ratio vs. temperature suggests MPI as a promising non-invasive temperature monitoring method for MFH. Our next goal is to improve the sensitivity of our receive coil to increase measurement accuracy. Next, we will implement simultaneous signal acquisition and heating.

ACKNOWLEDGEMENTS Cagla Deniz Bahadir and Mustafa Ütkür contributed equally to this work. This work was supported by the Scientific and Technological Research Council of Turkey through a TUBITAK 3501 Grant (114E167), by the European Commission through an FP7 Marie Curie Career Integration Grant (PCIG13-GA-2013-618834), and by the Turkish Academy of Sciences through TUBA-GEBIP 2015 program.

REFERENCES
[1] J.B. Weaver, A.M. Rauwerdink and E.W. Hansen. *Medical Physics* 36 (1822) 2009. doi:10.1118/1.3106342
[2] A. Jordan, R. Scholz, P. Wust, H. FaKhling, R. Felix. *Journal of Magnetism and Magnetic Materials*, 201: 413-419, 1999). doi: 10.1016/S0304-8853(99)00088-8
[3] A.M. Rauwerdink, E.W. Hansen and J.B. Weaver, *Phys. Med. Biol.*, 54: L51–L55,2009. doi: 10.1088/0031-9155/54/19/L01
[4] B. Gleich and J. Weizenecker. *Nature*, 435(7046):1217—1217, 2005. doi: 10.1038/nature03808.

First Results: Phantoms for MPI and Ultrasound Therapy

Ankit Malhotra*, Corinna Stegelmeier, Thomas Friedrich, Kerstin Lüdtke- Buzug and Thorsten M. Buzug*

Institute of Medical Engineering, University of Lübeck, Lübeck, Germany
*Corresponding author, email: {buzug, malhotra}@imt.uni-luebeck.de

INTRODUCTION Magnetic Particle Imaging (MPI) is an emerging modality for quantitative information regarding the spatial distribution of magnetic material inside a tissue matrix [1, 2]. In this work, research focus in MPI has been extended from imaging to therapy and monitoring with the help of specifically tailored superparamagnetic materials. Therapy and monitoring includes hyperthermia and magnetic manipulation, moreover there is current research going to combine other modalities such as High-Intensity Focused Ultrasound (HIFU) with MPI.

There are various phantoms developed for MPI to quantify the spatial and temporal resolution [3]. These phantoms usually consist of materials, which are hydrophobic in nature as the nanoparticles are water-soluble concerning biocompatibility. For the ultrasound phantoms, it is necessary that the phantoms used, mimic the properties of the tissues. Therefore, many different types of phantoms have been tested consisting of different materials such as Agar-Agar, Polyvinyl alcohol (PVA), Polyacrylamid (PAA) etc. [4-6]. PVA phantoms are able to mimic the acoustic and optical properties of tissues [5] and PAA phantoms are temperature sensitive hydrogels, which change their transparency and elastic modules when they reach a critical temperature known as the cloud point. On the other hand, the fabricated phantoms consist of matrices, which are formed by crosslinking of different chemical constitutes. Crosslinking makes them porous in nature and hence, there is a high probability that the nanoparticles will eventually disperse in the gel.

The aim of this research is to create a phantom consisting of the superparamagnetic nanoparticles in a gel-forming matrix, which could be used for different modalities. The phantoms should be mechanically stable as well as the nanoparticles should not disperse in the gel. Moreover, the future aim is to characterize these phantoms with HIFU and MPI for inducing cell ablation and hyperthermia therapy.

MATERIAL AND METHODS

Preparation of Agar-Agar Phantoms: Homogenous solutions of Agar in water (1.5-3 wt.%) were fabricated by heating to the boiling point and then transferred to appropriate molds to achieve rigidity. These phantoms were filled with nanoparticles ($\approx$17 mg/ml). Moreover, a magnetic gel-like structure was prepared by mixing superparamagnetic nanoparticles with the hot Agar-Agar solution.

Preparation of PVA Phantoms: The recipe for preparation of the PVA phantoms was based on ref. [7]. PVA with a hydrolysis degree higher than 99% was used. Two different phantoms were manufactured with two different solvents: distilled water and a mixture of dimethyl sulfoxide (DMSO) and water (80:20) were used for solutions containing 15wt.% of PVA at 80 °C.

RESULTS

Agar-Agar Phantoms: The first phantoms prepared lacked mechanical stability. The fragility decreased with higher concentration of Agar. Moreover, dispersion and distribution of nanoparticles in the gel was also observed (figure 1(A)). The mixing of the nanoparticles in Agar solution proved to be successful and we manufactured a magnetic gel (figure 1(B)).

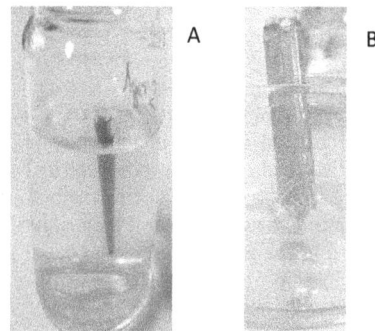

Figure 1: (A) Photograph of the Agar-Agar phantom filled with liquid nanoparticles visualizing the dispersion of particles into the gel. (B) Photograph of the magnetic gel.

PVA Phantoms: They do not suffer from mechanical stress and no dispersion of magnetic nanoparticles occurred. Moreover, the phantoms made in the water/DMSO mixture were highly transparent. However, the phantoms prepared with pure water were opaque and therefore of little use to our experiments.

CONCLUSION Agar phantoms lack stability but are optimal for making magnetic gel, which in combination could be used with other phantom materials. PVA proves to be an ideal material for construction of mechanically stable phantoms for MPI and ultrasound studies.

ACKNOWLEDGEMENTS We would like thank the German Federal Ministry of Education and Research (BMBF SAMBA-PATI 13GW0069A) for supporting this project.

REFERENCES
[1] B. Gleich and J. Weizenecker. *Nature*, 435(7046):1217—1217, 2005. doi: 10.1038/nature03808.
[2] T. Knopp and T. M. Buzug. Springer, Berlin/Heidelberg, 2012. doi: 10.1007/978-3-642-04199-0.
[3] J. Haegele et al. *International Workshop on Magnetic Particle Imaging IWMPI 2014*, 57-58.
[4] Lafon, C et al. *Ultrasonics Symposium*, 2:1295-1298 2001. doi: 10.1109/ULTSYM.2001.991957.
[5] Shieh, J. et al. *Applied Thermal Engineering 2014*, 62(2), 322-329. doi: 10.1016/j.applthermaleng.2013.09.021.
[6] Surry KJ, Austin HJ, Fenster A, Peters TM. *Phys Med Biol. 2004* Dec 21 ; 49(24):5529-46.
[7] A. Kharine et al. *Phys. Med. Biol. 2003*, 48 (2003) 357 – 370. Url: stacks.iop.org/PMB/48/357

Magnetic particle imaging in a mouse model of acute ischemic stroke

Peter Ludewig[1*], Nadine Gdaniec[2,3*], Jan Sedlacik[6], Sarah Behr[1], Scott J. Kemp[4], R. Matthew Ferguson[4], Amit P. Khandhar[4], Kannan M. Krishnan[4,5], Jens Fiehler[6], Christian Gerloff[1], Tobias Knopp[2,3], Tim Magnus[1]

[1] Department of Neurology, University Medical Center Hamburg-Eppendorf, Hamburg, Germany.

[2] Section for Biomedical Imaging, University Medical Center Hamburg-Eppendorf, Hamburg, Germany.

[3] Institute for Biomedical Imaging, Hamburg University of Technology, Hamburg, Germany.

[4] LodeSpin Labs, Seattle, Washington, USA

[5] Materials Science and Engineering Department, University of Washington, Seattle, USA

[6] Neuroradiology, University Medical Center Hamburg-Eppendorf, Hamburg, Germany.

*P.L. and N.G. contributed equally to this study.

Corresponding author: n.gdaniec@uke.de; p.ludewig@uke.de

INTRODUCTION With two million neurons dying every minute after stroke onset [1] fast and accurate diagnosis of ischemic stroke is fundamental for successful treatment of stroke patients. Ideally, stroke imaging should be able to distinguish between hemorrhage and ischemic stroke, but also provide further information on the tissue at risk, the so-called penumbra, and stenosis of extra- or intracranial vessels within a short period.

Magnetic particle imaging (MPI) is a new imaging modality acquiring 3D datasets combining high spatial resolution with short image acquisition times [2, 3] allowing to perform a broad range of functional cerebrovascular measurements such as cerebral perfusion to detect ischemic stroke. The purpose of this work was to visualize vasculature and vessel occlusion in a murine model of ischemic stroke with MPI.

MATERIAL AND METHODS The left common carotid artery (CCA) was ligated in a 12 week-old C57BL/6 mouse mimicking acute occlusion of this vessel. To assess anatomical information about the murine head and neck, a 7 T small animal MRI was used (Bruker Clinscan). Head position was marked with Resovist (Bayer Schering Pharma AG) filled fiducials for geometry planning and image registration of the MRI and MPI data during post-processing. Magnetic particle imaging was performed using a pre-clinical MPI scanner (Bruker/Philips). A 100 μL bolus with 46 mmol(Fe)/L (LS8, LodeSpin Labs) was injected via the mouse tail vein. MPI scans (3D data with 21.5 ms temporal resolution, a gradient strength of 1.5 $Tm^{-1}\mu_0^{-1}$ in two directions, and a drive-field amplitude of 14 $mT\mu_0^{-1}$) were acquired dynamically while administering the contrast-agent bolus.

RESULTS In the first milliseconds after injection, contrast-agent inflow could be detected only in the vascular tree of the right common carotid artery. As a result of the occlusion of the left CCA, no signal was detected contralaterally. In later phases, some signal was assessed in the left CCA territory probably due to collateral flow via the posterior and anterior communicating arteries.

DISCUSSION AND CONCLUSION For the first time, we show that MPI is capable of visualizing acute pathologies like occlusion of extracranial vessels in a mouse model of CCA ligation. Of course, this is only a small portion in stroke imaging, but the first step. Further works need to be done to provide information about tissue perfusion and functional parameters to assess cerebral tissue viability. MPI/ MRI hybrid systems might be the future of acute stroke imaging.

REFERENCES

[1] Saver, J.L., Time is brain--quantified. Stroke, 2006. 7(1):p. 263-6.

[2] Buzug, T.K.A.T.M., Magnetic Particle Imaging: An introduction to Imaging Principles and Scanner Instrumentation. 2012: Springer.

[3] Gleich, B. and J. Weizenecker, Tomographic imaging using the nonlinear response of magnetic particles. Nature, 2005. 435(7046): p. 1214-7.

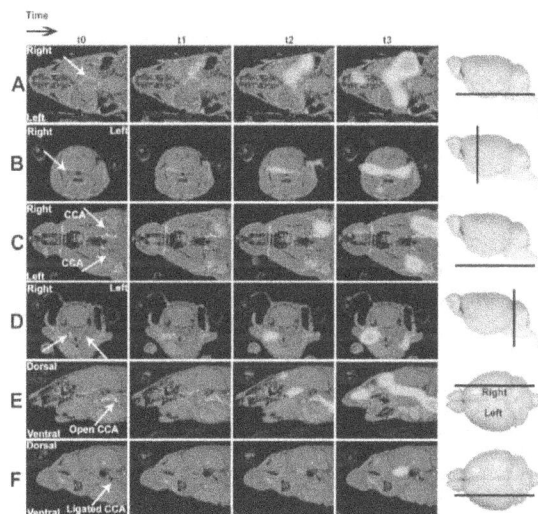

Figure 1: MPI shows reduced perfusion in the left CCA vascular territory after proximal ligation: The left CCA was ligated before the bifurcation of the CCA followed by injection of the contrast agent via a tail vein catheter. MPI scans were acquired dynamically after injection. Row A/B show transverse (A) and coronal section (B) in the region of the circle of Willis. Shortly after the injection (t1), the signal was detected in the right temporal lobe near the circle of Willis (arrow A/B t1), while no signal was detected contralaterally. At later points (t2/t3) signals were also seen in the left regions probably due to collateral flow via ACOM and PCOM. C/D show transverse (C) and coronal section (D) in the area of the bifurcation of the CCA, E/F show sagittal sections through the right (E) and left (F) hemispheres. Less signal was seen in the left CCA (arrows indicate the CCA). (CCA: common carotid artery, PCOM/ACOM posterior/anterior communicating artery)

First *in-vivo* Perfusion Imaging with MPI

Ryan Orendorff[a,*], Paul Keselman[a], Steven M. Conolly[a,b]

[a] Department of Bioengineering, University of California at Berkeley, Berkeley, CA, USA
[b] Department of Electrical Engineering, University of California at Berkeley, Berkeley, CA, USA
[*] Corresponding author, email: ryan.orendorff@berkeley.edu

INTRODUCTION: In order to diagnose stroke and other related diseases, doctors use Computed Tomography (CT), CT Perfusion (CTP), and CT Angiography (CTA) to determine blood flow rate and volume through the cerebral vasculature. While these techniques work well for determining large scale deficits like hemorrhagic stroke, they are fundamentally ill-equipped to track blood changes over time due to high background signal from the tissue surrounding the vasculature. Additionally, CTP and CTA have high radiation dose and use an iodinated contrast agent that is toxic for patients with Chronic Kidney Disease (CKD)---the majority of those who have a stroke have some form of CKD [1].

MPI is a promising new modality that images only a safe magnetic tracer, commonly SPIO nanoparticles [2, 3]. MPI contains no signal from tissue, and thus can see the cerebral vascular clearly if iron nanoparticles as they flow through the brain. Here we present, for the first time, perfusion imaging using MPI.

MATERIAL AND METHODS: MPI was performed on a Fisher-344 rat over a period of a few minutes; a brief description of this process follows. The rat was first anesthetized using 3% isoflurane. A catheter was then inserted into the tail vein of the rat. Then the rat was placed in the scanner with access to the catheter from outside the imaging bore. 0.2 mL of 25 mg Fe/mL Resovist iron nanoparticle solution was injected as a bolus into the animal over a period of six seconds. MPI was used to monitor the iron over time.

Imaging was performed using the Berkeley field free point (FFP) scanner. This MPI scanner has a gradient strength of 7 T/m by 3.5 T/m by 3.5 T/m in the x, y, and z directions, respectively. A single transverse, xy plane through the rat's brain was captured at a rate of one frame every 2.16 seconds, for a period of three and a half minutes. The field of view was 3.25 cm by 3.25 cm.

The time series images were then processed to produce a map of the relative cerebral blood flow (rCBF) through each pixel in the image. This calculation was done by finding the maximum rate of change of the iron signal through each pixel in the image, using a Savitzky–Golay filter to calculate the derivative of the signal. The maximum slope at each pixel was then plotted to produce an rCBF map.

RESULTS: Here we present the world's first MPI perfusion imaging. We have calculated, for the first time, physiologically relevant parameters (rCBF) from an MPI time course. This information is vital for the diagnosis and treatment of stroke and other diseases involving the cerebral vasculature. We have shown that MPI shows great promise to provide a safe, effective, and fast way in which to acquire this information.

CONCLUSION: Here we present the world's first MPI perfusion imaging. We have calculated, for the first time, physiologically relevant parameters (rCBF) from an MPI time course. This information is vital for the diagnosis and treatment of stroke and other diseases involving the cerebral vasculature. We have shown that MPI shows great promise to provide a safe, effective, and fast way in which to acquire this information.

ACKNOWLEDGEMENTS: We gratefully acknowledge funding from NIH R01 EB013689, CIRM RT2-01893, Keck Foundation 009323, NIH 1R24 MH106053 and NIH 1R01 EB019458, ACTG 037829, and NSF GRFP.

REFERENCES
[1] Krishna PR, et al. *Indian Journal of Nephrology*. 2009
[2] P.W. Goodwill et al. *IEEE Trans. Med. Imaging*. 2010
[3] T. Knopp et al. *Magnetic Particle Imaging*. 2012
[4] L. Axel, *Radiology*, 1980.

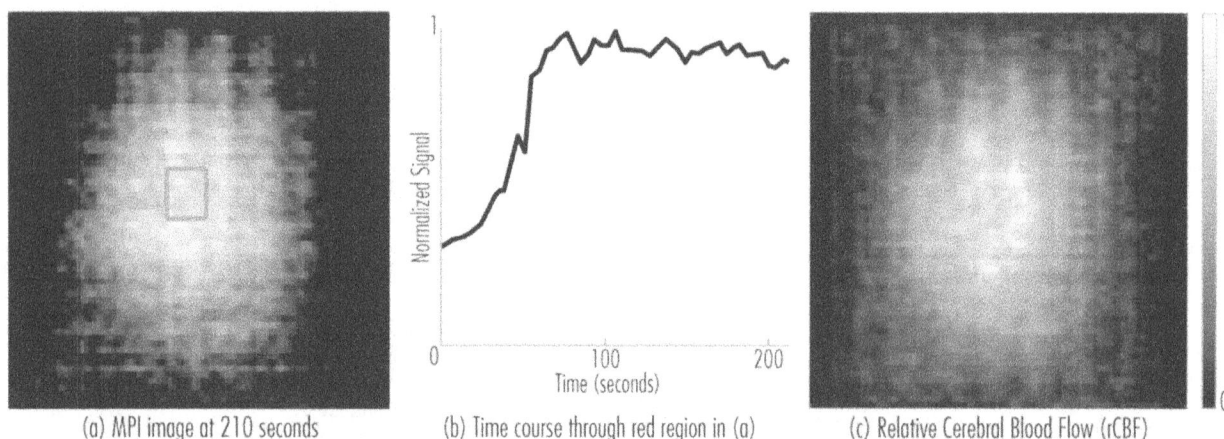

(a) MPI image at 210 seconds   (b) Time course through red region in (a)   (c) Relative Cerebral Blood Flow (rCBF)

Figure 1: MPI perfusion imaging. (a) A single frame at the end of the imaging sequence (t=210 seconds), showing iron signal in the brain of the rat. (b) The averaged signal in the red box shown in (a) over time. There is a clear rise in the signal as the iron bolus passes through the cerebral vasculature. (c) Using the data shown in (b) (but for each pixel), a map of the relative cerebral blood flow (rCBF) was created.

Long term *in vivo* biodistribution and clearance of tailored MPI tracers

Paul Keselman[a*], Bo Zheng[a], Patrick W. Goodwill[c], and Steven M. Conolly[a,b]

[a] University of California Berkeley, Department of Bioengineering, Berkeley, CA;
[b] University of California Berkeley, Department of Electrical Engineering and Computer Sciences, Berkeley, CA;
[c] Magnetic Insight, Inc., Newark, CA;
[*] Corresponding author, email: pkeselma@berkeley.edu

INTRODUCTION As the field of magnetic particle imaging (MPI) [1] matures, more emphasis is being put into the design and optimization of SPIOs specifically tailored to MPI physics [2,3]. This is driven both by the demand for higher resolution and adequate circulation time for *in vivo* applications [2,3]. In assessing these tailored tracers, it is imperative that we measure both the short term and long term fate of these nanoparticles. Here, we measure the *in vivo* long term clearance time of one tailored particle.

MATERIAL AND METHODS A Fisher 344 female rat (11 weeks old) was used for this long term biodistribution and clearance experiment. Briefly, PEG-coated 25 nm core diameter iron oxide nanoparticles (Senior Scientific Inc.), were diluted 2-fold in phosphate buffered saline (PBS) solution, and then injected as a bolus at a dose of 10 mgFe/kg into the tail vein of the animal. The rat was then imaged with respiratory triggering using a 3.5×3.5×7 T/m imager [4,5] with a drive field of 40 mTpp in z, and FOV of 3.5×3.5×10.5 cm. The animal was positioned in dorsal recumbency. Three time point scans (~7 min. duration) were acquired immediately after the injection (12 minutes), at day 13 after the injection, and finally at day 75 after the injection. All animal experiments were performed in compliance with the Animal Care and Use Committee (ACUC) at UC Berkeley. Half-life was calculated using non-linear least-squares exponential fit of the data.

RESULTS We imaged the animal immediately after the injection of the tracer, and then again on day 13 and day 75. All three time points are shown in Figure 1.

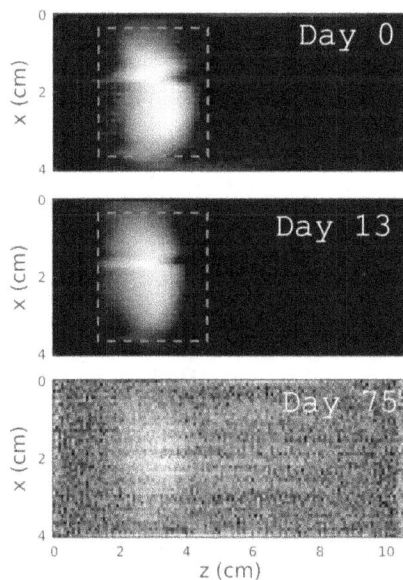

Figure 1: Ventral images of the animal at three time points after the injection of tracer.

It appears that the particles were immediately cleared from the blood into the liver via the reticuloendothelial system and then slowly cleared away, half-life of 30.2 days. In contrast to a half-life of 1-3 days for Resovist as measured in MRI [6]. To compute the half-life, we summed up all the signal in a rectangular region of interest (ROI) over the liver (shown by the dotted blue line) and then fitted a 1st order exponential to the results. Both the data and fit are shown in Figure 2.

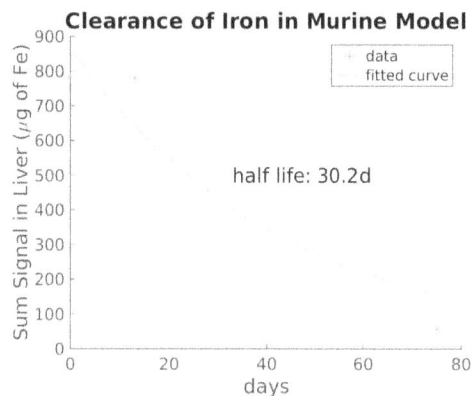

Figure 2: Graph of the total signal in the liver over 75 days.

CONCLUSION In this experiment we were able to successfully track long term biodistribution and eventual clearance of an MPI tracer. We are planning to conduct a more comprehensive study with a larger sample size, biweekly imaging time points, and several particle types including Resovist and MPI-specific particles synthesized by Lodespin Labs [3]. These experiments may guide optimal particle design for repeated MPI experiments or to optimize MPI signal persistence in longitudinal studies.

ACKNOWLEDGEMENTS The authors would like to acknowledge Dr. Erika Vreeland from Senior Scientific LLC for providing the particles, as well as the support from the following grants: NIH R01 EB013689, CIRM RT2-01893, Keck Foundation 009323, NIH 1R24 MH106053 and NIH 1R01 EB019458, and ACTG 037829.

REFERENCES
[1] B. Gleich and J. Weizenecker, *Nature*, 435(7046):1217—1217, 2005, doi: 10.1038/nature03808
[2] M. Ferguson, A.P. Khandar, S.J. Kemp, H. Arami, E.U. Saritas, L.R. Croft, et al, *IEEE TMI*, 34(5):1077-1084, 2015
[3] R. Hufschmid, H. Arami, M.R. Ferguson, M. Gonzales, E. Teeman, L.N. Brush, N.D. Browning, K.M. Krishnan, *Nanoscale*, 7(25):11142—11154, 2015. doi: 10.1039/C5NR01651G.
[4] P.W. Goodwill, L.R. Croft, J.J. Konkle, K. Lu, E.U. Saritas, B. Zheng, S.M. Conolly., *IWMPI*, 2013
[5] P.W. Goodwill & S.M. Conolly. *IEEE TMI*, 29(11):1851–1859, 2010
[6] R.C. Semelka and T.K. Helmberger, *Radiology*, 218(1):27–38, 2001

Stem cell tracking potential of Magnetic Particle Imaging compared with 19F Magnetic Resonance Imaging

Friso G. Heslinga [a,b,*], Steffen Bruns [a], Elaine Yu [a], Paul Keselman [a], Xinyi Y. Zhou [a], Bo Zheng [a], Sebastiaan Waanders [b], Patrick W. Goodwill [a], M. Wendland [a], Bennie Ten Haken [b], Steven M. Conolly [a,c]

[a] Department of Bioengineering, University of California at Berkeley, [b] Department of Neuroimaging, MIRA, University of Twente, [c] Department of Electrical Engineering & Computer Sciences, University of California at Berkeley, * Corresponding author, email: f.g.heslinga@student.utwente.nl

INTRODUCTION Stem cells have the potential to be used as treatment for regenerating cardiac and brain tissue, which both have limited self-healing capacities [1]. An important indicator for success of these type of treatments is cell fate: survival and engraftment of the stem cells. Localization of stem cells can be done by cell tracking; a process in which the cells are 'labeled' with a contrast agent or tracer material and a corresponding imaging technique is used to follow the stem cells, by scanning the body several times during several weeks.

The main modalities that are able to perform stem cell tracking deep inside the body (Single Photon Emission Computed Tomography, Positron Emission Tomography and Magnetic Resonance Imaging with T1 or T2 contrast agents) require at least 10,000 cells for detection, according to Nguyen (2014) [1]. Recently, fluorine-based MRI (19F MRI), has received more attention as a potential candidate for stem cell tracking even though sensitivity is still limited for clinical use [2]. Meanwhile, Magnetic Particle Imaging (MPI) has been shown to be useful for stem cell tracking as well [3]. A comparison between MPI and 19F MRI for stem cell tracking purposes has not yet been made. In this work, we aim to make a fair comparison between both techniques to determine which technique is most sensitive and therefore has the highest potential for future stem cell tracking research.

MATERIAL AND METHODS Sensitivity is here defined as the number of cells that is needed per voxel to be positively identified in an image. This detection limit depends on the physical sensitivity of the system and the amount of tracer that can be incorporated inside a cell through cell labeling.

For MPI, the detection limit of the system is tested by positioning 13 'point sources' with small amounts of Resovist® (Bayer Schering Pharma AG, Leverkusen, Germany; 0.5 mmol Fe/m), ranging from 225 to 3600 ng Fe inside 50 g of chicken breast, five samples at a time. Scans have been performed with a 7 T/m/μ_0 system, built at UC Berkeley. Signal to Noise Ratio (SNR) is determined for all samples and a linear fit is obtained for a better estimate of the detection limit of the system, which we define to be three times the standard deviation of the background noise.

For 19F MRI, similar SNR experiments are performed with 13 samples (0.1 to 6.4 mol/l KF) in chicken breast with a Bruker 7 T small animal MRI scanner. Since point sources are harder to make for MRI due to local field inhomogeneity artifacts, instead volumes of 100 μl of diluted KF are scanned with a 90° Spin Echo sequence. Due to the bigger volume, several voxels should now contain maximum signal (resolution = 1x1x1 mm^3), and averaging is done for four voxels containing the highest signal in one slice.

Cell uptake information was gathered for bone marrow-derived human mesenchymal stem cells (hMSCs; ATCC PCS-500-012, Manassas, VA). Uptake information of Resovist was already known [4], but labeling information of PFPE in hMSCs is limited and has been experimentally determined.

Normalization between experiments is done based on scanning parameters such as acquisition time and amount of voxels, which both affect the number of repetitions. Since cell labeling for MRI is done with an emulsified PFPE CS-1000 (Celsense Inc., Pittsburgh USA) with a shorter longitudinal relaxation time (T1) than KF, the detection limit was normalized for this factor as well. 1.5 x 10^6 hMSCs are co-incubated with CS-1000 at 5 mg/ml for 24 hours. Labeling efficiency is determined in a 400 MHz NMR system with a known quantity of KF as a reference.

RESULTS An overview of the results is given in table 1. For MPI and 19F MRI, mass refers to the iron content in Resovist and fluorine content in KF and CS-1000 respectively. Note that for 19F MRI the detection limit is given for 19F in KF while cell loading is given for 19F in the emulsified PFPE.

Table 1: Overview of detection limits and cell loading.

	MPI	19F MRI
Detection limit tracer (ng)	250-350	3800-5700
Cell loading (pg/cell)	78 [4]	41

CONCLUSION & DISCUSSION Loading of PFPE inside the hMSCs has been improved compared to existing literature [5], which has led to an improved detection limit of 19F MRI. Comparing sensitivity of MPI and 19F MRI is not straightforward. An interesting challenge is to normalize for scanning time, which is affected by repetitions, amount of voxels and T1 relaxtion time. Despite these challenges and being in early stages of its development, MPI is found to be at least an order of magnitude more sensitive for cell tracking of hMSCs than 19F MRI. This would imply that MPI has tremendous potential to become the leading modality for tracking stem cells and thereby improve research towards stem cell therapy.

REFERENCES

[1] P.K. Nguyen, J. Riegles, J.C. Wu. Stem cell imaging: from bench to bedside. Cell Stem cell. 2014; 14(4):431-44.
[2] M. Srinivas, A. Heerschap et al. 19F MRI for quantitative in vivo cell tracking. Trends in Biotechnology. 2010;28(7):363-70
[3] B. Zheng, T. Vazin, W. Yang, P.W. Goodwill et al. Quantitative stem cell imaging with Magnetic Particle Imaging. IWMPI 2013; 2013.
[4] B. Zheng, M.P. von See, E. Yu, B. Gunel, K. Lu, T. Vazin, D.V. Schaffer, P.W. Goodwill, S.M. Conolly. Quantative Magnetic Particle Imaging Monitors the Transplantation, Biodistribution, and Clearance of Stem Cells in Vivo. Theranostics. In press.
[5] E.J. Ribot, J.M. Gaudet, Y. Chen, K.M. Gilbert, P.J. Foster. In vivo MR detection of fluorine-labeled human MSC using the bSSFP sequence. International Journal of Nanomedicine. 2014;9(1):1731-9

Growth inhibition of *Pseudomonas Aeruginosa* by extremely low frequency Pulsed Magnetic Field (PMF)

Fadel M.Ali[1], Nermeen.Serag[2*], A. M. Khalil[3]

[1]Biophysics department, Faculty of Science, Cairo University, Egypt.
[2]Physics department, Faculty of Basic Science, German University in Cairo, Egypt
[3]Basic Sciences Dept. Faculty of Engineering, Pharos University in Alex. (PUA), Egypt.
*nermeen.serag@guc.edu.eg

INTRODUCTION P.aeruginosa opportunistic, dangerous and dreaded pathogen, as well as the principal cause of morbidity and mortality in cystic fibrosis patients. It has been demonstrated that bacterial exposure to ELF EMF can alter its viability and growth rate with frequency and amplitude dependency (**Moshe *et al.*, 2008**), antibiotic susceptibility (**Segatore *et al.*, 2012**), and *ultrastructural* shape (**Inhan *et al.*, 2011**).Biological systems, *in vivo, generate electric currents and fields that* associated with magnetic fields, as a result of the running physiological mechanisms. In these mechanisms ionic motions are involved which are responsible for all bioelectric signals generated. The flow rate, period and direction of flow of the ions will generate an electric impulse with a specific frequency, shape and amplitude (Fadel *et al., 2009*).

MATERIAL AND METHODS

Microorganism Culture Conditions:

- Reference strain of **P. aeruginosa** (Ref. 0353P, ATCC 27853, LOT 353604, OXOID) is used for all the present comparative experiments.
- The microorganism is plated on **MacConkey** agar and incubated at **37 °C, 24 hrs** Thereafter, several colonies plated again on a fresh agar plate and also incubated for 24 hrs at 37 °C.
- For maintaining a fresh strain, this procedure is repeated weekly before running the experiment

Exposure System:

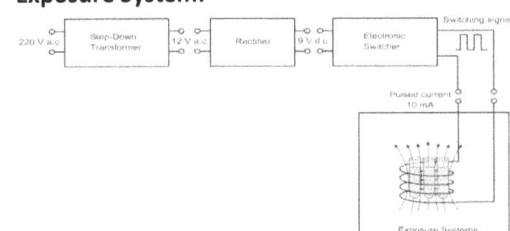

RESULTS This work study the effect of PMF at resonance frequencies on the activity of P. aeruginosa. For this the growth characteristics, molecular structure, antibiotic sensitivity, and TEM for the microorganism are studied. The results indicated inhibition of the microbial growth at 0.5 Hz and 0.7 Hz for the influencing field And Maximum inhibition of microorganism cellular growth occurred after one hour exposure to PMF. The exposure of microorganism to different exposure systems for different periods indicated that the most effective time gave maximum inhibition for bacterial growth was 1 hr. On the other hand, the exposed cells showed remarkably morphological changes and impacts on disrupted cells. The severity of exposure to 0.5 Hz and 0.7 Hz leads to that almost the majority of the cells are disintegrated, fragmented, and destroyed. pores are formed with irreversible cell membrane breakdown and significant biomass loss.

DISCUSSION There are several parameters involved cellular division mechanism for the microorganism; such as its cellular membrane phospholipid packing properties and structure, synthesis, intracellular constituents and DNA. The changes in permeability or in structure of cellular membrane may lead to change in the ion channel properties, to loss of the cellular inter-constituents, and/or the permeation of extracellular components to come inside the cell, which may interact with DNA. The TEM images shows the severity of exposure to PMF, that almost the majority of the cells are disintegrated, fragmented, and destroyed and indicated that pores are formed with irreversible cell membrane breakdown and significant biomass loss because of the extrusion of cytoplasmic contents from the cell wall and almost completely dissolution of the cytoplasm. This may be due to the fact that both hydrophobic and electrostatic interactions are involved in maintaining the organization of the membrane *Ingram L. and Buttke T. (1984),* it is reasonable to speculate that PMF has altered the electrostatic balance of the membrane components in a manner similar to that caused by cationic molecules, producing loss of integrity and/or disorganization and thus triggering growth arrest or death of bacterial cell. This understanding is logic since the cellular membranes play an important role in cell division mechanisms. **As a conclusion** - It may be presumed from the results that maximum inhibition on microbial growth can arise when resonance interference of the applied Field with the bioelectric signals generated from physiological functions of bacterial cells occurs.

ACKNOWLEDGEMENTS I wish to acknowledge Mr. Asem Abd El-Monem Ali (Physics Specialist) from German University in Cairo for helping on manufacturing and making the exposure facilities.

REFERENCES

[1] Fadel M. Ali, M. A. Ahmed and M. A. El Hag (2009): "Control of Sclerotium cepivorum (Allium White Rot) activities by electromagnetic waves at resonance frequency." Aust. J. Basic& Appl. Sci. 3:1994-2000.

[2] Inhan-Garip A., Aksu B., Akan Z., Akakin D., Ozaydin A.N., and San T. (2011): "Effect of extremely low frequency electromagnetic fields on growth rate and morphology of bacteria", Int J Radiat Biol. 87(12):1155-61.

[3] Moshe G., Yaara P., Alexandra B., Yoram W., *etal., (2008) "Microbial Growth Inhibition by* Alternating Electric Fields"

[4] Ingram L. and Buttke T. (1984): "Effects of alcohols on microorganisms" In: Rose A., Tempest D., Editors. "Advances in microbial physiology". 25, Academic Press, New York, USA. 25: 280–282.

Compression of FFP System Matrix with a Special Sampling Rate on the Lisssajous Trajectory

Marco Maass[a,*], Klaas Bente[b], Mandy Ahlborg[b], Hanne Medimagh [b], Huy Phan[a], Thorsten M. Buzug[b], and Alfred Mertins[a]

[a] Institute for Signal Processing, University of Lübeck, Germany
[b] Institute of Medical Engineering, University of Lübeck, Germany
[*] Corresponding author, email: maass@isip.uni-luebeck.de

INTRODUCTION The tracer based imaging method magnetic particle imaging (MPI) allows reconstructing the distribution of superparamagnetic iron oxid nanoparticles [1]. For the calculation of the spatial particle concentration c from the voltage signal u, a system-matrix based reconstruction is widely used. The system matrix S, which maps between the particle concentration and voltage signal, has to be known [2]. Since the system matrices can be very large in size and thus consume a huge amount of memory in workspace, a compression method was introduced [3]. The idea was to transform the system matrix via well-known transforms into another space where it becomes sparse.

In this paper, we study the relationship between the sampling pattern and the compressibility of the system matrix and introduce a new compression method for MPI system matrices using a field-free point (FFP) on a Lisssajous trajectory.

MATERIAL AND METHODS The signal equation in time-domain is described by $u = Sc$, where $u \in \mathbb{R}^N$, $S \in \mathbb{R}^{N \times M}$ and $c \in \mathbb{R}_+^M$. With the discrete Fourier transform matrix $\mathcal{F}$ we obtain the frequency-domain representation $\hat{u} = \mathcal{F}u = \mathcal{F}Sc = \hat{S}c$, which is the standard representation in MPI. In this work, we consider the two–dimensional case for the field-of-view (FOV). In [3], the rows of $\hat{S}$ were transformed with a transformation matrix T with respect to the dimensions of the FOV. Hence, the signal equation was re-written as $\hat{u} = \hat{S}TT^{-1}c = \hat{S}_T c_T$. The work in [3] also showed that the discrete Chebychev transform (DTT) and discrete cosine transform of type two (DCT-II) are able to compress system matrices significantly.

For the compression scheme proposed in this work, the frequency ratio is chosen to be rational with $f_x/f_y = N_y/N_x = N_y/(N_y - 1)$ where $N_x, N_y \in \mathbb{N}$. We can show that, if we sample the trajectory equidistantly in time at $N_s = k \cdot N_x \cdot N_y$ sampling points in one period, where $k \in \mathbb{N}$, we obtain k separable Euclidian-like grids. The sampling points are shown for $N_y = 5$ and two different k in Fig. 1. Some equivalent observation results were recently also confirmed in [4]. In this paper, we show that the orthogonal transform on the FOV should have symmetric and antisymmetric basis functions to result in a maximally sparse representation of the system matrix. In fact, for simulated system matrices, the use of a spatial transform with the above mentioned symmetry properties results in a large number of coefficients in $\hat{S}_T$ being exactly zero. For measured system matrices, these values are approximately zero. While the DCT-II and the DTT automatically satisfy the symmetry requirements, they are not yet optimal to compress the system matrix. We show that a further compression can be achieved by applying a secondary orthogonal transform to the matrix $\hat{S}_T$ that can be composed of optimized rotation matrices.

Figure 1: Sampling pattern for a Lissajous trajectory with the ratio $f_x/f_y = 4/5$ and $k = 4$ (left) and $k = 8$ (right).

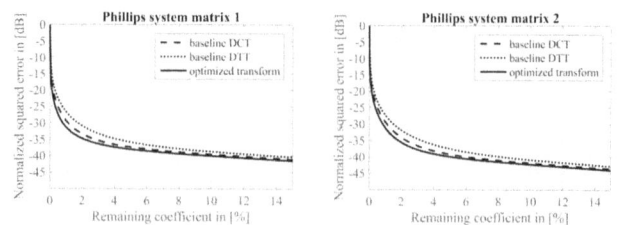

Figure 2: Normalized squared error as function of the percentage of remaining coefficients after hard-thresholding in the range of 0% to 15%. The highest possible error is 0 dB.

For testing, we use the system matrices from the Philips dataset [2]. The frequency ratio in this dataset is $f_x/f_y = 33/32$, and the above mentioned time-domain sampling condition was originally fulfilled when this matrix was acquired. In a second step, several frequencies have been deleted so that only the first 1268 frequency components were available.

RESULTS As can be seen in Fig. 2, our method is performing with a better compression ratio in terms of normalized mean squared error than the standard approach by [3] on the Philips system matrix dataset. One can see that gains of up to 2 dB can be achieved.

CONCLUSIONS We showed that the number of samples per Lissajous trajectory has to obey a certain sampling rule and that the applied spatial transform should obey certain symmetries to ensure sparsity for the transform coefficients. Experimentally, we were able to verify that under these conditions, better compacting transforms can be found than the DCT or DTT.

ACKNOWLEDGEMENTS This work was supported by the German Research Foundation under Grant No. ME 1170/7-1 and BU 1436/7-1.

REFERENCES
[1] B. Gleich and J. Weizenecker. *Nature*, 435(7046):1217—1217, 2005. doi:10.1038/nature03808.
[2] T. Knopp and T. M. Buzug. Springer, Berlin/Heidelberg, 2012. doi: 10.1007/978-3-642-04199-0.
[3] J. Lampe et al, *Physics in Medicine and Biology*, 57(4):1113—1134, 2012. doi:10.1088/0031-9155/57/4/1113.
[4] W. Erb et al, *Numerische Mathematik* , 2015. doi: 10.1007/s00211-015-0762-1

Investigation and Removal of Artifacts Due to Particles Located Outside the Field-Free-Point Trajectory

A. Weber[a,b,*], F. Werner[c,d], J. Weizenecker[e], T.M. Buzug[b], T. Knopp[c,d]

[a] Bruker Biospin MRI GmbH, Rudolf-Plank-Str. 23, 76275 Ettlingen, Germany
[b] Institute of Medical Engineering, University of Lübeck, Ratzeburger Allee 160, 23562 Lübeck, Germany
[c] University Medical Center Hamburg Eppendorf, Section of Biomedical Imaging, Martinistr. 52, 20246 Hamburg, Germany
[d] Hamburg University of Technology, Institute for Biomedical Imaging, Schwarzenbergstr. 95, 21073 Hamburg, Germany
[e] University of Applied Sciences, Moltkestr. 30, 76133 Karlsruhe, Germany
* Corresponding author, email: alexander.weber@bruker.com

INTRODUCTION Magnetic Particle Imaging (MPI) is an imaging modality that allows to visualize the distribution of superparamagnetic iron oxide (SPIO) nanoparticles. For signal generation the field-free-point (FFP) of a gradient field is moved along a given trajectory covering the field-of-view (FOV) [1]. If SPIOs are passed by the FFP, they induce a signal in the receive coils due to their magnetization change. However, because of the non-sharpness of the FFP and rotation effects, particles that are not located within the FOV also generate a MPI signal. If these signals are not handled appropriately, artifacts appear in the reconstructed images. The problem especially arises in *in-vivo* experiments when the FOV is smaller than the particle distribution within the object [2]. Furthermore, signals generated by particles outside the FOV are relevant for multi-patch reconstruction [3]. In this work, we therefore investigate these artifacts on experimental MPI data. Furthermore, we describe a method to handle the signals during reconstruction such that artifacts are strongly suppressed.

MATERIAL AND METHODS To investigate the influence of particles that are not covered by the FFP trajectory, a two-point phantom consisting of two 20 µl delta probes of undiluted Resovist is moved in 1mm steps out of the FOV. The 2D experiments are carried out using a preclinical MPI scanner (Bruker/Philips) at a drive-field amplitude of 14 mT ×14 mT and a gradient strength of 1.25 Tm^{-1} in both imaging directions. This results in a FOV of 22.4×22.4 mm^2. The system matrix is calibrated with a 1×1×1 mm^3 delta probe of undiluted Resovist and covers an area of 42×42 mm^2 with a grid size of 42×42 pixel. Hence, the system matrix is acquired at a larger area than the FOV. By cropping the system matrix prior to reconstruction, different sizes of the so-called overscan can be adjusted. The reconstruction is performed using the Kaczmarz algorithm in combination with Tikhonov regularization.

RESULTS In Figure 1, the reconstruction results of the two-point phantom located 6 mm beyond the FOV are presented. The actual x-position is marked by the white arrow and the white dashed box represents the FOV. It can be seen, that a significant signal is generated, but the intensity is mapped to the exterior border of the FOV. In Figure 2, the reconstruction results for varying overscan sizes, but constant phantom position (4 mm to the left of the FOV) are shown.

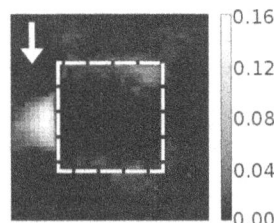

Figure 1: Reconstruction result of a two-point phantom, which is located 6 mm to the left of the FOV border.

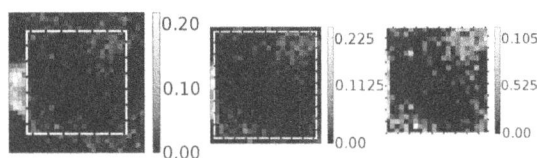

Figure 2: Reconstruction results of a two-point phantom, which is located 4 mm to the left of the FOV border with varying overscan.

Using the system matrix without overscan, artifacts at the interior border of the FOV occur. These artifacts can be removed by employing a system matrix with overscan. For the considered data an overscan of one voxel is sufficient.

CONCLUSION Particles that are not covered by the FFP trajectory also generate a significant signal contribution, which causes artifacts at the inner border of the FOV if the system matrix only covers the FOV. By using a system matrix that is larger than the FOV, these artifacts are mapped outside the FOV such that they can be easily removed by cropping the final image to the FOV.

ACKNOWLEDGEMENTS The authors thankfully acknowledge the financial support by the German Federal Ministry of Education and Research (BMBF, grant number 13N11088) and the German Research Foundation (DFG, grant number AD 125/5-1).

REFERENCES
[1] B. Gleich and J. Weizenecker. *Nature*, 435(7046):1217—1217, June 2005.
[2] J. Weizenecker et al, *Phys. Med. Biol.*, 54:L1-10, 2009
[3] T. Knopp et al, *Phys. Med. Biol.*, 60:L15, 2015

MMSE MPI Reconstruction Using Background Identification

Hanna Siebert[a], Marco Maass[a,*], Mandy Ahlborg[b], Thorsten M. Buzug[b], and Alfred Mertins[a]

[a] Institute for Signal Processing, University of Lübeck, Germany
[b] Institute of Medical Engineering, University of Lübeck, Germany
[*] Corresponding author, email: maass@isip.uni-luebeck.de

INTRODUCTION Magnetic particle imaging (MPI) is an imaging method for the determination of the distribution of superparamagnetic iron oxide nanoparticles (SPIONs) injected into the human body [1]. Since the SPIONs often do not spread within the whole field of view (FOV) due to anatomical structures, areas without SPIONs are likely to exist in the FOV. As each column of the system matrix represents the impact of one position in the FOV, we propose to estimate which columns of the system matrix correspond to the positions in the FOV without SPIONs. The advantage of identifying and removing the columns of the system matrix belonging to positions without particle concentration is studied. Several methods to find these positions are introduced and evaluated. Furthermore, a method for reconstruction based on the minimum mean-square error (MMSE) estimator is applied.

MATERIAL AND METHODS The particle concentration c is reconstructed from the voltage signal u with the help of the system matrix S by the signal equation $u = Sc$, where $u \in \mathbb{R}^N$, $S \in \mathbb{R}^{N \times M}$ and $c \in \mathbb{R}_+^M$. The reconstruction based on the MMSE estimator uses $\hat{c}(u) = \left[R_{cc}^{-1} + S^T R_{nn}^{-1} S \right]^{-1} S^T R_{nn}^{-1} u$, where $R_{cc} \in \mathbb{R}^{M \times M}$ represents a matrix used for regularization and $R_{nn} \in \mathbb{R}^{N \times N}$ is the correlation matrix of additive zero-mean noise n [2]. Here, white Gaussian noise is assumed, which leads to $R_{nn} = \sigma_n^2 I$. The matrix $R_{cc} = E\{cc^T\}$ is the correlation matrix for the particle concentration c. It can be generated based on a priori knowledge of the typical particle concentrations. The applied system matrix S is obtained by simulation of a Lissajous sequence [3].

To identify background pixels with the SPION concentration zero, an iterative threshold method and a linear classifier are investigated. For the experiments, phantoms showing different lower case letters with the values zero (background) and one (foreground) were designed and divided up into training and test data. With the knowledge of the identified background pixels, the corresponding columns of the system matrix can be eliminated in the signal equation.

The iterative thresholding method detects pixels with the value zero in several iterations. In each iteration, a reconstruction based on the MMSE estimator is performed, followed by an application of a threshold value deciding which pixels are assigned with the value zero. Afterwards, the corresponding columns of the system matrix are deleted and the modified system matrix is used for the reconstruction in the following iteration. The algorithm terminates as soon as the threshold value stops identifying pixels with the value zero.

The linear classifier identifies pixels with the value zero based on its neighborhood. By applying Fisher's linear discriminant analysis, a linear function that allows us to differentiate between pixels of the background and foreground is determined.

As a method to prevent pixels of the foreground to be classified as background, the classification is followed by a correction step that assigns every pixel in a four-pixel neighborhood of a foreground pixel as foreground as well.

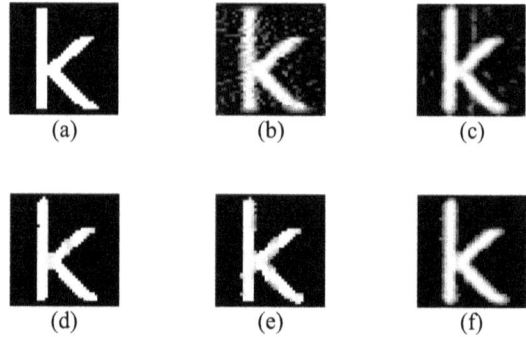

Figure 1: Phantom (a) and reconstruction results (b-f) for SNR=30dB.

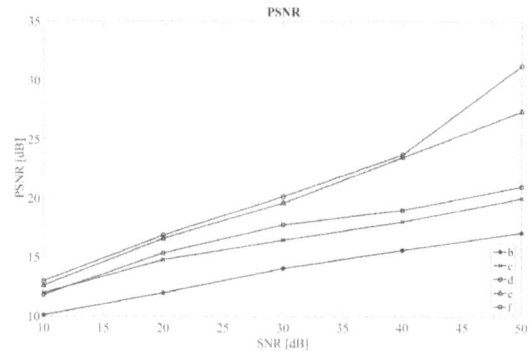

Figure 2: PSNR for SNR of 10-50dB.

RESULTS Fig. 1 shows a comparison of the reconstruction results for different cases. (b) MMSE estimator with $R_{cc_\lambda} = \lambda I$; (c) MMSE estimator with proper R_{cc}; (d) MMSE estimation after background pixel identification by iterative thresholding method and deletion of columns of the system matrix; (e) MMSE estimation after background-pixel deletion based on a linear classifier; (f) MMSE estimation after background deletion based on a linear classifier combined with post processing. As can be seen in Fig. 2, the peak signal-to-noise ratio (PSNR) of (d-f) with identified background pixels is higher than the PSNR of (b-c) without identified background pixels.

CONCLUSION We were able to show that a proper correlation model R_{cc} improves the reconstruction result. Furthermore, removing the columns of the system matrix corresponding to background pixels identified by our introduced methods can improve the result.

REFERENCES
[1] B. Gleich and J. Weizenecker. *Nature*, 435(7046):1217—1217, 2005. doi: 10.1038/nature03808.
[2] S. M. Kay. Statistical Signal Processing – Estimation Theory. Prentice Hall, New Jersey, 1993.
[3] T. Knopp and T. M. Buzug. Springer, Berlin/Heidelberg, 2012. doi: 10.1007/978-3-642-04199-0

Optimizing the Coil Setup for a Three-Dimensional Magnetic Particle Spectrometer

Xin Chen[a,*], André Behrends[a], Matthias Graeser[a], Alexander Neumann[a], Thorsten M. Buzug[a]

[a] Institute of Medical Engineering, University of Lübeck, Lübeck, Germany
[*] Corresponding author, email: {chen, buzug}@imt.uni-luebeck.de

INTRODUCTION Magnetic particle imaging (MPI) is a novel tomographic imaging technique, which measures the distribution of superparamagnetic iron oxide nanoparticles (SPIONs) with a high sensitivity and resolution [1]. A magnetic particle spectrometer (MPS) can be used to analyze the magnetic response of the SPIONs and determine their usability for imaging [2]. For the particle synthesis this is a fast and easy way to investigate the synthesis parameters. Due to the known spatial distribution the selection field can be excluded from the spectrometer. The MPS applies oscillating magnetic fields to the SPIONs and measures their response. Additionally, homogeneous offset fields can be applied to measure the magnetization for asymmetric excitation. Recently, a spectrometer has been introduced that is advancing the use of MPS to achieve the system matrix for a given combination of trajectory and particle type [3]. However, this spectrometer is limited to measurements in two dimensions. The work presented here optimizes the coil setup for a three-dimensional magnetic particle spectrometer in terms of power consumption. Moreover, experimental fabrication molds and assembly frames are presented to proof the manufacturability and feasibility of the setup. Furthermore, a cooling system has been designed and has been tested to prove the applicability of implementation.

MATERIAL AND METHODS The aim of this work is to optimize the coil setup in order to reduce power consumption. To ensure that the magnetization of the nanoparticles is driven in saturation, the magnetic field of $20\,mT/\mu_0$ should be achieved [4]. The main aspect of the coil setup optimization is to find a proper geometry for the coils. Four different coil setups have been compared, i.e. flat circular coils, curved circular coils, flat rectangular coils and curved rectangular coils. In the simulation study, flat rectangular coils show the best result. In addition, the closer the pair of coils is, the lower the power consumption. As the bore of the coil setup should be easily accessed, a quasi-rectangular coil setup has been proposed (Figure 1).

After the coils being fabricated, a magnetometer is used to measure the magnetic field in order to adjust the current, which is then compared to the simulation result. Furthermore, a coil support frame is designed for assembling the coil setup, and two fans are used for cooling the coil setup.

RESULTS The adjusted current amplitude for the inner and outer pair of coils are 18.4 A and 22.4 A, the measured resistances are 45.2 mΩ / 44.3 mΩ and 60.4 mΩ / 64.7 mΩ (@ 25 kHz), respectively. The total power consumption for this coil setup is 50.6 W. In comparison with the former designed two-dimensional coil setup [3] with power consumption of 163 W, this is a considerable improvement. However, the former coil setup is optimized in terms of field homogeneity for particle parameter estimation. As this setup is planned for recording hybrid system matrices [5], some inhomogeneity can be tolerated. The heating test of the two-pair coil setup with the cooling system shown above reaches a stable temperature of 63 c after 1 hour for the desired field strength. By air guidance, this can be further improved.

CONCLUSION In this work, an optimized coil setup for a three-dimensional magnetic particle spectrometer is presented. It shows not only the improvement of power consumption by the factor of three, but also verifies the applicability of future technical implementation.

ACKNOWLEDGEMENTS This work was supported by the Federal Ministry of Education and Research, Germany (BMBF) under Grant 13GW0069A.

REFERENCES

[1] B. Gleich, J. Weizenecker. Tomographic Imaging Using the Nonlinear Response of Magnetic Particles. Nature, 435(7046):1214- 1217, 2005. doi: 10.1038/nature03808.

[2] S. Biederer, T. Knopp, T. F. Sattel, K. Lüdtke-Buzug, B. Gleich, J. Weizenecker, J. Borgert, T. M. Buzug. Magnetization Response Spectroscopy of Superparamagnetic Nanoparticles for Magnetic Particle Imaging. Journal of Physics D: Applied Physics, 42(20), 205007, 2009. doi: 10.1088/0022-3727/42/20/205007.

[3] M. Graeser, M. Ahlborg, A. Behrends, K. Bente, G. Bringout, et al. A Device for Measuring the Trajectory Dependent Magnetic Particle Performance for MPI. 2015 International Workshop on Magnetic Particle Imaging (IWMPI), 2015. doi: 10.1109/IWMPI.2015.7107078.

[4] S. Biederer, T. Knopp, T. F. Sattel, K. Lüdtke-Buzug, B. Gleich, J. Weizenecker, J. Borgert, T. M. Buzug, Estimation of Magnetic Nanoparticle Diameter with a Magnetic Particle Spectrometer. Springer IFMBE Series, 25/3:61-64, 2009. doi: 10.1007/978-3-642-03887-7_17.

[5] M. Graeser, et al. Reconstruction using a Hybrid System Matrix for 2D Imaging, 2016 International Workshop on Magnetic Particle Imaging (IWMPI), (same proceeding), 2016.

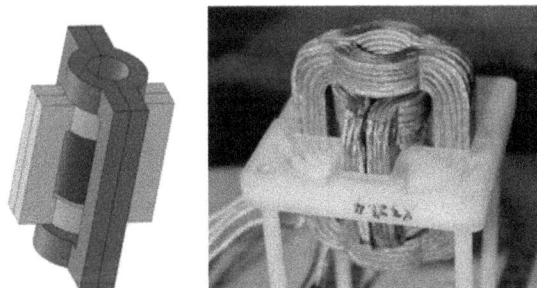

Figure 1: The quasi-rectangular coils setup in simulation (left) and implementation (right).

Development and Testing of Magnetic Nanoparticle-Gel Materials for Magnetic Particle Imaging Phantoms

R. Sandig[a], A. Mattern[a], D. Baumgarten[a], O. Kosch[b], F. Wiekhorst[b], A. Weidner[a], S. Dutz[a,*]

[a] Institut für Biomedizinische Technik und Informatik (BMTI), Technische Universität Ilmenau, Germany
[b] Physikalisch-Technische Bundesanstalt, Berlin, Germany

* Corresponding author, silvio.dutz@tu-ilmenau.de

INTRODUCTION A promising approach to directly determine the spatial distribution of magnetic nanoparticles (MNP) within tissue is Magnetic Particle Imaging (MPI) [1]. In the past 10 years, there was a rapid progress in development of MPI scanners and nowadays the very first preclinical devices are commercially available from Bruker BioSpin (Ettlingen, Germany). To evaluate performance of commercial as well as various custom-made scanners in several laboratories, dedicated phantoms with defined MNP distributions are required. MPI directly detects the magnetic response of MNP and is not influenced by any tissue or background signals. Therefore, conventional phantoms used for MRI have very limited suitability for MPI imaging. Thus, we aim at developing and evaluating test phantoms for MPI that enable their assessment in present MPI scanners. Prerequirement for the development of the phantoms is the establishment of suitable MNP-matrix combinations, which enable homogeneous MNP distributions at defined concentrations combined with immobilization of the MNP within the matrix to guarantee long-term stability of magnetic behavior and high mechanical stability of the matrix material. As matrix materials, gels are used which show similar imaging behavior like body tissue in MRI and MPI.

MATERIAL AND METHODS Magnetic multicore iron oxide nanoparticles of about 50 nm core diameter were prepared as described before [2], coated with different shells (Dextran, CM-Dextran, and DEAE-Dextran) and embedded in 6 various matrix materials (agar-agar: "AA", agarose: "Ag", ballistic gel: "BG", gelatin: "Gel", and two different gelatin-oil mixtures: "GelOl" and "GelOlF").
The obtained MNP-matrix combinations were tested for their mechanical stability by means of mechanical load tests. The homogeneity of MNP distribution and immobilization within the matrix was determined by optical investigation of the samples with a microscope and by investigation of the magnetic particle properties measured by vibrating sample magnetometry (VSM).
The most promising MNP-matrix combination was used to manufacture measurement objects of different shape (spheres, cubes) and different size (5, 10, 20 mm) embedded in a phantom matrix with an overall geometry of a cylinder with D = 50 mm and H = 60 mm. The resulting test phantoms were evaluated for their suitability to simulate MNP loaded areas within a nonmagnetic matrix by means of MRI (Bruker Icon) and MPI (Bruker BioSpin preclinical MPI-scanner).

RESULTS The coercivity served as a measure for inhomogeneity of the MNP distribution in the MNP-matrix combinations, caused by agglomeration of the MNP within the matrix (cf. Fig 1). We assume that a high coercivity is caused by unwanted agglomeration of the particles. Low coercivity can be caused by not sufficiently immobilized single particles as well as by mobile agglomerates. To distinguish between these both cases, the absence of agglomerates in the matrix was checked by

microscopy. Microscopic and magnetic investigations revealed that bare and CM-Dextran coated MNP show the best results regarding homogeneous distribution of MNP within a ballistic gel matrix.

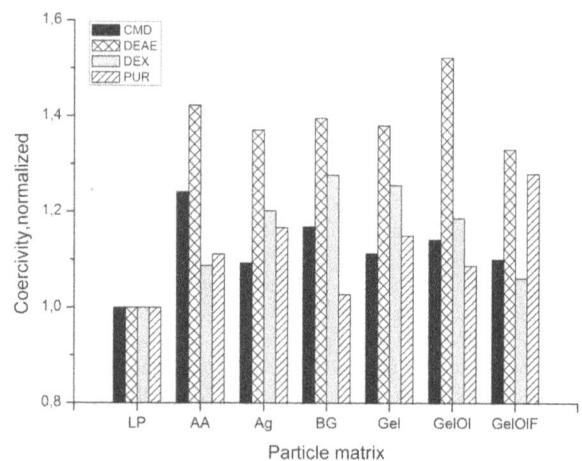

Figure 1: Coercivity of different MNP-matrix combinations, normalized to the coercivity of dried MNP ("LP").

Resulting from these investigations, the combination of CM-Dextran coated MNP within ballistic gel was chosen as the most promising one of the investigated combinations for test phantom building. Several strategies for preparation of measurement objects and their embedding in the phantom matrix were tested and will be discussed together with the results of the test measurements in our presentation.

CONCLUSION We found suitable combinations of coated magnetic nanoparticles and matrix materials for the buildup of MPI phantoms, which guarantee a fixation of the MNP within the matrix without agglomeration of the particles. From these materials, MPI phantoms were designed, produced, and tested by means of MRI and MPI.

ACKNOWLEDGEMENTS The authors thank M.Sc. Mathias Fritz (TU Ilmenau) for assistance during microscopy.

REFERENCES
[1] Gleich B, Weizenecker J. Tomographic imaging using the nonlinear response of magnetic particles. nature 2005; 435: 1214–1217.
[2] Dutz S, Clement JH, Eberbeck D, Gelbrich Th, Hergt R, Müller R, Wotschadlo J, Zeisberger M. Ferrofluids of magnetic multicore nanoparticles for biomedical applications. J Magn Magn Mater 2009; 321/10: 1501–1504.

Dynamic Magnetization of Immobilized Magnetic Nanoparticles for Cases with Aligned and Randomly Oriented Easy Axes

Takashi Yoshida [a,*], Thilo Viereck [b], Teruyoshi Sasayama [a], Keiji Enpuku [a], Meinhard Schilling [b], and Frank Ludwig [b]

[a] Department of Electrical Engineering, Kyushu University, Japan
[b] Institute of Electrical Measurement and Fundamental Electrical Engineering, TU Braunschweig, Germany
[*] Corresponding author, email: t_yoshi @ees.kyushu-u.ac.jp

INTRODUCTION In magnetic particle imaging (MPI), nonlinear dynamic magnetization signals from magnetic nanoparticles (MNPs) are detected. Therefore, the performance of MPI strongly depends on the dynamic magnetization properties of the MNPs [1]. In this study, we investigated the dynamic magnetization properties via the Néel mechanism of immobilized MNPs by a numerical simulation. In order to clarify how the easy-axis direction of the MNPs with respect to an AC excitation field affects the dynamic properties, we studied the case of aligned and randomly oriented easy axes.

METHODS The behavior of MNPs via the Néel mechanism can be described by the Fokker–Planck equation [2]. In this numerical simulation, the characteristic Néel relaxation time τ_{N0} is set to be $\tau_{N0} = 10^{-9}$. By performing numerical calculations of the Fokker–Planck equation, we obtain the k-th harmonic of the complex magnetization, M_k, in the direction of the AC excitation field. In the calculation, an AC excitation field $H(t) = H_{ac} \cos 2\pi f t$ is applied with an angle β relative to the easy-axis direction. For the case where the easy axes of all the MNPs align along the direction of the excitation field, the magnetization is given by performing a numerical simulation with $\beta = 0$. On the other hand, when the easy axes of the MNPs are randomly orientated, the magnetization can be calculated by averaging the magnetization over $0 \leq \beta \leq \pi/2$.

RESULTS In Fig. 1, the frequency dependence of real part M_1' and imaginary part M_1'' of the fundamental component of the magnetization are shown for the case of $\xi_{ac} = \mu_0 m H_{ac}/k_B T = 5$ and $\sigma = K V_C / k_B T = 13$. As shown, the magnetization M_1' at very low frequencies in the aligned case is approximately 1.8 times of that in the randomly orientated case. The imaginary part M_1'' has a peak value at a specific frequency f_p. In Fig. 2, the dependence of f_p on ξ_{ac} is shown. We can see that f_p in the random case increases significantly faster with ξ_{ac} than that for the aligned case. Since the effective Néel relaxation time is given by $\tau_{Neff} = 1/2\pi f_p$, this result indicates that the effective Néel relaxation time in a large ac field is much shorter for the random case than the aligned case. We note that the analytical expression for the field-dependent Néel relaxation time in the case of aligned MNPs is given by [2]

$$\tau_N(\xi, \sigma) = \frac{\sqrt{\pi}}{2\sqrt{\sigma}} \tau_{N0} \frac{\exp[\sigma(1+h^2)]}{(1-h^2)(\cosh\xi - h\sinh\xi)}, \quad \text{with } h = \xi/2\sigma. \quad (1)$$

As shown in Fig. 2, eq.(1) explains well the simulation results, where we set $\xi = \xi_{ac}/\sqrt{2}$ in eq. (1). We will also discuss the difference in the amplitude of the harmonic magnetization spectra from MNPs between the cases with aligned and randomly orientated easy axes.

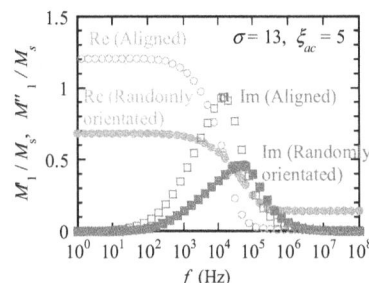

Figure 1: Frequency dependence of the real part M_1'/M_s and imaginary part M_1''/M_s of the magnetization of MNPs.

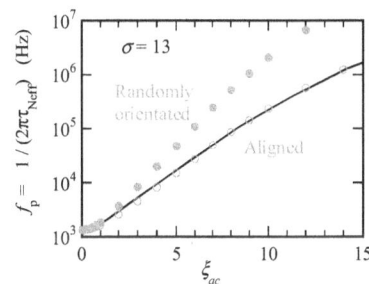

Figure 2: Relationship between f_p and ξ_{ac}. Symbols show simulation results. The solid line is obtained from eq. (1).

CONCLUSION We investigated the dynamic magnetization properties of immobilized MNPs for the cases of aligned and randomly oriented easy axes. A numerical simulation showed that the dynamic magnetization properties strongly depend on the easy-axis orientation. Therefore, the easy-axis direction is also an important parameter for MNP dynamics.

ACKNOWLEDGEMENTS This work was supported by the JSPS KAKENHI (15H05764 and 26820159) and by the European Commission Framework Programme 7 under NANOMAG Project (NMP-LA-2013-604448).

REFERENCES
[1] B. Gleich and J. Weizenecker. Nature, 435(7046):1217—1217, 2005. doi: 10.1038/nature03808.
[2] W. T. Coffey, P. J. Cregg, Y. P. Kalmykov, in: I. Prigogine, S. A. Rice (Eds.), Advances in Chemical Physics, vol.83, Wiley, New York, 1993, p.263. doi: 10.1002/SERIES2007.

61

Elevator speeches 2:

Methodology

A Novel Magnetic Particle Imaging Scanner with Lower Amplitude of an Excitation Field

Xingming Zhang[ab], Tuấn Anh Lê[a], and Jungwon Yoon[*a]

[a] School of Mechanical and Aerospace Engineering, Gyeongsang National University, Jinju 660-701, Republic of Korea
[b] School of Naval Architecture and Ocean Engineering, Harbin Institute of Technology at Weihai, Weihai, Shandong, China.
[*] Corresponding author, email: jwyoon@gnu.ac.kr

INTRODUCTION Magnetic Particle Imaging (MPI) is a promising technique to provide a fast and sensitive imaging modality to measure the spatial distribution of magnetic nanoparticles (MNPs). Currently, a drive field with a frequency of 25 kHz and amplitude of a 20 mT/μ_0 has been widely used in MPI[1]. However, the amplitude of the drive field should be reduced significantly for the medication safety[2]. Since a signal resource usually has harmonic distortions, MPI needs high quality passive filters after signal processing processes by the power amplifier (PA) and receive coil. The filter used after PA can reduce harmonic distortions of drive currents with high currents circuit, while the filter used after receive coil removes the excitation signal. These two filters make the system very complex. Moreover, the amplitude of particle signal at the first harmonic provides main information of a carrier component, which is removed by the passive filter.

Thus, this study presents a novel MPI scanner using a Lower Amplitude Excitation Field (LAEF) and using RMS of signal at the first harmonic to reconstruct an image.

MATERIAL AND METHODS Fig.1 shows the coil configuration of the proposed MPI. An excitation coil can change the magnetization of MNPs and induce signals with a high frequency LAEF. Due to LAEF, Field of View (FOV) shrinks to a point, which coincides with Field Free Point (FFP). Thus, the concentration of MNPs (c_p) at FFP becomes uniform which can only cause the amplitude change of signal. Empty coil signal can be reduced by cancellation coils[3]. Then, c_p can be obtained after demodulating signal. The selection coils load AC currents of a lower frequency with a DC offset to achieve and move the FFP in a whole workspace. The signal RMS is linearly proportional to the c_p at FFP. Fig. 2 shows signal flow diagram. Experiments were carried out to verify the proposed new concept of a single axis MPI (Fig.1). The amplitude of LAEF was reduced to a 68 μT/μ_0 with 40kHz, then 1mL MNPs with 60~70nm core sizes with c_p =8mg(Fe)/mL were placed in the center of the receive coil.

RESULTS The Fig.3 shows the 1D image reconstruction of the proposed MPI system. MNPs' samples in different positions were used for image reconstruction. The more detailed results and explanations will be presented later.

CONCLUSION Because the proposed MPI system only requires a low amplitude and low RF power, this can be easily scaled up to a human size range due to LAEF, and it also allows the reconstruction for MPI to be more simplified. These possible advantages will make the proposed MPI more feasible with high possibility of widespread use.

Cancellation and recieve coils / Low amplitude excitatoin coil / Low amplitude excitatoin coil

Figure 1: The configuration of an MPI Scanner with LAEF

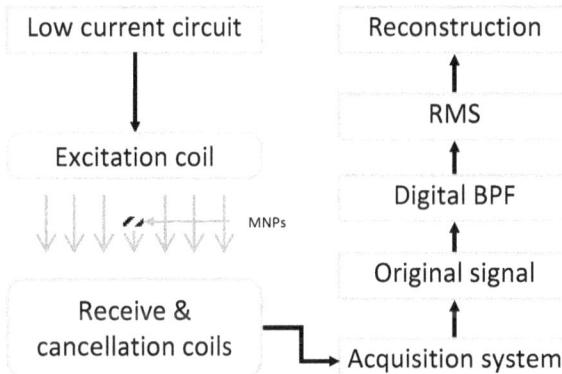

Figure 2: Signal flow diagram of the realized setup

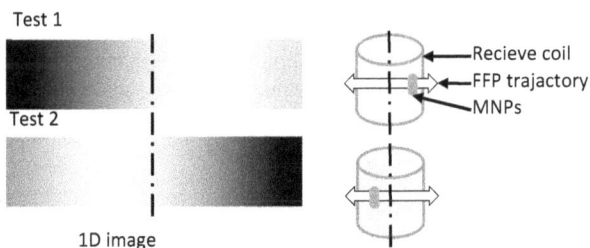

Figure 3: 1D image reconstruction. Each test spends 5 sec acquisition time.

REFERENCES
[1], R. M., Ferguson, K. R., Minard &, K. M. Krishnan. Journal of magnetism and magnetic materials, 2009. . doi:10.1016/j.jmmm.2009.02.083
[2] I. N. Weinberg,, P. Y. Stepanov, S. T. Fricke, R. Probst, M. Urdaneta, D. Warnow, & J. P. Reilly,. Medical physics, 2012. doi: 10.1118/1.3702775
[3] V. Schulz, M. Straub, M. Mahlke, S. Hubertus, T. Lammers, & F. Kiessling, *IEEE Transactions on*, 2015. doi: 10.1109/TMAG.2014.2325852

RDS Toolbox – Simulation of 3D Rotational Drift

A. Vilter[*], M. A. Rückert, T. Kampf, V. J. F. Sturm, V. C. Behr

University of Würzburg, Experimental Physics V, Würzburg, Germany
[*] Corresponding author, email: avilter@physik.uni-wuerzburg.de

INTRODUCTION The optical detection of rotational drift of individual magnetic micro-particles in liquid suspension driven by an external magnetic field B has been published in [1]. In this work B was a field rotating in a plane (hereafter referred to as the xy-plane) with constant amplitude B and rotating frequency ω_0. Due to constraints imposed by the experimental setup the motion of particles was restricted to the xy-plane (2D system). Depending on the ratio $\beta = \Omega_c/\omega_0$ of the driving frequency ω_0 and a critical frequency Ω_c (comprising field strength and parameters of particles and environment) two different types of rotational behavior are observed. For $\beta \geq 1$ the magnetization vector of the particle follows the external field with a defined characteristic phase lag (*lock case*). For $\beta < 1$ the magnetization vector performs a non-linear rotational drift at periodically oscillating angular velocity $\Omega[t]$ and period of oscillation T (*drift case*). The average angular velocity $\bar{\Omega} = 2\pi T^{-1}$ shows a functional dependency on the external field and the observed system (particles and environment). Therefore, it is a characteristic of the system under observation – comparable to the Larmor frequency in magnetic resonance. This triggered the development of the concept of rotational drift spectroscopy (RDS) [2], which aims at measuring inductively the behavior of magnetic particle ensembles as well as their interaction with their environment based on the effect of rotational drift. For further development of the RDS concept mathematical description and simulation of this effect including systems with no motion restriction (3D system) is fundamental.

MATERIAL AND METHODS In contrast to the description of the magnetization behavior of SPIOs based on the Langevin theory of paramagnetism the dynamics of the particles considered in [1] can be described by the time evolution of a magnetic moment m. This moment is assumed to be locked relatively to particle geometry and has a constant amplitude m. Neglecting inertial terms [2], the corresponding equation of motion is given by

$$\frac{d\boldsymbol{m}}{dt} = \frac{1}{\zeta}(\boldsymbol{m} \times \boldsymbol{B}) \times \boldsymbol{m} \tag{1}$$

For further considerations, B was chosen as a vector rotating in the xy-plane with constant angular velocity ω_0 and amplitude B following the assumptions in [1].
First numerical RDS simulations [3] base on a discretized version of this equation. It is solved by a standard time domain explicit Euler method with successive normalization step to account for the conservation of m. Additionally, a semi analytical two-step procedure in a coordinate system rotating with B was implemented. In this rotating frame of reference the time evolution of m can be described by a rotation of m around a current axis $\boldsymbol{\Omega}_r[t] = \Omega_c(\hat{n}_m[t] \times \hat{e}_x) - \omega_0 \hat{e}_z$. In the first step, this axis $\boldsymbol{\Omega}_r[t_i]$ and corresponding angle of rotation $\Delta\vartheta[t_i] = |\boldsymbol{\Omega}_r[t_i]| \Delta t$ were calculated. In the second step, the rotation of m around this current axis was executed. Being norm-conserving in itself this saves renormalization. Additionally, an analytical analysis was performed.

RESULTS For the assumed field geometry, a complete analytical solution can be found. It shows that in the rotating frame of reference, the amplitude of Ω_r varies over time while its direction is preserved. In a similar manner to the effects described in [1] also for 3D systems two different types of rotational behavior occur. Corresponding trajectories are shown in Fig. 1. For $\beta \geq 1$ the magnetization relaxes to the xy-plane resulting in the same time constant phase lag as described for the *lock case* of the 2D system. For $\beta < 1$ m performs a precessing motion on a cone around Ω_r resulting in circular trajectories in the rotating frame. Hereby, $\Omega_r[t]$ oscillates periodically resulting in the same $\bar{\Omega}_r$ as in *drift case* of the 2D system.

Comparing the numerical implementations shows that the two-step procedure overcomes issues of the original explicit Euler method and yields more stable results, which agree with the analytical solution.

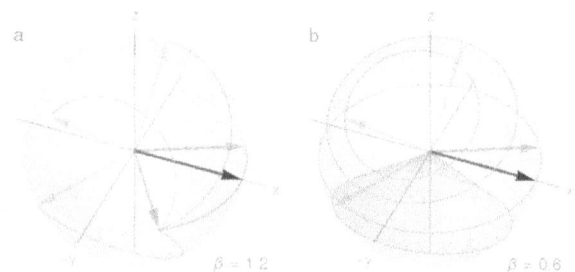

Figure 1: Trajectories of the time evolution of m in the rotating frame for three random starting orientations (gray arrows): Two rotational behaviors can be observed (a: lock case b: drift case).

CONCLUSION Even for a motion of magnetic particles not restricted to the plane of the rotating field a rotational drift occurs, which is the basis for the RDS concept. Both, numerical simulations as well as the analytical solution describe this effect of 3D rotational drift and serve as toolbox for further RDS simulation. The analytical approach provides a complete solution for the problem considered in this work while the numerical approaches offer the option to easily extend the simulations to more complex questions covering parameter distributions for particle systems, thermal noise and more complex field geometries.

ACKNOWLEDGEMENTS This work is funded by the DFG (BE 5293/1-1) and the Elite Network of Bavaria of the Bavarian State Ministry of Education, Science and the Arts.

ACKNOWLEDGEMENTS
[1] B.H. McNaughton, K.A. Kehbein, J.N. Anker, R. Kopelman, J. Phys. Chem. B, 2006. doi: 10.1021/jp060139h.
[2] M.A. Rückert, P. Vogel, A. Vilter, W.H. Kullmann, P.M. Jakob, V.C. Behr, IEEE Trans. Magn. 51(2):6500604, 2015. doi: 10.1109/TMAG.2014.2334138.
[3] M.A. Rückert, P. Vogel, T. Kampf, W.H. Kullmann, P.M. Jakob, V.C. Behr, IEEE Trans. Magn. 51(2):6500704, 2015. doi: 10.1109/TMAG.2014.2335536.

Magnetic signal detection method based on active vibration of magnetic nanoparticles

Akihiro Matsuhisa [a,*], Tomoki Hatsuda [b], Tomoyuki Takagi [b], Masahiro Arayama [a], Yasutoshi Ishihara [a]

[a] School of Science and Technology, Meiji University
[b] Graduate School of Science and Technology, Meiji University
* Corresponding author, email: akn0611@i.softbank.jp

INTRODUCTION Magnetic particle imaging (MPI) [1] is a medical imaging technology suitable for noninvasive diagnostics. In the conventional MPI method, image artifacts and blurring appear when the magnetic-field gradient is not sufficiently steep. Therefore, huge coils and multiple power supplies are required to generate steep magnetic-field gradients and alternating magnetic fields. We proposed a new signal detection method based on vibrating magnetic nanoparticles (MNPs) that does not require any huge hardware system [2]. Although the same concept system was reported [3], we built a prototype system introducing a shaker to control oscillation accurately and evaluated the influence of oscillation parameters on the detection sensitivity of MNPs.

METHODS Figure 1 shows the principle of the proposed method. A phantom containing MNPs is located within the magnetic-field gradient generated by a pair of Maxwell coil, and it is vibrated by the shaker. The position of MNPs varies when they are vibrated within the magnetic-field gradient. Consequently, the magnetic field generated from the MNPs changes. An oscillating magnetization signal is detected when MNPs exist at the field-free point (FFP), while only a small magnetization signal is detected when MNPs exist at regions with strong magnetic field. Magnetization signals are detected as induced electromotive force (EMF) by a receiver coil. In this study, the FFP position is scanned by moving the Maxwell coil, and EMF is measured at each point. The data acquired through the averaging process is discrete Fourier-transformed. By extracting the intensity of the first harmonic component at each position, the distribution of MNPs is estimated.

Figure 2 schematically shows the experimental system. A magnetic-field gradient of 1.13 [T/m] is applied along the x-axis to MNPs in the direction perpendicular to the vibration direction of the shaker. The FFP position is adjusted using a 2-axis stage and scanned every 2 mm in the range of ±30 [mm] along the x-axis direction. The amplitude of oscillation is set as ±0.95 [mm], and the frequency is set as 30 [Hz] at each FFP. Cylindrical phantoms (length of 15 [mm], diameter of 13 [mm]), which included MNPs made of polymethyl methacrylate (PMMA) with a particle diameter of 19 [nm], are used. Two such phantoms are arranged 10 [mm] apart along the x-axis. The receiver coil (length of 60 [mm], diameter of 20 [mm]) with 570 turns is set along the x-axis.

RESULTS The normalized profile of one-dimensional MNP distribution based on the first harmonic component is shown in Fig. 3. The distribution of MNPs reflects the position and lengths of two phantoms. Signal acquisition with averaging over 500 points was required in this system.

CONCLUSION In this study, we proposed a new signal detection method based on vibrating MNPs that does not require the huge hardware. From the obtained result, it was confirmed that the

signal generated from MNPs could be detected by vibrating them. Further experiments should be conducted in the future to determine the optimum oscillating frequency and amplitude corresponding to the gradient of the magnetic-field intensity.

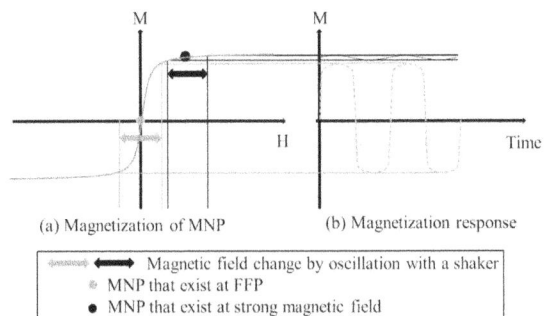

(a) Magnetization of MNP   (b) Magnetization response

Magnetic field change by oscillation with a shaker
● MNP that exist at FFP
● MNP that exist at strong magnetic field

Figure 1: Principle of the proposed method.

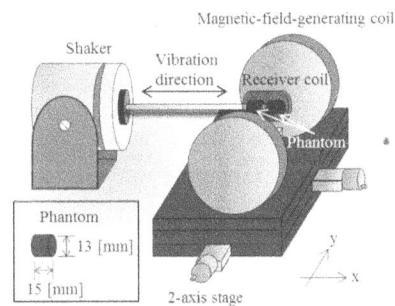

Figure 2: Schematic of experimental system.

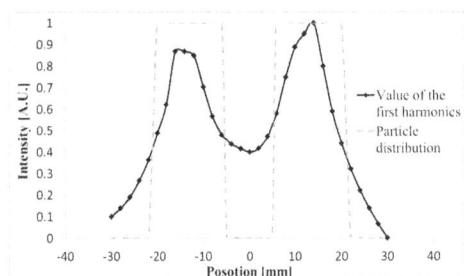

Figure 3: Normalized one-dimensional MNP distribution.

REFERENCE

[1] B. Gleich and J. Weizenecker. *Nature*, 435(7046):1214–1217, 2005.

[2] Y. Ishihara, "Improvement of the MPI image resolution by using spatial property of magnetization," FY2012 Procedures for Grant-in-Aid for Specially Promoted Research, 2011.

[3] P. Vogel, M. A. Rückert, P. M. Jakob, and V. C. Behr. *IEEE Trans. Mag.*, 51(2):6502104.1–6502104.4, 2015.

The Influence of Discretization of DC Field on Magnetic Nanothermometer

Le He [a,b], Shiqiang Pi [b], Qingguo Xie [a], Wenzhong Liu [b,*]

[a] Wuhan national laboratory for optoelectronics, Huazhong University of Science and Technology
[b] School of Automation, Huazhong University of Science and Technology
[*] Corresponding author, email: lwz7410@hust.edu.cn

INTRODUCTION In the recent decades, magnetic nanoparticles (MNPs) has been attracted increasing attention in biomedical research such as magnetic hyperthermia [1, 2] and noninvasive nanothermometer [3, 4]. Magnetothermometers have been proposed by using the first-order Langevin function under DC applied field. [4] In this study, discrete magnetization curves induced by two discretization strategies of DC applied field are utilized to probe temperature by simulation. Moreover, the accuracy of temperature probing for these two strategies of DC applied field is discussed in the present study.

MATERIAL AND METHODS The static magnetization of superparamagnetic nanoparticles is described by:

$$M = nm\left(\coth\left(\frac{mH}{kT}\right) - \frac{kT}{mH}\right) \qquad (1)$$

where n is the concentration of MNPs, m is the average effective moment, H is the applied magnetic field, k is the Boltzmann constant, and T is the absolute temperature.

From magnetization measurement, discrete points of magnetization M_i and corresponding applied field H_i are obtained to constitute a series of equations shown in [4]. We can then obtain information on temperature from the equations in [4]. According to non-linearity of first-order Langevin function, the problem of discretization of magnetization curve has influence on solving the equations, which ultimately impacts the accuracy of temperature probing.

Therefore, discretization strategy of magnetization curve is one of important impact factors on the precision of temperature probing. Moreover, the discretization quality of magnetization curve is determined by discretization of applied field. [5] Finally, the discrete quality of DC applied field influence on temperature probing precision. Lloyed-Max algorithm is employed to optimize discrete DC applied field in this study, which is a common method in optimal quantization. [5]

RESULTS To compare precision of temperature probing under above two ways of DC field applying, the uniform discrete DC field points are set from 20 to 200 Oe with a step of 10 Oe. According to Lloyed-Max algorithm, the uniform discrete DC field points have been optimized into non-uniform ones based on iterative methods in [5]. Figure 1 shows the uniform and optimized discrete magnetization curves and temperature errors. In the range of our investigation, the maximum error decreases from 0.25 K under uniform discrete DC field to 0.08 K under optimized discrete DC field, and corresponding standard deviation decreases from 0.12 K to 0.05 K as well. The temperature probing precision increases nearly by a factor of 3. Therefore, the accuracy can be improved by optimizing the discrete points of DC applied field.

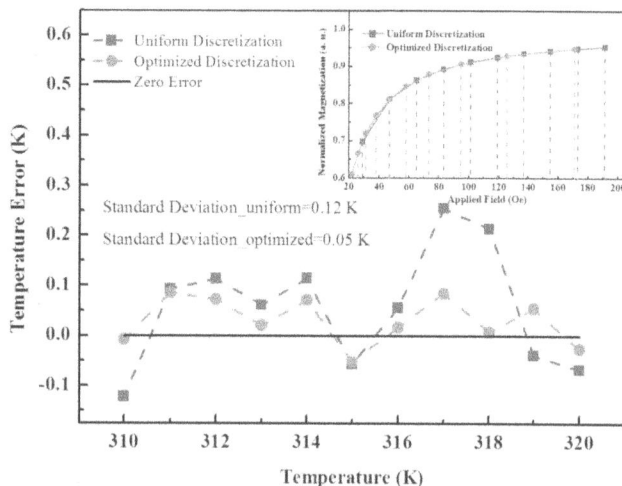

Figure 1: Temperature probing errors under uniform and optimized discretization of DC field and discrete strategies of magnetization curves embedded. The effective magnetic moment is 4.4×10^{-19} emu. And Signal to Noise Ratio of DC magnetization is 80 dB.

CONCLUSION In the present study, we investigated the influence of discretization strategy of DC applied field on precision of temperature probing by simulation. It indicated that the discrete quality of DC applied field influence on temperature probing accuracy. It is not the most appropriate way that temperature is probed under uniform discrete points of DC field. Because of the non-linearity of the first-order Langevin function, non-uniform discretization by optimization method can improve the accuracy of noninvasive magnetonanothermometer.

ACKNOWLEDGEMENTS This study was supported by projects 61571199 (NSFC) and 2014AEA048 (HBSTD).

REFERENCES
[1] R. E. Rosensweig, Journal of Magnetism and Magnetic Materials, 252(1-3): 370-374, 2002. doi: 10.1016/S0304-8853(02)00706-0.
[2] A. Jordan, R. Scholz, P. Wust, H. Fähling and R. Felix, Journal of Magnetism and Magnetic Materials, 201(1-3): 413-419, 1999. doi: 10.1016/S0304-8853(99)00088-8.
[3] J. B. Weaver, A. M. Rauwerdink and E. W. Hansen, Medical Physics, 36(5): 1822-1829, 2009. doi: 10.1118/1.3106342.
[4] J. Zhong, W. Z. Liu, L. Jiang, M. Yang and P. C. Morais, Review of Scientific Instruments, 85(9): 094905-1, 2014. doi: 10.1063/1.4896121.
[5] W. Z. Liu, J. Zhong, Q Xiang, G. Yang and M, Zhou, IEEE Transactions on Nanotechnology, 10(6): 1231-1237, 2011. doi: 10.1109/TNANO.2010.2089697

The effect of dc field strength on the performance of a magnetic nanothermometer

Jing Zhong[a,*], Frank Ludwig[a], Meinhard Schilling[a]

[a] Institut für Elektrische Messtechnik und Grundlagen der Elektrotechnik, TU Braunschweig, Germany
[*] Corresponding author, email: jingzhonghust@gmail.com

INTRODUCTION A novel method of noninvasive and precise temperature probing is crucial for biomedical applications, such as hyperthermia for cancer therapy [1] and drug delivery [2]. A suitable technique to access the local temperature of a target tissue contributes to improve the treatment efficiency of hyperthermia and drug delivery for cancer therapy. However, the existing thermometry cannot meet these requirements of noninvasive and precise temperature probing in the stated applications.

In recent years, several approaches on magnetic nanothermometry for noninvasive and precise temperature probing were proposed [3-5]. They employ the temperature sensitivity of magnetic nanoparticle (MNP) magnetization to achieve remote and noninvasive temperature probing, which has the potential for *in vivo* temperature measurements. In a large amplitude ac magnetic field, the first and third harmonics of MNP magnetization can be used to calculate temperature. In ac and dc magnetic fields, the first and second harmonics allows to measure temperature with a higher accuracy due to the higher signal-to-noise ratio (SNR) of second harmonic than that of third harmonic [5]. However, the performance of magnetic nanothermometry in different dc fields has not yet been explored.

In this paper, we study the performance of magnetic nanothermometry using the first and second harmonics under ac and dc magnetic fields by simulation, especially the effect of dc field strength on the performance of magnetic nanothermometry is considered.

MATERIAL AND METHODS The magnetization of ideal MNPs exposed to ac and dc magnetic fields with a low frequency can be described by the Langevin function with no need to take into account MNP dynamics. The magnetization M is

$$M = \phi M_s \left[\coth\left(\frac{\mu_0 M_s VH}{kT} \right) - \frac{kT}{\mu_0 M_s VH} \right], \quad (1)$$

where ϕ is the volume fraction of MNPs, M_s is the saturation magnetization, V is the particle volume, $H=H_0 \cdot \sin(\omega t)+H_{dc}$ is the applied magnetic field with ac field amplitude H_0 and dc field strength H_{dc}, ω is the angular frequency, k is the Boltzmann constant and T is the absolute temperature. When H_0 and H_{dc} are small enough, the MNP magnetization can be described by the Fourier expansion, and the fundamental and second harmonic A_1 and A_2, are

$$\begin{cases} A_1=f(\phi,T) \\ A_2=g(\phi,T) \end{cases}. \quad (2)$$

Temperature probing is based on solving Eqs. (2) knowing the magnetic moment of the particles.

RESULTS In our simulation, the first and second harmonics of MNP magnetization under ac and dc magnetic fields were used for temperature probing. The ac magnetic field has an amplitude of 20 Oe and a frequency of 70 Hz whereas the strength of dc magnetic field varies from 10 to 60 Oe. The simulation results are shown in Figure 1. Herein, the temperature range is from 310 to 320 K. The simulations were performed for monodisperse MNPs with 20 nm core diameter and a saturation magnetization of 7.6×10^6 A/m. The SNR is set to 70 dB.

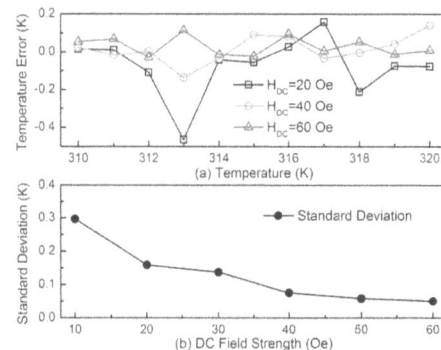

Figure 1: Simulation results of temperature probing error in different applied dc magnetic fields. (a) shows the temperature probing error whereas (b) shows the standard deviation.

Figure 1 shows the simulation results of temperature probing errors in different dc magnetic fields. For an applied dc magnetic field of 10 Oe, the standard deviation of temperature probing error is about 0.30 K whereas it is about 0.05 K for an applied dc magnetic field of 60 Oe. It shows that the temperature probing error decreases about 6 times with an increase in dc magnetic field of about a factor of 6.

CONCLUSION In this study, we report on the effect of dc magnetic field strength on the temperature probing accuracy by using the first and second harmonics of MNP magnetization under ac and dc magnetic fields. Our simulation result indicates that in a certain dc field range from 10 to 60 Oe an increase in dc magnetic field improves the performance of a magnetic nanothermometer.

ACKNOWLEDGEMENTS Financial support from the Alexander von Humboldt Foundation is acknowledged.

REFERENCES
[1] L. C. Branquinho, M. S. Carriao, A. S. Costa, N. Zufelato, M. H. Sousa, R. Miotto, R. Ivkov and A. F. Bakuzis. *Scientific Reports* 3: 2887(10pp), 2013. doi: 10.1038/srep02887.
[2] T. Hoare, J. Santamaria, G. F. Goya, S. Irusta, D. Lin, S. Lau, R. Padera, R. Langer and D. S. Kohane. Nano Letters 9: 3651-3657, 2009. doi: 10.1021/nl9018935.
[3] J. B. Weaver, A. M. Rauwerdink and E. W. Hansen. *Medical Physics* 36: 1822-1829, 2009. Doi: 10.1118/1.3106342.
[4] A. M. Rauwerdink, E. W. Hansen and J. B. Weaver. *Physics in Medicine and Biology* 54: L51-L55, 2009. doi: 10.1088/0031-9155/54/19/L01.
[5] M. Zhou, J. Zhong, W. Liu, Z. Du, Z. Huang, M. Yang, and P. C. Morais. *IEEE Transactions on Magnetics*, 51: 9(6pp), 2015. doi: 10.1109/TMAG.2015.2434322.

67

Magnetic signal separation using independent component analysis

Masahiro Arayama[a,*], Tomoyuki Takagi[b], Tomoki Hatsuda[b], Akihiro Matsuhisa[a], Hiroki Tsuchiya[b], Yasutoshi Ishihara[a]

[a] School of Science and Technology, Meiji University
[b] Graduate School of Science and Technology, Meiji University
[*] Corresponding author, email: keisoku.arayama@gmail.com

INTRODUCTION In magnetic particle imaging (MPI) [1], image artifacts and blurring appear in a reconstructed image when the strength of a gradient magnetic field is insufficient, and the magnetization signals generated from magnetic nanoparticles (MNPs) existing outside a field free point (FFP) are detected. In order to overcome this drawback, we have proposed a reconstruction method using the orthogonal basis obtained using singular value decomposition (SVD) [2]. However, a reconstructed image quality is inadequate. Therefore, we attempt to apply the concept of independent component analysis (ICA) [3] to discriminate the signal generated from inside and outside of the FFP region. ICA is a method of separating independent signal components from a signal, which is mixed with many individual original signals. In addition, it is expected to separate signals generated from inside and outside of the FFP region. In this study, we performed numerical simulation to investigate our proposed method.

METHODS ICA is a computational method for separating a multivariate signal. A schematic of the ICA is shown in Fig. 1. It is possible to estimate the original independent signals by only using the information of the mixed signals even when the mixture ratio of the original signals is unknown. In the ICA, the mixed signals are previously standardized and whitened, and the non-Gaussianity of the mixed signals is maximized. If the non-Gaussianity of the mixed signals is maximized, the mixed signals become independent of each other. Consequently, if the mixed signals are independent of each other, the mixed signals are similar to the original signals.

In this proposed method, first, the observed signals are decomposed using the SVD since the ICA requires that the original signal consists of the independent components,

$$\mathbf{M} = \mathbf{U}\,\mathbf{\Sigma}\,\mathbf{V}^{T}, \qquad (1)$$

where $\mathbf{M}$ is the observed signal, $\mathbf{U}$ is the left-singular vector, $\mathbf{\Sigma}$ is the singular value, $\mathbf{V}$ is the right-singular vector, and $\mathbf{V}^{T}$ is the transpose of the matrix $\mathbf{V}$. Here, the left-singular vectors $\mathbf{U}$, which are decomposed as orthogonal basis, represent the mixture information generated from each MNP. Therefore, the left-singular vectors $\mathbf{U}$ are independently separated as $\mathbf{U}'$ using the ICA. By performing the ICA, it is possible to extract a component only inside the FFP region from the mixed components inside and outside the FFP region. Then, the separated signals are calculated using $\mathbf{U}'$, $\mathbf{\Sigma}$, and $\mathbf{V}^{T}$:

$$\mathbf{M}' = \mathbf{U}'\,\mathbf{\Sigma}\,\mathbf{V}^{T}, \qquad (2)$$

where $\mathbf{M}'$ is the separated signal.

In order to confirm the effectiveness of the proposed method, we performed numerical experiments. In the numerical experiments, a field of view (FOV) was set to 30 mm × 30 mm with a matrix size of 5 × 5. A gradient magnetic field of 2.5 T/m was generated, and an alternating magnetic field of 20 mT was applied. An MNP (particle diameter: 30 nm) was located at the center of the FOV.

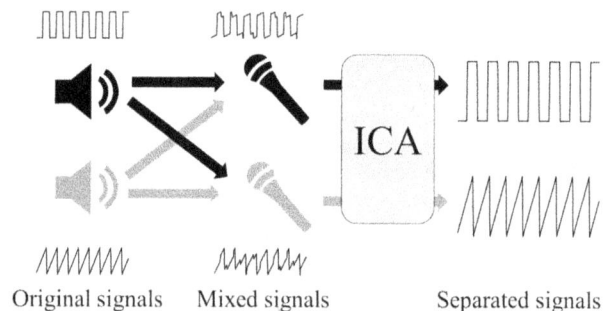

Figure 1: Schematic of the independent component analysis.

RESULTS Figure 2 shows the normalized separated signals that are extracted using ICA and SVD. In this figure, the largest signal and the second largest signal, which were the part of signals separated as 25 components, are shown. Although the largest signal is considered to correspond to the signal generated from an MNP arranged at a central matrix, the source location of each separated signal cannot be determined on each matrix using the analysis of this study. This is because there is a permutation problem (e.g., an ambiguity in the order of the rows of the output signal) in the ICA [3].

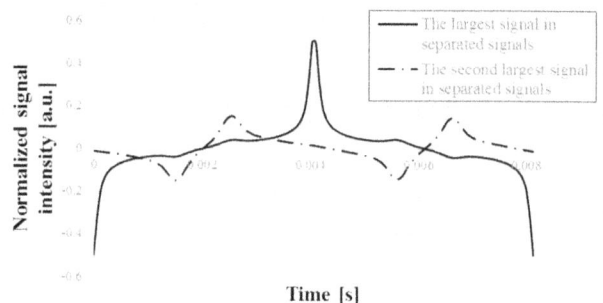

Figure 2: Separated signals by using ICA and SVD.

CONCLUSION We proposed a new magnetic signal separation method using the ICA. It was confirmed that a signal generated from an MNP existing at the FFP region can be separated. However, the permutation problem was not solved. Hence, in the future, we will attempt to solve the permutation problem, and investigate the separation of various MNP distributions.

REFERENCES

[1] B. Gleich and J. Weizenecker. Nature, 435 (7046):1214—1217, 2005.
[2] T. Takagi, S. Shimizu, H. Tsuchiya, T. Hatsuda, T. Noguchi, and Y. Ishihara. Image reconstruction method based on orthonormal basis of observation signal by singular value decomposition for magnetic particle imaging. In 5th International Workshop on Magnetic Particle Imaging, Istanbul, 2015.
[3] A. Hyvärinen, J. Karhunen, and E. Oja, Independent component analysis. John Wiley & Sons, 2001.

MPI meets MRI – simultaneous measurement of MPI and MRI signals

P. Vogel [a,*], T. Kampf [a], M.A. Rückert [a], A. Vilter [a], P.M. Jakob [a], V.C. Behr [a]

[a] Department of Experimental Physics 5 (Biophysics), University of Würzburg, Würzburg
* Corresponding author, email: Patrick.Vogel@physik.uni-wuerzburg.de

INTRODUCTION Magnetic Particle Imaging (MPI) is a promising tomographic method, which can determine the spatial distribution of superparamagnetic iron-oxide nanoparticles (SPIOs) directly [1]. Unlike Magnetic Resonance Imaging (MRI), it cannot provide anatomical background information.

To overcome this issue, different approaches like fusion-MPI/MRI imaging, where both datasets are co-registered retrospectively or hybrid-MPI/MRI scanners, where both measurements can be performed within the same system [2] have been shown. Due to the different magnetic field configurations required for MPI and MRI, the experiments have to be executed one after the other.

A novel concept based on the traveling wave MPI-hardware [3] is theoretically investigated to perform MPI and the acquisition of nuclear induction signal simultaneously.

MATERIAL AND METHODS For MPI a field free point (FFP) with a strong gradient is moved through the sample to scan it point-by-point. At each FFP-position the nonlinear changing of the magnetization of the SPIOs is inductively measured and a full 3D reconstruction can be performed [1].

The behavior of the nuclear magnetization vector during a MPI-measurement was numerically analyzed using the Bloch equations and a simplified model of the TWMPI-system. Contrary to typical MRI-measurements a coil generates an oscillating main magnetic field which cancels out the nuclear magnetization for typical relaxation times and driving frequencies.

However, applying a constant offset field perpendicular to the main field with only 5 percent of its field strength, a permanent nuclear magnetization arises. During the zero crossing of the main magnetic field the nuclear magnetization vector starts to precess around the offset field direction. This generates a transverse magnetization component in relation to the oscillating main magnetic field. With increasing main field strength this transverse component precesses around the main field direction and an induction signal is generated. Thus for every period of the main field the spin system undergoes an excitation. In Fig. 1 (a) the simulated signal is shown, which increases according to the longitudinal relaxation time of the material. A zoom into the signal reveals the frequency sweep (fig. 1 (b)), which can also be seen in the spectrum (Fig. 1 (c)). Fig. 1 (d) shows a 3D plot of a magnetization vector over a short time span of two periods. During the phase of maximum main magnetic field the system's polarization is increased.

It must be emphasized that the zero crossing of the main magnetic field corresponds to the passage of the FFP in a MPI-experiment. Hence signal can be reconstructed similar to a MPI-measurement [2]. Fig. 1 (e) shows the reconstructed 2D signal of a simulated point sample. Due to the varying magnetic field the signal is widely spread out in the spectrum increasing to the Larmor frequency of about 1.9 MHz representing the maximum magnetic field in the TWMPI-scanner. The ring shaped image in Fig. 1 (e1) is the result of using only a small area of the spectrum for the reconstruction. Higher partial frequency bands corresponds to a

different size of the ring due to higher frequency components encoding areas farther away from the sample (see Fig. 1 (e2)-(e4)).

The simulation suggests that the strength of the expected nuclear induction signal is about one-third of a corresponding NMR-signal using a constant B_0 field. The signal depends only on the main field strength and a corresponding offset field, but not on the oscillating frequency of the main field.

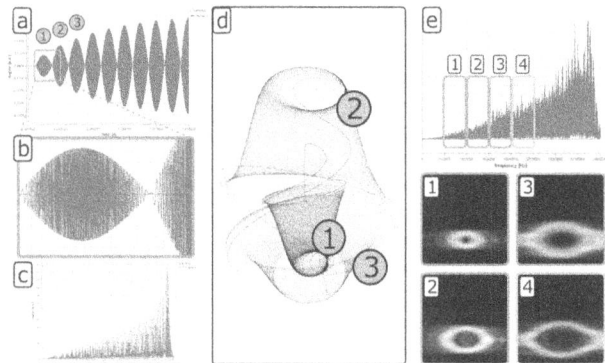

Figure 1: (a) the increasing induced signal is the result of single frequency sweeps, which is enlarged in (b), and the corresponding spectrum (c). A tracking of the magnetization vector gives an idea of the magnetization effect (d). In (e) a spectrum of a point sample simulated in a TWMPI scanner driving in slice-scanning mode is shown. The reconstruction shows a ring shaped figure, which grows up the higher frequency band for the reconstruction is selected.

CONCLUSION A novel approach for acquiring a nuclear induction signal is introduced. Contrary to NMR it uses a strong sinusoidal magnetic field and a small constant offset field for magnetization and signal generation. Thus it can be used in a TWMPI-scanner for simultaneously signal acquisition. A closer look at the underlying effect gives an idea about the dynamics of the spin system, which can yield access to measuring new system parameters and may constitute the basis for a new tomographic technique using dynamic magnetic fields.

ACKNOWLEDGEMENTS This work was supported by the DFG (BE-5293/1-1).

REFERENCES

[1] B. Gleich and J. Weizenecker, *Nature*, 435:1214-1217, 2005. doi: 10.1038/NATURE03808.
[2] P. Vogel and S. Lother, et al., *IEEE TMI*, 33(10):1954-1959, 2014. doi: 10.1109/TMI.2014.2327515.
[3] P. Vogel, et al., *IEEE TMI*, 33(2):400-407, 2013. doi: 10.1109/TMI.2013.2285472.

Effects of Safety Limits on Image Quality in MPI

Ecem Bozkurt [a,b], Omer Burak Demirel [a,b], Damla Sarica [a,b], Yavuz Muslu [a,b], Emine Ulku Saritas [a,b*]

[a] Department of Electrical and Electronics Engineering, Bilkent University, Ankara, Turkey
[b] National Magnetic Resonance Research Center (UMRAM), Bilkent University, Ankara, Turkey
* saritas@ee.bilkent.edu.tr

INTRODUCTION Drive field in magnetic particle imaging (MPI) has to abide by safety restrictions due to the risk of peripheral nerve stimulation [1-4]. Accordingly, the maximum allowable drive field amplitude goes down at high frequencies. Recent work has shown that the resolution of MPI improves at low drive field amplitudes [5-6]. However, how the resolution varies across the safety limits is not yet known. In this work, we investigate the impact of safety limits on image quality for x-space MPI. The results are significant while determining optimal frequency for x-space MPI, complying safety restrictions.

MATERIAL AND METHODS The simulation parameters in this work depend on a recent work [6], where nanoparticle relaxation times were measured for Resovist at various frequencies (4.5 kHz to 25 kHz) and drive field amplitudes (5 mT to 30 mT). Here, we particularly concentrate at the lowest and highest frequencies investigated in [6]. The magnetostimulation thresholds [1] at these frequencies for the human torso are listed in Table 1, together with the corresponding relaxation time constants for Resovist, as measured in [6].

Table 1: For 4.5 kHz and 25 kHz, drive field safety limits [1], and related relaxation time constant [6] for Resovist.

	Safety Limit		Additional Test Field	
	B_{peak} (mT)	τ (μsec)	B_{peak} (mT)	τ (μsec)
4.5 kHz	9.9	6.2	30	3.9
25 kHz	7.6	1.2	30	0.92

The x-space MPI images at these drive field amplitudes and frequencies were compared via simulations. An x-space MPI system was simulated in 2D using a custom MPI simulation toolbox in MATLAB. Here, the relaxation effect was modeled as a convolution of nanoparticle's Langevin response with an exponential function defined by a relaxation time constant [7]. Selection fields were chosen as G_{xx}=7 T/m and G_{zz}=3.5 T/m. For all images, partial field-of-view overlap along the z-direction was chosen as 60%. For fair comparison, the total scan-time was kept constant at 60ms. Additive noise was also kept uniform across all simulations.

RESULTS Our simulations confirm that better resolution is obtained at low drive fields (see Fig. 1.b,c). As seen in Fig. 2.b, the 1D cross-sections of reconstructed images at the safety limits at 4.5kHz and 25kHz are almost identical, with the image at 4.5 kHz displaying few artifacts due to reduced signal at that frequency. Our simulations at 9.3 kHz and 12.2 kHz safety limits also exhibit the same trends (results not shown).

CONCLUSION In this work, we showed that at drive field safety thresholds, resolution remains almost identical across frequencies. While it was known that the resolution improved at lower drive fields, it is quite unexpected for the resolution to be independent of the operating frequency, as long as one operates at the safety limits. This implies that by increasing the frequency,

one can not only maintain resolution but also enhance SNR. Whether this remains valid for frequencies extending to 150 kHz remains as future work.

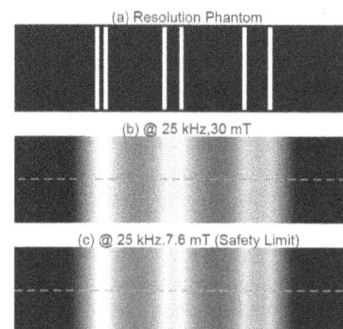

Figure 1: Comparison at 25 kHz. (a) Resolution phantom, line sources with 1-mm width are separated by 2 mm ,4 mm, 6 mm. (b) Image at B_{peak}=30 mT. (c) Image at B_{peak}=7.6 mT (safety limit at 25kHz). 2x12cm^2 FOV (2x8cm^2 central region shown).

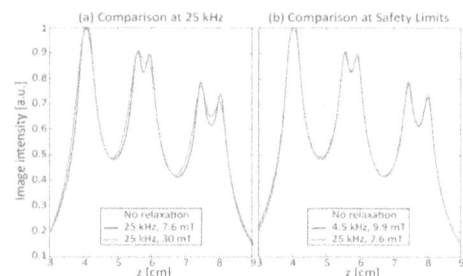

Figure 2: 1D central cross-sections. (a) Cross-sections from Fig. 1.b,c. Resolution is visibly better at lower drive field. (b) Safety limits at 4.5 kHz and 25 kHz yield almost identical images.

ACKNOWLEDGEMENTS This work was supported by the Scientific and Technological Research Council of Turkey through a TUBITAK 3501 Grant (114E167), by the European Commission through an FP7 Marie Curie Career Integration Grant (PCIG13-GA-2013-618834), and by the Turkish Academy of Sciences through TUBA-GEBIP 2015 program.

REFERENCES
[1] E. U. Saritas, *et al.*, *IEEE Trans. Med. Imaging*, 32(9): 1600–1610, 2013. doi: 10.1109/TMI.2013.2260764
[2] I. Schmale, *et al.*, *IEEE Trans. on Magn.*,51(2): 18–21, 2015. doi:10.1109/TMAG.2014.2322940
[3] E. U. Saritas, *et al.*, *Med Phys* 42(6):3005-3012, 2015. doi: 10.1118/1.4921209.
[4] I. Schmale, *et al.*, *2013 Int. Work. Magn. Part. Imaging, IWMPI 2013*, 7965, 2012: 796510, 2013. doi: 10.1109/IWMPI.2013.6528346
[5] A. Weber et al. Int. Work. Magn. Part. Imaging, 2015. doi: 10.1109/IWMPI.2015.7107020.
[6] L. Croft, *et al.*, 'Low Drive Field Amplitude for Improved Image Resolution in Magnetic Particle Imaging', *Med Physics*. In Press.
[7] L. Croft, *et al.*, *IEEE Trans. Med Imaging*. 2012; 31(12):2335-42. doi: 10.1109/TMI.2012.2217979.

Lissajous Node Points for a Sytem Matrix based MPI Image Reconstruction Approach

Christian Kaethner[a,*], Mandy Ahlborg[a], Wolfgang Erb[b], Thorsten M. Buzug[a]

[a] Institute of Medical Engineering, Universität zu Lübeck, Lübeck, Germany
[b] Institute of Mathematics, Universität zu Lübeck, Lübeck, Germany
[*] Corresponding author, email: {kaethner,buzug}@imt.uni-luebeck.de

INTRODUCTION In Magnetic Particle Imaging (MPI) [1], the acquisition of the particle signal can be achieved by a field free point (FFP) traveling along specific trajectories. The Lissajous trajectory is a possibility for such a path [2,3]. It features a dense but inhomogeneous coverage of the respective field of view (FOV), which means that the regions near the FOV edges are sampled denser than the region towards the center of the FOV. In this contribution, a characterization of the Lissajous trajectory is performed by trajectory specific node points that precisely reflect these characteristics. In addition to this, a system matrix (SM) based reconstruction procedure based on these node points is proposed.

MATERIAL AND METHODS The Lissajous trajectory used to sample the FOV can be described by

$$\gamma(t) = \left(A_x \sin(2\pi n_y t/T_R), A_y \sin(2\pi n_x t/T_R)\right)$$

with A_x, A_y being the drive field amplitudes, n_x, n_y being the frequency dividers, t representing the time, and T_R the repetition time. According to [4], the characteristic Lissajous node points for such a trajectory can be achieved, when the trajectory is sampled at the time points

$$t_k = \frac{k}{4n_x n_y T_R} \quad \text{with} \quad k = 1, \dots, 4n_x n_y.$$

An example of these node points as well as the respective trajectory are shown in Fig. 1.

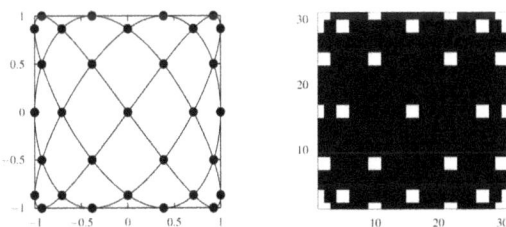

Figure 1: *(left)* Visualization of a Lissajous trajectory based on the frequency dividers $n_x = 3$ and $n_y = 4$ and the respective Lissajous node points. *(right)* Potential point sample positions for the SM acquisition based on Lissajous node points.

The idea of the here proposed reconstruction approach including Lissajous node points is based on a SM acquisition. Instead of an equidistantly spaced positioning of the point sample during the SM acquisition, the locations of the Lissajous node points are used (see Fig. 1). Afterwards, a reconstruction using an iterative solver, for example a Kaczmarz approach, is used. Based on the modified SM acquisition grid, the resulting reconstructions are

located only at the positions of the Lissajous node points. In order to get a full image representation, the reconstructed particle signals are interpolated [4].

The MPI simulations are carried out under the assumption of ideal magnetic fields. The FOV has a size of 2.9 cm × 2.9 cm and is located in the xy plane. The simulated gradient strength of the selection field is assumed to be $1\,\mathrm{Tm}^{-1}$. With a base frequency $f_B = 2.5\,\mathrm{MHz}/3$ and frequency dividers $n_x = 31$ and $n_y = 32$, the repetition time is given by $T_R = (n_x n_y)/f_B = 0.0012$ s. The particle characteristics are modeled via Langevin theory assuming a particle size of 30 nm.

RESULTS The initial results of the proposed reconstruction approach that includes the Lissajous node points are shown in Fig. 2. It can be seen that in comparison to the used phantom, the reconstruction results are very promising.

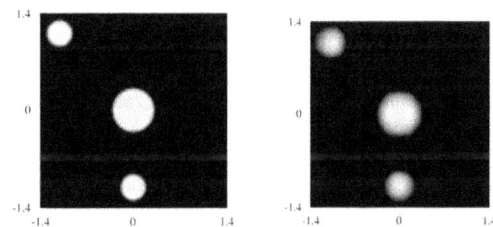

Figure 2: Visualization of the used simulation phantom and the reconstruction based on the Lissajous node points.

CONCLUSION The use of Lissajous node points is a possibility to characterize the data acquisition path of specific MPI scanners, i.e. the Lissajous trajectory. It can be seen in the results that an SM based reconstruction procedure using these node points leads to promising first results.

ACKNOWLEDGEMENTS The authors gratefully acknowledge the financial support of the German Federal Ministry of Education and Research (BMBF, grant number 13N11090), the German Research Foundation (DFG, grant number BU 1436/9-1 and ER 777/1-1), and the European Union and the State Schleswig-Holstein (EFRE, grant number 122-10-004).

REFERENCES
[1] B. Gleich and J. Weizenecker. *Nature*, 435(7046):1217—1217, 2005. doi: 10.1038/nature03808.
[2] J. Borgert et al. Biomedizinische Technik/Biomedical Engineering, 2013. doi: 10.1515/bmt-2012-0064.
[3] N. Panagiotopoulos et al. International Journal of Nanomedicine, 2015. doi: 10.2147/IJN.S70488.
[4] W. Erb, C. Kaethner, M. Ahlborg, and T. M. Buzug. Numerische Mathematik, 2015. doi: 10.1007/s00211-015-0762-1.

Basic Study of Image Reconstruction Method Using Neural Networks with Additional Learning for Magnetic Particle Imaging

Tomoki Hatsuda[a,*], Tomoyuki Takagi[a], Akihiro Matsuhisa[b], Masahiro Arayama[b], Hiroki Tsuchiya[a], Yasutoshi Ishihara[b]

[a] Graduate School of Science and Technology, Meiji University
[b] School of Science and Technology, Meiji University
* Corresponding author, email: ikuta.keisoku@gmail.com

INTRODUCTION In magnetic particle imaging (MPI) [1], image blurring and artifacts occur in a reconstructed image because the magnetization signals generated from magnetic nanoparticles (MNPs) at the field free point (FFP) are similar to those around the FFP regions. In order to overcome these problems, we proposed a new reconstruction method using neural networks [2]. In this method, a data set of magnetization signals and MNP location pairs is used for learning in neural networks. If all possible combinations of the data sets are learned, an accurate estimated result is obtained. However, it is difficult to learn all the combinations in a reasonable period of time. In this study, the number of data sets learned in the first stage is minimized, and additional learning using the appropriate data sets, which reduces the error between observed signals and estimated signals, is performed. By learning the minimum number of required data sets, it is expected that image blurring and artifacts will be suppressed even when the MNP's magnetization is insufficient, *e.g.*, when an applied alternative magnetic field and/or a gradient magnetic field are/is weak. We performed numerical experiments to confirm the effectiveness of our proposed method.

METHODS In the learning process, a magnetization signal (Fig. 1 (a)) is calculated analytically from a known MNP location (Fig. 1 (b)). In the first learning process, some randomly selected data sets from combinations of all the magnetization signals and MNP location pairs are used. After the first learning process, the iterative process is executed as described below. First, the observed signal (Fig. 1 (c)) measured from an unknown MNP location is input to the neural network, and the MNP location (Fig. 1 (d)) is estimated using the neural network (Fig. 1 (1)). Second, the estimated signal (Fig. 1 (e)) is calculated analytically from the estimated MNP location (Fig. 1 (2)). Third, the error (Fig. 1 (f)) between the observed signal and estimated signal is calculated (Fig. 1 (3)), and the error region (Fig. 1 (g)) depending on the error intensity at each FFP is imaged (Fig. 1 (4)). Fourth, some probability mappings (Fig. 1 (h)) where the MNPs possibly exist are generated considering the error region and the estimated MNP location (Fig. 1 (5)). Fifth, data sets of the magnetization signal and MNP location pairs (Fig. 1 (i)) are randomly created according to the probability mappings (Fig. 1 (6)). Finally, additional learning is performed using the newly created data sets (Fig. 1 (7)). The above procedures are performed iteratively until the error between the observed signal and estimated signal is sufficiently reduced.

We performed numerical experiments to confirm the effectiveness of the proposed method. In the experiments, the field of view was set as 11 mm × 11 mm with a matrix size of 11 × 11, and the MNPs (particle diameter: 20 nm) were located as shown in Fig. 2 (a). A gradient magnetic field of 3.0 T/m was generated, and an alternating magnetic field of 10 mT was applied. In the first learning process, 10000 data sets were input

to neural networks, and 5000 additional data sets were created in every iterative process.

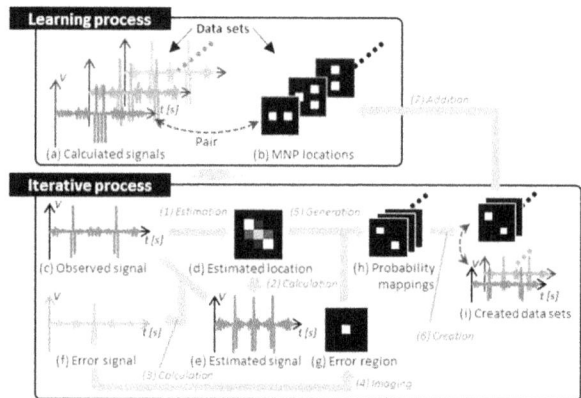

Figure 1: Concept of proposed method.

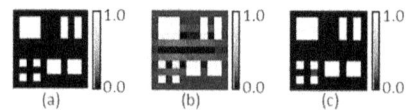

Figure 2: Numerical experiment results with (b) inverse-matrix solution and (c) proposed method for (a) MNP location.

RESULTS Figures 2 (b) and (c) show the reconstructed images after using the inverse-matrix solution [3] and the proposed method, respectively. In the proposed method, the error between the observed signal and estimated signal was converged in 20 iterations, and the number of data sets used for learning was suppressed to 110000 although the combination of data sets reached an order of 10^{36} for an 11 × 11 matrix size. These results show that the MNP location is reconstructed more precisely with the proposed method than with the inverse-matrix solution.

CONCLUSION We proposed a new reconstruction method using neural networks with additional learning. This method was able to reconstruct the image more clearly even when the MNP's signals were not detected sufficiently. However, it is difficult to reconstruct an accurate image when appropriate data sets are not selected for learning in neural networks. Hence, in the future, we will improve the method for selecting the data sets.

REFERENCES
[1] B. Gleich and J. Weizenecker. *Nature*, 435 (7046):1214—1217, 2005.
[2] T. Hatsuda, S. Shimizu, H. Tsuchiya, T. Takagi, T. Noguchi, and Y. Ishihara. A basic study of an image reconstruction method using neural networks for magnetic particle imaging. In 5th International Workshop on Magnetic Particle Imaging, Istanbul, 2015.
[3] J. Weizenecker, J. Borgert, and B. Gleich. *Phys. Med. Biol.*, 52 (21):6363—6374, 2007.

A new 3D MPI model using realistic magnetic field topologies for algebraic reconstruction

Wolfgang Erb [a,*], Gael Bringout [b], Jürgen Frikel [c], Thorsten M. Buzug [b]

[a] Institute of Mathematics, [b] Institute of Medical Engineering, Universität zu Lübeck, Germany
[c] DTU Compute, Technical University of Denmark, Denmark
[*] Corresponding author, email: erb@math.uni-luebeck.de

INTRODUCTION One of the main advantages using field free lines for the acquisition of data in Magnetic Particle Imaging (MPI) is the availability of efficient Radon-based reconstruction algorithms [1]. However, due to inhomogeneities and the rotation movement of the generated magnetic fields, the FFL-MPI data is not acquired along parallel straight lines but on rotating curved regions. Using the classical Radon inversion for the reconstruction leads to geometrical distortions and rotational artefacts in the reconstructed images. Goal of this work is to avoid these shortcomings by using a new 3D MPI model for the reconstruction that includes realistic magnetic field topologies.

MATERIAL AND METHODS In a configuration with K receive coils with sensitivities $\vec{\rho}_i(\vec{r})$, $i = 1, \ldots, K$, K time-dependent voltage signals $u_i(t)$ are measured. The goal is to reconstruct the tracer distribution $c(\vec{r})$ from the measured voltage signals. In contrast to classical field free points or lines [1], low field regions are considered. To this end, a model based on realistic magnetic field topologies is used in which all involved magnetic fields are represented in terms of spherical harmonics [2]. Based on this representation, the low field regions $F(t) \subset \mathbb{R}^3$ at time t in which the absolute value of the magnetic field is smaller than a given threshold are computed. As can be observed in Fig 1 a) and b), realistic magnetic fields generate curved low field regions rather than straight ones. Using Faraday's law of induction and the Langevin model for the magnetization, a velocity vector $\vec{v}(\vec{r}, t)$, describing the change in the magnetization through the time-dependent movement of the involved magnetic fields, is modelled. In order to discretize the problem, a series expansion $c(\vec{r}) = \sum_{n=1}^{N} c_n p_n(\vec{r})$ with respect to voxel basis functions $p_n(\vec{r})$ is used. Given discretizations $\vec{u}_i = (u_i(t_1, \ldots, t_M)$ and $\vec{c} = (c_i, \ldots, c_N)$, the modelled system matrix $S^i = (S^i_{m,n})$ can be constructed as follows:

$$S^i_{m,n} = \int_{F(t_m)} \langle \vec{\rho}_i(\vec{r}), \vec{v}(\vec{r}, t_m) \rangle \, p_n(\vec{r}) d\vec{r}.$$

Then, the system equation for each receive channel i reads

$$S^i \vec{c} = \vec{u}_i.$$

Combining the information from the different receive channels in a single system of equation, the reconstructed tracer distribution is computed by using the Matlab implementation of the LSQR algorithm. The termination step for the iteration serves as a regularization parameter. Since the low field regions $F(t)$ are small compared to the whole field of view, the matrices S^i are in general sparse. This sparse structure can be used for a time-efficient implementation of the reconstruction.

RESULTS The phantom shown in Fig. 1 c) is used to simulate the MPI signal in two different scanners. They are set up with either an ideal or a realistic magnetic field topology. Moreover, rotation frequencies of 100 and 1000 Hz are used. As shown in Fig. 2, the filtered backprojection (FBP) algorithm introduces a rotational artefacts and geometric distortions. Rotational artefacts become more severe with increasing rotational frequency while geometric

distortions seem to be independent from the rotation frequency. The observations suggest that the source for the geometric distortions are the inhomogeneities of the magnetic fields. Since both effects are captured by the new model, they are corrected with the presented method.

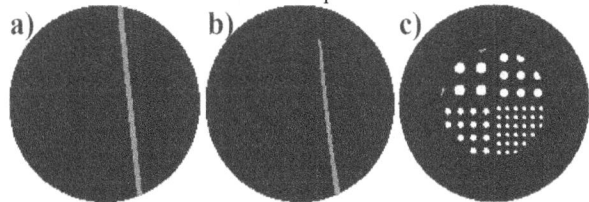

Figure 1: (a) and (b) Areas (in gray) with a field amplitude smaller than 2 mT for an ideal and realistic field topology. (c) Phantom with circles of 4, 6, 8 and 10 mm diameter limited to a 100 mm diameter circle.

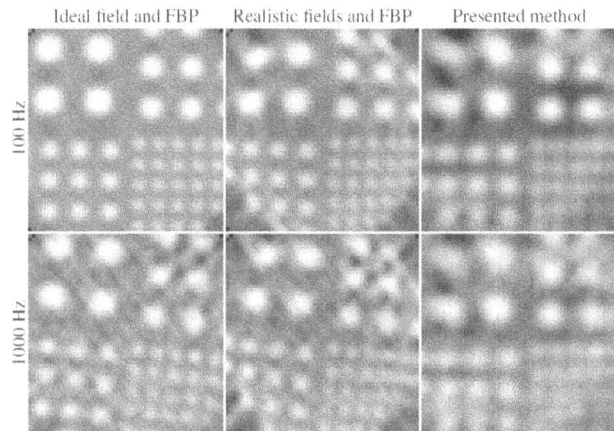

Figure 2: Reconstruction using either a FBP or the presented methods for 3 different rotation frequencies and 2 field types. The results for the presented method are cropped to match the area obtained via the FBP.

CONCLUSION An algebraic reconstruction based on a model including realistic descriptions of magnetic field topologies is adequate to correct geometric distortions and rotational artefacts which are generated in a FBP reconstruction.

ACKNOWLEDGEMENTS The first three authors contributed equally to this work. The authors gratefully acknowledge the financial support of the German Research Foundation (DFG, grant number ER 777/1-1), the German Federal Ministry of Education and Research (BMBF, grant number 13N11090), the European Union and the State Schleswig-Holstein (Programme for the Future – Economy, grant number 122-10-004) and of the HC Ørsted Postdoc programme, co-funded by Marie Curie Actions.

REFERENCES
[1] T. Knopp and al., *Inverse Problems*, 27(9):095004, 2011. doi: 10.1088/0266-5611/27/9/095004.
[2] G. Bringout and T. M. Buzug. *Biomedical Engineering / Biomedizinische Technik*, 59(s1):675—649,

Nonlinear Scanning in X-Space MPI

Ahmet Alacaoglu [a], Ali Alper Ozaslan [a], Omer Burak Demirel [a,b], Emine Ulku Saritas [a,b*]

[a] Department of Electrical and Electronics Engineering, Bilkent University, Ankara, Turkey
[b] National Magnetic Resonance Research Center (UMRAM), Bilkent University, Ankara, Turkey
* Corresponding author, email: saritas@ee.bilkent.edu.tr

INTRODUCTION Linear trajectories are typically used with x-space reconstruction in MPI, where the field free point (FFP) scans a field-of-view (FOV) by covering horizontal line segments in space [1-3]. Nonlinear trajectories, on the other hand, are widely used in MPI methods that utilize system matrix reconstruction [4-6]. The application of nonlinear scanning in x-space MPI, while argued to be feasible, has not yet been presented. In this work, nonlinear scanning ability of x-space MPI is demonstrated via a Lissajous trajectory.

MATERIALS AND METHODS According to the 3D x-space theory [1], the MPI image is the convolution of the particle distribution with collinear and transverse point spread functions (PSF), the orientations of which are defined with respect to the FFP velocity vector (Fig. 1a). The transverse PSF is significantly wider and hence deteriorates the image quality. To capture the collinear component only, one can choose one of two options: (1) use a linear trajectory with drive and receive coils aligned, or (2) use a nonlinear trajectory with two perpendicular receive coils, virtually rotating the coils to align with the instantaneous velocity vector. Here, we undertake the second approach to demonstrate the nonlinear scanning capability of x-space MPI.

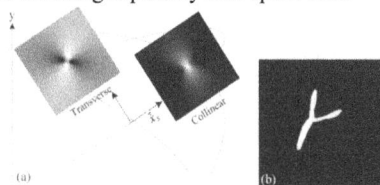

Figure 1: (a) The orientations of the collinear and transverse PSFs, shown with respect to the velocity vector. Figure adapted from [1]. **(b)** Phantom (2x2 cm^2) used in this simulation study.

We developed an x-space MPI simulation toolbox in MATLAB. A nanoparticle diameter of 25 nm was assumed. Selection field of $[3, 3, -6]$ $T/m/\mu_0$ in the [x,y,z] axes was utilized. A Lissajous trajectory with a frequency ratio of 24/25 (24 kHz & 25 kHz), 30 mT drive field amplitude was generated with 10 MHz sampling frequency. Particle relaxation effects were ignored.

Figure 2: Schematic of the proposed x-space image reconstruction scheme for Lissajous trajectory.

Figure 2 summarized the proposed x-space image reconstruction scheme. First, the Lissajous trajectory is decomposed into two partitions containing non-overlapping scanning directions. Next, MPI signal for each partition is gridded to the instantaneous position of the FFP. The resulting images are then combined to form the final image.

RESULTS Figures 3a-b show the resulting MPI images from individual receive coils in x and y. Note that these images contain the effects of both transverse and collinear PSFs. Figure 3c, on the other hand, shows the combined-coil image, where the signal from x- and y-coils are combined to virtually align with the instantaneous velocity vector. As a result, this combined-coil image only contains the effect of the collinear PSF (note the relatively higher resolution when compared to Fig. 3a-b). Finally, Fig. 3d shows the effect of filtering the fundamental frequencies (HPF with $f_c = 36$ kHz). This last image mainly had a DC loss as explained in [7], and the corners of the image were corrupted (not shown). As the FFP speed approaches zero at the corners, it is expected that these locations have unreliable signal levels. Hence, a windowing based on the FFP speed is performed to reach the final image in Fig. 3d.

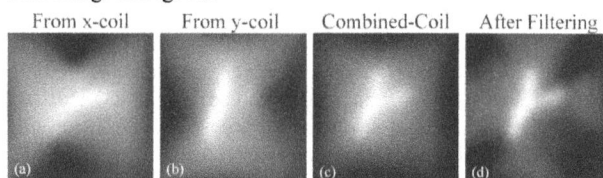

Figure 3: Simulation results. **(a)** & **(b)** Images from individual receive coils. Combined-coil image, **(c)** without filtering and **(d)** after fundamental frequency filtering.

CONCLUSION We have demonstrated nonlinear scanning ability of x-space MPI via a Lissajous scanning scheme. The resulting images are comparable to ones obtained with linear trajectories, demonstrating the versatility of x-space MPI. The effects of relaxation and partial FOVs (i.e., overlapping/non-overlapping patches) remain to be shown as future work.

ACKNOWLEDGEMENTS A. Alacaoglu and A. A. Ozaslan contributed equally to this work. This work was supported by the Scientific and Technological Research Council of Turkey through a TUBITAK 3501 Grant (114E167), by the European Commission through an FP7 Marie Curie Career Integration Grant (PCIG13-GA-2013-618834), and by the Turkish Academy of Sciences through TUBA-GEBIP 2015 program.

REFERENCES
[1] P. W. Goodwill, S. M. Conolly, *IEEE Trans Med Imag,* 30 (9):1581-1590, 2011. doi: 10.1109/TMI.2011.2125982.
[2] E. U. Saritas *et al.*, *J Magn Reson*, 229:116-126, 2013. doi: 10.1016/j.jmr.2012.11.029.
[3] P. W. Goodwill *et al.*, *Adv Mat,* 24:3870-3877, 2012. doi:10.1002/adma.201200221.
[4] B. Gleich and J. Weizenecker. *Nature*, 435(7046):1214-1217, 2005. doi: 10.1038/nature03808.
[5] T. Knopp *et al.*, *Phys Med Biol*, 54:385-397, 2009. doi:10.1088/0031-9155/54/2/014
[6] J. Rahmer *et al.*, *BMC Med Imaging*, 9 (4), 2009. doi:10.1186/1471-2342-9-4.
[7] K. Lu *et al.*, *IEEE Trans Med Imag*, 32 (9):1565–1575, 2013. doi: 10.1109/TMI.2013.2257177.

X-Space and Chebyshev Reconstruction in Magnetic Particle Imaging: A First Experimental Comparison

Tobias Knopp[a,b,*], Chrisitan Kaethner[c], Mandy Ahlborg[c] and Thorsten M. Buzug[c]

[a] University Medical Center Hamburg-Eppendorf, Section for Biomedical Imaging, 20246 Hamburg, Germany
[b] Hamburg University of Technology, 21073 Hamburg, Germany
[c] Institute of Medical Engineering, Universität zu Lübeck, 23562 Lübeck, Germany
[*] Corresponding author, email: t.knopp@uke.de

INTRODUCTION Magnetic particle imaging (MPI) allows for the determination of the spatial distribution of magnetic nanoparticles [1]. For the direct reconstruction of 1D MPI data two different algorithms are known in literature. The one uses the time domain representation of the induced voltage signal [2] while the other considers the voltage in frequency space [3]. Recently, it has been shown that both approaches are mathematically equivalent when considering a continuous model of the signal chain [4]. In a discrete setting there are numerical differences though. In this work, we compare the x-space and the Chebyshev reconstruction approach for the first time on experimental MPI data and confirm their equivalence up to a negligible numerical error.

MATERIAL AND METHODS The reconstruction via x-space and Chebychev approach are performed based on the formulas described in [4] assuming the imaging sequence

$$H(x,t) = Gx - A\cos(2\pi tf)$$

with G being the gradient strength, x the spatial position, A the drive field amplitude, t the time, and f the drive-field frequency. It holds for the Chebyshev scheme that the particle concentration can be determined by

$$\bar{c}(x) = \frac{iT}{\pi\mu_0 pA} \sum_{k=1}^{\infty} \hat{u}_k^{\mathrm{TF}} \frac{\sin\left(k\arccos\left(\frac{G}{A}x\right)\right)}{\sin\left(\arccos\left(\frac{G}{A}x\right)\right)}$$

and for the x-space technique by

$$\bar{c}(x) = \frac{T}{2\pi\mu_0 pA} \frac{u^{\mathrm{TF}}\left(\frac{1}{2\pi f}\arccos\left(\frac{G}{A}x\right)\right)}{\sin\left(\arccos\left(\frac{G}{A}x\right)\right)}.$$

Here, $T = f^{-1}$ denotes the acquisition time, μ_0 the permeability of free space, p the receive coil sensitivity, and u^{TF} the transfer function corrected time signal. Further, the signal used for the reconstruction is corrected for relaxation artifacts by setting the real part of the complex valued spectrum to zero. In addition to this, a correction of the missing fundamental frequency, i.e. the DC offset, is performed after image reconstruction.

Experimental data has been acquired with a pre-clinical MPI scanner (Bruker/Philips Preclinical MPI System, Ettlingen, Germany). With a gradient strength of $1\,\mathrm{Tm}^{-1}\mu_0^{-1}$ and a drive field strength of $14\,\mathrm{mT}\mu_0^{-1}$, the resulting drive-field FOV is $28\,\mathrm{mm}$. For the determination of the transfer function, we applied the linear regression approach outlined in [5]. The experiments were carried out using a phantom that consists of a tube with 4 mm length and 2 mm diameter. The tube is filled with undiluted Resovist (Bayer Schering Pharma AG, Berlin, Germany).

RESULTS In Fig. 1, the reconstruction results using the Chebyshev and the x-space method are shown. As one can clearly see, both methods yield nearly identical reconstruction results. Even at those positions where no particles where located a very similar noise pattern is reconstructed with both methods. The normalized root mean square deviation between the Chebyshev and the x-space reconstruction is below 0.9 %.

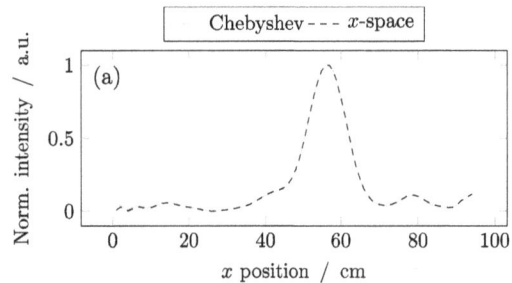

Figure 1: Reconstruction results for the Chebyshev and the x-space approach.

CONCLUSION In this contribution, we have shown that x-space and Chebyshev reconstruction are not only equivalent in a continuous mathematical setting but also yield highly similar results when applied to experimental 1D MPI data.

ACKNOWLEDGEMENTS TK gratefully acknowledges funding and support of the German Research Foundation (DFG, grant number AD 125/5-1). CK, MA, and TMB gratefully acknowledge the financial support of the German Federal Ministry of Education and Research (BMBF, grant number 13N11090), the German Research Foundation (DFG, grant number ER 777/1-1), the European Union and the State Schleswig-Holstein (EFRE, grant number 122-10-004).

REFERENCES
[1] B. Gleich and J. Weizenecker. *Nature*, 435(7046):1217—1217, 2005. doi: 10.1038/nature03808.
[2] P. W. Goodwill and S. M. Conolly. *IEEE Trans. Med. Imag.*, 29(11):1851–1859, 2010. doi: 10.1109/TMI.2010.2052284
[3] J. Rahmer et al. *BMC Med Imaging*, 9, 2009. doi: 10.1186/1471-2342-9-4
[4] M. Grüttner et al. *Biomed. Tech./Biomed. Eng.*, 58(6):583–591, 2013. doi: 10.1515/bmt-2012-0063
[5] T. Knopp et al. *IEEE Trans. Med. Imag.*, 29(1):12-18, 2010. doi: 10.1109/TMI.2009.2021612

Self Calibration for Relaxation- and System-Induced Delays in X-space MPI

Baturalp Buyukates[a], Damla Sarica[a,b], Emine Ulku Saritas[a,b,*]

[a] Department of Electrical and Electronics Engineering, Bilkent University, Ankara, Turkey
[b] National Magnetic Resonance Research Center (UMRAM), Bilkent University, Ankara, Turkey
[*] Corresponding author, email: saritas@ee.bilkent.edu.tr

INTRODUCTION Both system and relaxation induced time delays complicate image reconstruction in x-space magnetic particle imaging (MPI). Relaxation reduces the signal-to-noise ratio (SNR) and blurs the image along the scan direction [1-4]. System delays, on the other hand, introduce unwanted image artifacts. Hence, determining the delays accurately is crucial for obtaining artifact-free images. Here, we propose a method to estimate the effective time delay for relaxation- and system-induced sources, without any a priori knowledge.

MATERIAL AND METHODS X-space image reconstruction in the presence of time delays can be represented as:

$$\text{IMG}\big(x_s(t)\big) = \frac{s(t + \Delta t_r - \Delta t_s)}{\dot{x}_s(t)}$$

Here, IMG is the reconstructed image, $s(t)$ is the received signal, Δt_r is the effective relaxation-induced delay, Δt_s is the delay in the system, $x_s(t)$ is the instantaneous position of the field-free point (FFP). Typically, Δt_s is measured through a prior phase calibration. A reasonable value for Δt_r was suggested as $\tau/2$, where τ is the relaxation time constant of the nanoparticle [1], which can be measured on a relaxometer setup. Here, we lump both delays into one estimated delay, Δt_{est}, which we calculate *directly* from the MPI signal.

We developed a custom x-space MPI simulation toolbox in MATLAB. Simulation parameters were as follows: 25 nm nanoparticle diameter, $\tau = 2.9\mu s$, 20 kHz and 25 mT-pp drive field, 2.3 T/m selection field (similar to values as in [1]). Scan axis was partitioned into 80% overlapping partial field-of-views (pFOVs). The first harmonic of the signal was filtered out [5] and noise added. 1D cross-sections of images were analyzed.

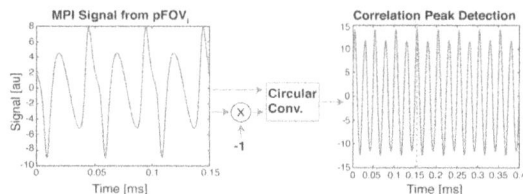

Figure 1: Phase shift estimation for one partial field-of-view.

As outlined in Fig. 1, to determine Δt_{est}, the signal from each pFOV was circularly convolved with its mirror symmetric version. A subsequent peak detection yields twice the value of Δt_{est} (after some further processing). From this time delay, the phase shift of the signal can be computed.

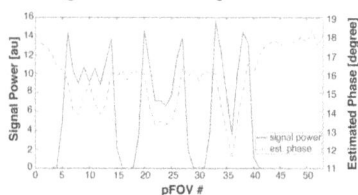

Figure 2: Signal power and estimated phase shift for each pFOV.

Finally, an overall phase shift was computed via weighted mean of the estimated shifts from all pFOVs. The weight for each pFOV was its total signal power, effectively penalizing low signal regions of the image. Fig. 2 shows both the corresponding signal power and estimated phase shifts for each pFOV.

RESULTS Figure 3 compares the proposed technique with calibration-based methods. As seen in Fig. 3a, 45° phase error causes significant artifacts in the image. With prior system phase calibration (Fig. 3b), artifacts are not fully alleviated. If τ of the nanoparticle is known (Fig. 3c), applying an additional $\tau/2$ time shift further improves the image. Finally, Fig. 3d shows the results of the proposed method. Even though our method does not require any calibration or a priori knowledge, it provides results closest to the ideal image.

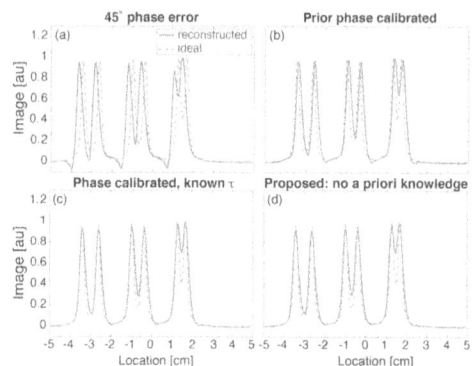

Figure 3: Comparison of the proposed method with calibration-based techniques. In each plots, dashed line is the ideal image without relaxation, solid line is the reconstructed MPI image.

CONCLUSION In this work, we propose a new method to estimate the effective system- and relaxation-induced time delays in MPI. This method does not require any a priori knowledge about the system or the nanoparticle. Our results showed that this practical approach outperforms calibration-based methods. Experimental validation of this technique remains as future work.

ACKNOWLEDGEMENTS This work was supported by the Scientific and Technological Research Council of Turkey through a TUBITAK 3501 Grant (114E167), by the European Commission through an FP7 Marie Curie Career Integration Grant (PCIG13-GA-2013-618834), and by the Turkish Academy of Sciences through TUBA-GEBIP 2015 program.

REFERENCES
[1] L. R. Croft *et al.*, *IEEE Trans Med Imag*, 31(12):2335-2342, 2012. doi: 10.1109/TMI.2012.2217979.
[2] P. W. Goodwill and S. M. Conolly, *IEEE Trans Med Imag*, 30 (9):1581- 1590, 2011. doi: 10.1109/TMI.2011.2125982.
[3] E. U. Saritas *et al.*, *J Magn Reson*, 229:116-126, 2013. doi: 10.1016/j.jmr.2012.11.029.
[4] P. W. Goodwill *et al.*, *Adv Mat*, 24:3870-3877, 2012. doi:10.1002/adma.201200221.
[5] K. Lu *et al.*, *IEEE Trans Med Imag*, 32 (9):1565–1575, 2013. doi: 10.1109/TMI.2013.2257177.

Spatial Resolution in MPI: Modeling the Role of Harmonic Number

H. Bagheri* and M.E. Hayden

Department of Physics, Simon Fraser Univeristy, Burnaby BC Canada V5A 1S6
* Corresponding author; email: hbagheri@sfu.ca

INTRODUCTION Spatial resolution in Magnetic Particle Imaging (MPI) improves as the harmonic number n is increased. A demonstration of this effect is presented in an accompanying contribution [1]. Of course this enhancement does not increase indefinitely; signal-to-noise ratio (SNR) limitations, or other considerations, invariably limit the maximum achievable resolution. Here we examine the role of n on resolution in the context of a simple analytic model of particle magnetization based on the Langevin function: $L(x) = \coth(x) - 1/x$.

MATERIAL AND METHODS The problem we consider consists of a point-like magnetizable sample at position z that is exposed to a magnetic field $\mathbf{H} = H_0\hat{z} + H_1 \sin \omega t \, \hat{z}$, where $\hat{z}$ is a unit vector. We assume that the bias field H_0 is locally associated with a linear magnetic field gradient G_z, such that $H_0 = G_z z$. In MPI, analogous conditions can be encountered in the vicinity of a Field Free Point (FFP); see [1] for an example. Changing z, and hence the distance between the sample and the FFP (at $z = 0$), changes the bias field to which the sample is exposed. Note also the close analogy between the configuration considered here and that of an infinite sample exposed to a uniform bias field [2].

We further assume that the magnetic moment $\mathbf{m}$ of the sample responds instantaneously to the oscillating (or excitation) component of the applied field, such that $\mathrm{m} \propto L(\mathrm{H}(z,t)/\mathrm{H}^*)$. Here, the parameter H^* characterizes the manner in which saturation is attained. We then compute the Taylor expansion of m about H_0 in terms of the variable H_1 and determine $\partial \mathrm{m}/\partial t$ to infer a quantity proportional to the emf that would be induced in a nearby pickup coil [3]. This analysis reveals that the leading order contribution to the emf induced at each harmonic n of the angular drive frequency ω scales as

$$\varepsilon_n \propto \frac{D_n \xi^n}{2^{n-1}(n-1)!} \frac{d^n L}{dx^n}\bigg|_{x=\chi} \qquad (1)$$

where $\xi = H_1/H^*$ and $\chi = H_0/H^* \equiv G_z z/H^*$ are dimensionless measures of the fields H_1 and H_0, and $D_{2p} = -D_{2p-1} = (-1)^p$.

RESULTS Figure 1 shows anticipated induced signal amplitudes as a function of the bias parameter χ; or equivalently, as a function the distance z between the sample and the FFP. An obvious narrowing of the peak at $\chi = 0$ is evident as n is increased, particularly when the phase of the induced signal is considered. This is accompanied by a progressive enhancement of the relative contrast between the central peak and the adjacent minimum. Both trends increase the precision to which measurements of ε_n as a function of z can be used to localize (or resolve) the source, as n is increased. Importantly, agreement is observed between the curves in Fig. 1 and the data in [1] when appropriate values for G_z (14 MA/m^2) and H* (7 kA/m) are used.

Figure 2 shows anticipated relative peak signal amplitudes (cf. $\chi = 0$ in Fig. 1) as a function of n and ξ. It is clear that ε_n decreases rapidly as n is increased, and so the resolution enhancement discussed above cannot be increased indefinitely. SNR limitations (or other factors) will always limit the minimum detectable emf, and hence the minimum detectable feature size.

Insofar as the scaling of ε_n with ξ is concerned, Eq. 1 predicts $\varepsilon_n \propto \xi^n$, which is valid for $(\xi/\pi)^2 \ll 1$. For larger ξ, ε_n begins to saturate. This effect, which can be seen through careful inspection of Fig. 2, leads to apparent scaling laws $\varepsilon_n \propto \xi^\alpha$ with $\alpha < n$, as reported in [1].

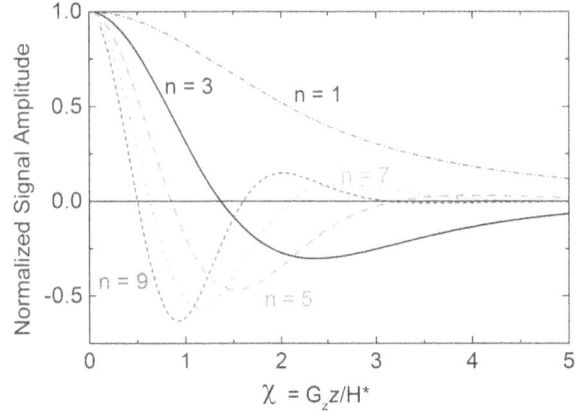

Figure 1: Normalized signal amplitude at odd harmonics of the excitation frequency, plotted as a function of bias parameter χ or distance z between the sample and the FFP. Results are shown for the case where $\xi \ll 1$.

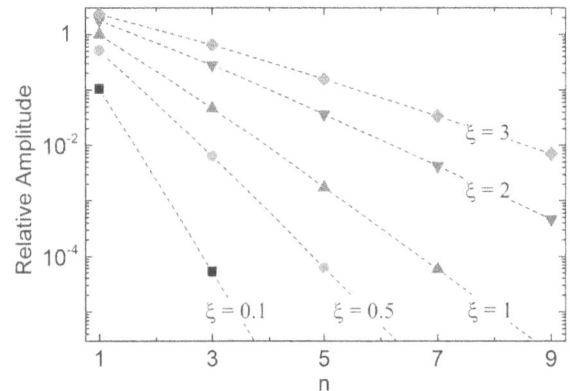

Figure 2: Relative peak signal amplitudes at odd harmonics of the excitation frequency, normalized to the response at $n = 1$ and $\xi = 1$. The dashed lines are intended as guides for the eye.

CONCLUSION The one-dimensional model of harmonic generation in a non-uniform magnetic field outlined here provides insight into the resolution enhancement that can accompany increases in MPI harmonic number [1]. A full description of this model and comparisons to experiment will be reported elsewhere.

ACKNOWLEDGEMENTS This work is funded by the Natural Sciences and Engineering Research Council of Canada.

REFERENCES
[1] Bagheri and Hayden, these proceedings; 6th *IWMPI* (2016).
[2] Weaver et al., *Med. Phys.* **35**, 1988 (2008).
[3] Hoult and Richards, *J. Magn. Reson.* **24**, 71 (1976)

Rapid Scanning in X-Space MPI: Impacts on Image Quality

Omer Burak Demirel[a,b], Damla Sarica[a,b], Emine Ulku Saritas[a,b,*]

[a] Department of Electrical and Electronics Engineering, Bilkent University, Ankara, Turkey
[b] Natioal Magnetic Resonance Research Center (UMRAM), Bilkent University, Ankara, Turkey
[*] saritas@ee.bilkent.edu.tr

INTRODUCTION Initially, piecewise constant focus fields were used in x-space MPI to cover the field-of-view (FOV) by dividing it into overlapping partial FOVs (pFOV) [1-2]. Since this step-by-step coverage proved to be time consuming, a linear focus field was utilized in more recent works [3-4]. Previously, effects of rapid scanning were analyzed for system matrix reconstruction [5], but have not been researched in x-space MPI. Here, we investigate the effects of rapid scanning on image quality for x-space MPI images.

MATERIAL AND METHODS In case of piecewise constant focus field, filtering the fundamental frequency results in a DC loss term (Fig. 1.a) [3]. Here, we utilize a linear focus field:

$$H_{focus}(t) = \frac{2B_f}{\mu_0 T_{Ramp}} t - \frac{B_f}{\mu_0}$$

Here, $B_f \triangleq \mu_0 G \cdot FOV/2$ [T] is the maximum focus field, G [T/m/μ_0] is the selection field, FOV [m] is the total coverage, and T_{Ramp} [s] is the ramp time of the focus field. We also define $S_r \triangleq 2B_f/T_{Ramp}$ [T/s] as the slew rate of the focus field. When S_r is relatively high, fundamental filtering not only causes a DC loss, but also distorts the edges of the reconstructed pFOVs (Fig. 1.b). This complicates x-space image reconstruction.

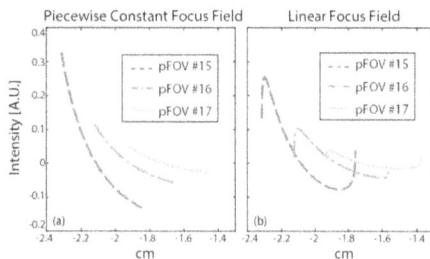

Figure 1: Reconstructed images for (a) piecewise constant focus field, and (b) linear focus field (shown for S_r = 150 T/s).

Here, we implemented an SNR-optimized reconstruction algorithm that robustly calculates the effective DC loss and stitches images from all pFOVs, even for high S_r values. In this algorithm, each point in a pFOV is weighted by the corresponding squared-velocity of the field-free point (FFP). We developed a custom MPI simulation toolbox in MATLAB to test the effect of S_r on image quality. 7mT-peak drive field at 25 kHz was utilized in accordance with magnetostimulation limits [6]. $G_{xx}=G_{zz}=3$T/m selection field, 10x4cm^2 FOV, 25nm nanoparticle diameter was used, ignoring relaxation effects.

RESULTS According to our simulations, there are two effects of fast scanning on image quality: overall signal loss and slight resolution enhancement. Figure 2 shows example images from 20 T/s (the safety limit for focus field, adopted from MRI [6]) and 150 T/s. As shown in Fig. 2d, the signal losses at 20 T/s and 150 T/s were 3.5% and 19%, respectively. Interestingly, higher slew rates resulted in slightly better resolution (see Fig. 3.b): approximately 1% improvement at 20 T/s and 10% at 150 T/s. It should be noted that this secondary effect may not be valid if

relaxation effects are taken into account. Additionally, we observed that linearity and shift invariance holds for slew rates up to 180T/s. At higher slew rates the image quality suffered considerably (results not shown).

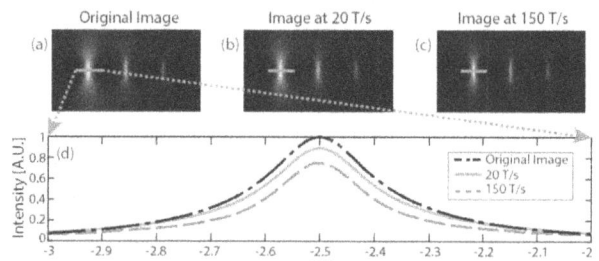

Figure 2: Images at various scanning speeds. (a)Original image for piecewise constant focus field. (b) Image for S_r = 20 T/s. (c) Image for S_r = 150 T/s. (d) Zoomed in 1D cross-section.

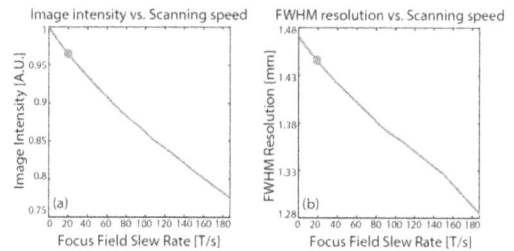

Figure 3: Impact of fast scanning on (a) image intensity and (b) FWHM resolution, given as a function of focus field slew rate.

CONCLUSION In this work, we showed that image quality is mainly preserved when slew rates are within the human-subject safety limits of MPI. When slew rate is further increased, the dominant effect is an overall signal loss. Investigating the effects of nanoparticle relaxation and SNR remains as future work.

ACKNOWLEDGEMENTS This work was supported by the Scientific and Technological Research Council of Turkey through a TUBITAK 3051 Grant (114E167), by the European Commission through an FP7 Marie Curie Career Integration Grant (PCIG13-GA-2013-618834), and by the Turkish Academy of Sciences through TUBA-GEBIP 2015 program.

REFERENCES
[1] P.W. Goodwill and S.M. Conolly. *IEEE Trans Med Imaging*, 30(9):1581-90, 2011. doi: 10.1109/TMI.2001.2125982.
[2] K. Lu et al. , *IEEE Trans Med Imaging*, 32(9):1565-1575, 2013. doi: 10.1109/TMI.2013.2257177.
[3] P.W. Goodwill *et al.*, *IEEE Trans Med Imaging*, 31(5):1076-85, 2012. doi: 10.1109/TMI.2012.2185247.
[4] J.J. Konkle *et al.*, *Biomed Tech (Berl)*, 58(6):565-576, 2013. doi: 10.1515/bmt-2012-0062.
[5] J. Rahmer *et al.*, *Proc of International Workshop on Magnetic Particle Imaging 2015*. doi: 10.1109/IWMPI.2013.6528353.
[6] E.U. Saritas *et al.*, *IEEE Trans Med Imaging*, 32(9):1600-10, 2013. doi: 10.1109/TMI.2013.2260764.

Influence of Particle Size Distribution of Magnetic Nanoparticles on the Spatial Resolution of Magnetic Particle Imaging

Xiuying Wang[a], Shiqiang Pi[a], and Wenzhong Liu[a,*]

[a] School of Automation, Huazhong University of Science and Technology, Wuhan, China
* Corresponding author, email: lwz7410@hust.edu.cn

INTRODUCTION Magnetic particle imaging (MPI) is a new tomographic imaging technique, which can be used in biomedical imaging. Based on the physical model and experiment of MPI, the particle size distribution is one of the key factors affecting the spatial resolution of MPI and also the precision [1-3]. However, the current researches just provide us a brief introduction of the influence of particle size distribution. In this paper, we go into more details on the influence of the particle size distribution, especially the standard deviation on the spatial resolution of MPI.

MATERIAL AND METHODS Statistically, the particle size distribution of magnetic nanoparticles is a gamma distribution or lognormal distribution. Generally, it is described by a lognormal distribution:

$$f(D) = \frac{1}{D\sigma\sqrt{2\pi}} \exp\left(-\frac{(\ln(D)-\mu)^2}{2\sigma^2}\right) \qquad (1)$$

where $\mu = \ln(E(D)) - \frac{1}{2}\ln\left(1 + \frac{\mathrm{Var}(D)}{\mathrm{E}^2(D)}\right)$, $\sigma = \sqrt{\ln\left(\frac{\mathrm{Var}(D)}{\mathrm{E}^2(D)}+1\right)}$, $E(D)$ is the mathematic expectation value of the particle size distribution, and $\sqrt{\mathrm{Var}(D)}$ is the standard deviation.

The magnetization response of the magnetic nanoparticles under the time varying magnetic field can be written as:

$$M = \int_0^\infty \phi \frac{\pi M_s D^3}{6} L(D,H) \cdot f(D) \cdot dD$$
$$L(D,H) = \coth(\xi) - \frac{1}{\xi}, \xi = \frac{\pi M_s D^3 H}{6k_B T} \qquad (2)$$

where ϕ is the nanoparticle density, M_s is the saturation magnetization, H is the excitation magnetic field, k_B is Boltzmann constant, and T is the absolute temperature.

The full width at half maximum (FWHM) of point spread function (PSF) is used to calculate the spatial resolution of MPI:

$$\Box x = \frac{k_B T}{\pi M_s D^3 G} \Box \xi_{FWHM} \qquad (3)$$

where G is the gradient strength.

RESULTS Based on the above formulas, several simulations ($G = 1.87$ T/m) were performed to analyze the influence of the particle size distribution on spatial resolution of MPI. In the case of a certain Var(D), the influence of E(D) on the resolution is shown in Fig. 1. Line 1 shows that when Var(D) is small, the spatial resolution goes to better with the increase of E(D). However, Line 2 shows that when Var(D) is large, the spatial resolution first goes to worse and then gradually improves with the increase of E(D). The decreasing E(D) in Line 2 improves the spatial resolution, but also results in severe image artifacts and low strength of

magnetic response. In the case of a certain E(D), the influence of Var(D) on the resolution is shown in Fig. 2. The greater the Var(D), the higher the resolution. The reason for this is that the percentage of large-size particles increases.

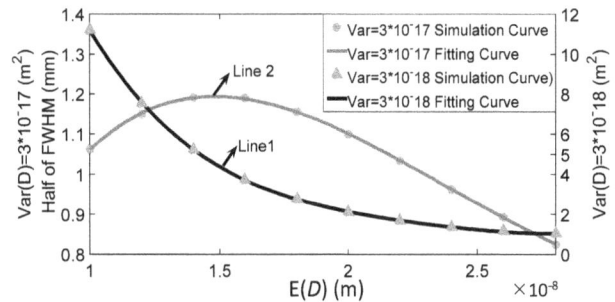

Figure 1: Simulation results of spatial resolution using magnetic nanoparticles with Var(D) of 3×10^{-17} (right axis) and 3×10^{-18} (left axis) m^2, respectively.

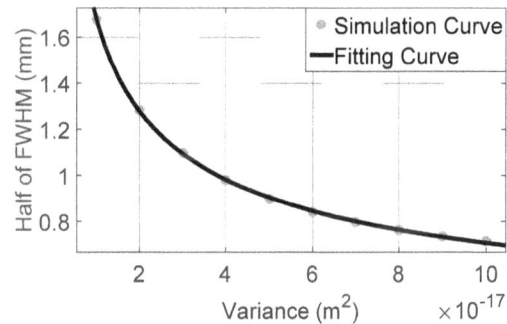

Figure 2: Simulation results of spatial resolution using magnetic nanoparticles with E(D) of 20 nm.

CONCLUSION Increasing E(D) and/or Var(D) can improve the spatial resolution as well as the strength of the magnetic response. However, when E(D) increases to a certain extent, the improvement of spatial resolution is not obvious. Taking account into both spatial resolution and magnetic relaxation, magnetic fluids ($Ms = 477$ kA/m) with E(D) and $\sqrt{\mathrm{Var}(D)}$ of about 20 and 6 nm respectively is enough.

ACKNOWLEDGEMENTS This work was supported by 61571199 (NSFC) and Hubei Provincial project of 2014AEA048.

REFERENCES
[1] T. Knopp, S. Biederer and TF. Sattel. *IEEE Trans Med Imaging*, 30(6):1284—1292, 2011. doi: 10.1101/TMI.2011.2113188.
[2] S. Biederer and T. Knopp. *J.Phys.D*: Appl. Phys. 42(20): 205007-205007(7), 2009. doi: 10.1088/0022-3727/42/20/205007.
[3] J. Rahmer, J. Weizenecker and B. Gleich. *BMC Medical Imaging*, 9:4, 2009. doi: 10.1186/1471-2342-9-4.

79

DC Shift Imaging for X-Space MPI Reconstruction

Damla Sarica [a,b], Omer Burak Demirel [a,b], Yavuz Muslu [a,b], Emine Ulku Saritas [a,b*]

[a] Department of Electrical and Electronics Engineering, Bilkent University, Ankara, Turkey
[b] National Magnetic Resonance Research Center (UMRAM), Bilkent University, Ankara, Turkey
* saritas@ee.bilkent.edu.tr

INTRODUCTION Direct feedthrough filtering in x-space magnetic particle imaging (MPI) [1-3] results in a constant (DC) loss in the image [4-5]. Following a DC recovery algorithm that utilizes overlapping regions of partial field-of-views (pFOVs), one can reconstruct an x-space MPI image that matches the ideal image [5]. Here, we propose an alternative x-space image reconstruction scheme that we name "DC shift imaging", which takes advantage of the lost DC components.

MATERIAL AND METHODS The lost first harmonic term, S_1, can be expressed as [4-5]

$$S_1 = \int_{-W/2}^{W/2} \hat{\rho}(x) \sqrt{1 - \left(\frac{2x}{W}\right)^2} \, dx$$

where $\hat{\rho}(x)$ is the ideal MPI image (i.e., particle distribution convolved with the Langevin-based point spread function), $\sqrt{1 - (2x/W)^2}$ is a space variant velocity term, and W is the total extent of the pFOV. If this equation is considered for each pFOV, it can be written as

$$S_{1j} = \int_{-W/2}^{W/2} \hat{\rho}(x_{0j} + x) \sqrt{1 - \left(\frac{2x}{W}\right)^2} \, dx$$

where S_{1j} and x_{0j} represent the DC loss and the center location for the jth p-FOV image, respectively. Using the symmetry of this formulation, it can be expressed as a convolution:

$$S_1(x_{0j}) = \hat{\rho}(x) * \left. \left(\sqrt{1 - \left(\frac{2x}{W}\right)^2} \right) \right|_{x=x_{0j}} = \hat{\rho}(x) * h(x) \, |_{x=x_{0j}}$$

Effectively, the DC loss experienced by the jth pFOV is the convolution of the ideal MPI image with $h(x)$, evaluated at the center of that pFOV. When DC shifts for all pFOVs are calculated, the result can be shown as an image sampled at all pFOV centers (see Fig. 1 for 1D schematic). Accordingly, the DC shift image is a blurred version of the ideal image. For smaller pFOV sizes, $h(x)$ gets narrower and the DC shift image converges to the ideal MPI image. Note that $h(x)$ is known for given scan parameters. Hence, one can deconvolve the DC shift image to obtain the underlying ideal MPI image.

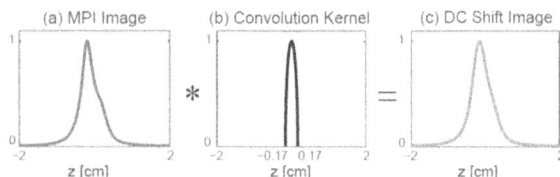

Figure 1: Convolution of (a) the ideal MPI image and (b) the convolution kernel $h(x)$ results in (c) the DC shift image

To demonstrate this new image reconstruction technique, we performed 2D simulations on a custom MPI simulation toolbox that we developed in MATLAB. The parameters for the simulation were: $G_{xx} = G_{zz} = 3$T/m, 5mT drive field at 25kHz, 3.3mm pFOV size, pFOV overlap = 95%, 25nm nanoparticle

diameter. The DC shift values were computed by comparing the overlapping portions of the images from neighboring pFOVs, as explained in [5]. A regular MPI image was then reconstructed. Additionally, the DC shift values were turned into an image by interpolating them on a finer grid. Finally, the resulting DC shift image was deconvolved by the computed $h(x)$, using Lucy-Richardson deconvolution with 30 iterations.

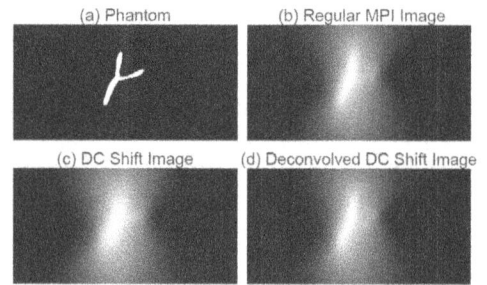

Figure 2: Simulation results for DC shift imaging. (a) Phantom, (b) regular x-space MPI Image, (c) DC Shift Image, and (d) deconvolved DC shift image. 4x2 cm^2 FOV.

RESULTS Figure 2 shows the results of the proposed method in comparison with regular x-space MPI reconstruction. DC shift image (Fig 2.c) is considerably blurred when compared to the regular x-space MPI image (Fig. 2.b). A subsequent deconvolution recovers the underlying ideal image perfectly (Fig. 2.d). Note that due to the finite and relatively small extent of $h(x)$, deconvolution results do not show any visible artifacts.

CONCLUSION In this work, we presented an alternative x-space image reconstruction scheme: DC shift imaging. This technique takes advantage of the lost DC components in each pFOV to reconstruct the underlying ideal MPI image. The next step will be to evaluate the robustness of this technique against noise and nanoparticle relaxation effects.

ACKNOWLEDGEMENTS This work was supported by the Scientific and Technological Research Council of Turkey through a TUBITAK 3501 Grant (114E167), by the European Commission through an FP7 Marie Curie Career Integration Grant (PCIG13-GA-2013-618834), and by the Turkish Academy of Sciences through TUBA-GEBIP 2015 program.

REFERENCES
[1] P. W. Goodwill, S. M. Conolly, *IEEE Trans Med Imaging*, 30 (9):1581-1590, 2011. doi: 10.1109/TMI.2011.2125982.
[2] E. U. Saritas *et al.*, *J Magn Reson*, 229:116-126, 2013. doi: 10.1016/j.jmr.2012.11.029.
[3] P. W. Goodwill *et al.*, *Adv Mat*, 24:3870-3877, 2012. doi:10.1002/adma.201200221.
[4] J. Rahmer *et al.*, *BMC Medical Imaging*, 9:4, 2009. doi: 10.1186/1471-2342-9-4.
[5] K. Lu *et al.*, *IEEE Trans Med Imaging*, 32(9):1565-1575, 2013. doi: 10.1109/TMI.2013.2257177.

Limitations of Magnetic Particle Imaging Resolving Large Contrasts

Nadine Gdaniec[a,b,*], Martin Hofmann[a,b], Tobias Knopp[a,b]

[a] Section for Biomedical Imaging, University Medical Center Hamburg-Eppendorf, Hamburg
[b] Institute for Biomedical Imaging, Hamburg University of Technology, Hamburg
[*] Corresponding author, email: n.gdaniec@uke.de

INTRODUCTION Magnetic Particle Imaging (MPI) is an imaging technique that allows determining the spatial distribution of superparamagnetic nanoparticles *in-vivo*. The nanoparticles are injected and distribute via the blood system. Depending on the perfusion of the different tissue types as well as the preceding time after injection of the tracer, the concentration of the nanoparticles can vary over a large range. The aim of this work is to investigate the maximum contrast that can be resolved with MPI and show that it is closely linked to the spatial resolution that can be achieved.

MATERIAL AND METHODS The spatial resolution was calculated for monosized particles in MPI [1,2]. The derived resolution expression was based on a 1D imaging model with the imaging process formulated as a convolution. However, the expressions were restricted to objects of the same concentration so far. In this work, we generalized them to the case of two objects with different concentration. In order to verify the theoretical findings, experiments were carried out with a preclinical MPI system (Bruker/Philips). Two samples were imaged at 5 mm and 10 mm separation, while the concentration ratio of one of the samples varied between 1 and 16.

RESULTS The theoretical dependence of the spatial resolution ratio and the concentration ratio is shown in Figure 1. The reconstructed images from the experiments for a separation of 5 mm are shown in Figure 2. The right sample in the images is made of pure Resovist, while the concentration ratio of the left is 16 (a), 8 (b), 4 (c), and 1 (d). While the samples with equal concentration are both visible (d), it is hard to distinguish the samples in (a,b,c) from the background. The reconstructed images for a separation of 10 mm are shown in Figure 3. The samples with concentration ratio 4 (c) and 8 (b) are separable from the background in this case. From the theoretical curve (Figure 1) a doubling of the sample distance corresponds to an increased resolvable contrast of factor 6.5, which agrees well with the experimental findings.

CONCLUSION It was shown that the resolvable contrast increases with increasing separation of the objects. The findings are especially relevant for future medical investigations since they allow to predict weather it is feasible to see perfusion in the vicinity of thick vessels that typically carry a much larger MPI signal.

ACKNOWLEDGEMENTS We gratefully acknowledge funding and support of the German Research Foundation (DFG, grant number AD 125/5 - 1).

REFERENCES
[1] J. Rahmer, J.Weizenecker, B.Gleich, and J.Borgert, BMC Med. Imag.,vol. 9, 2009
[2] T. Knopp, S. Biederer, T.F. Sattel, M. Erbe, and T. Buzug, IEEE Transactions on Medical Imaging, Vol.30, No.6, 2011

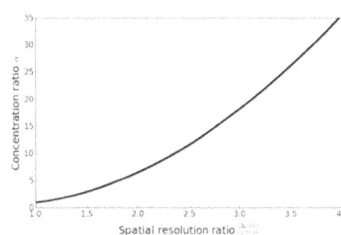

Figure 1: Resolvable contrast as a function of the spatial resolution

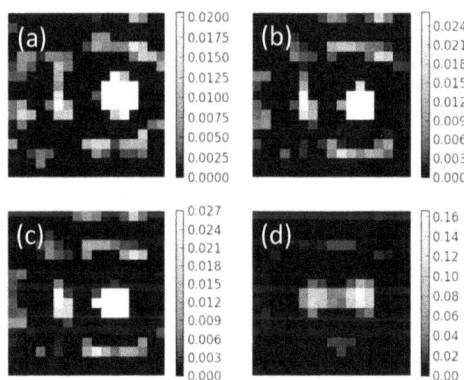

Figure 2: Reconstructed images for two samples with 5 mm separation. The right sample is pure Resovist, while the dilution factor of the left is 16 (a), 8 (b), 4 (c), and 1 (d).

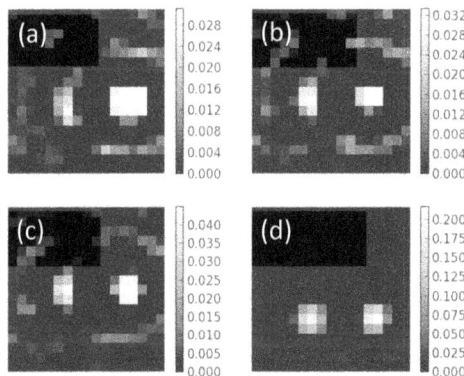

Figure 3: Reconstructed images for two samples with 10 mm separation. The right sample is pure Resovist, while the dilution factor of the left is 16 (a), 8 (b), 4 (c), and 1 (d).

Deconvolving Relaxation Effects in Multi-Dimensional X-space MPI

Gamze Onuker [a], Omer Burak Demirel [a,b], Damla Sarica [a,b], Yavuz Muslu [a,b], Emine Ulku Saritas [a,b,*]

[a] Department of Electrical and Electronics Engineering, Bilkent University, Ankara, Turkey
[b] National Magnetic Resonance Research Center (UMRAM), Bilkent University, Ankara, Turkey
[*] Corresponding author, email: saritas@ee.bilkent.edu.tr

INTRODUCTION Improving image resolution is one of the most important goals in x-space magnetic particle imaging (MPI) technique [1-2]. However, both the nanoparticles' Langevin response as well as relaxation effects blur the image significantly [3]. Recently, we proposed a technique to estimate and deconvolve the relaxation effects from the MPI signal, without any a priori information about the nanoparticles or their distribution in space [4]. Here, we extend this technique to multi-dimensional imaging using x-space MPI reconstruction.

MATERIAL AND METHODS Theoretically, relaxation is modeled as an exponential function $r(t) = (1/\tau)\exp(-t/\tau)u(t)$, where $r(t)$ is then convolved with the adiabatic MPI signal [3]. In [4], we proposed a technique to blindly estimate the relaxation time constant, τ, and deconcolve the relaxation effect from the MPI signal. Effectively, this technique yields the underlying adiabatic MPI image. Here, we validate that this method can be applied to multi-dimensional x-space MPI.

An in-house MPI simulation toolbox was utilized in MATLAB to test the proposed method. Selection field of $G_{xx} = G_{zz} = 3$ T/m, drive field at 25 kHz and 20 mT, 25 nm nanoparticle diameter, $\tau = 1.25\mu s$ were used. Fundamental filtering and additive noise were applied to the simulated MPI signal. A 2D space was scanned by dividing the field-of-view (FOV) into smaller, overlapping partial FOVs. Image from each pFOV was reconstructed individually, and stitched using the method described in [5]. The relaxation time constant was blindly estimated for each pFOV and an effective overall time constant was computed. Next, relaxation effect was deconvolved via Wiener deconvolution. The resulting signal was processed using the abovementioned reconstruction scheme. Finally, the resulting 2D MPI image was compared with the ideal and the relaxation blurred images.

RESULTS The relaxation time constant estimated from the MPI signal was $\tau = 1.15\mu s$, a close match to the actual value. The corresponding relaxation kernel was deconvolved from the MPI signal to yield the corrected image. As seen in Figs. 1 and 2, relaxation blurred images have visibly lower resolution than the ideal image. The corrected image, on the other hand, closely matches the ideal image. In the 1D cross-section given in Fig. 2, the relaxation blurred image cannot resolve the two separate peaks, while the corrected image can.

CONCLUSION In this work, we presented a multi-dimensional validation of our previously proposed simultaneous relaxation estimation and image reconstruction technique for x-space MPI. The relaxation time constant is estimated blindly, without any a priori information. Deconvolving the relaxation effects from the MPI signal improves the image quality visibly. Using this technique, we can recover the underlying adiabatic MPI image almost fully. Our future goal is to validate this technique experimentally.

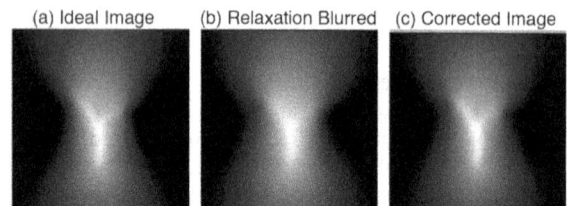

Figure 1: 2D simulation results for the proposed method. (a) Ideal MPI image (i.e., without relaxation), and (b) relaxation blurred image. (c) Deconvolution with the blindly estimated relaxation kernel significantly improves image resolution. (2×4 cm^2 FOV, 2×2 cm^2 central part shown)

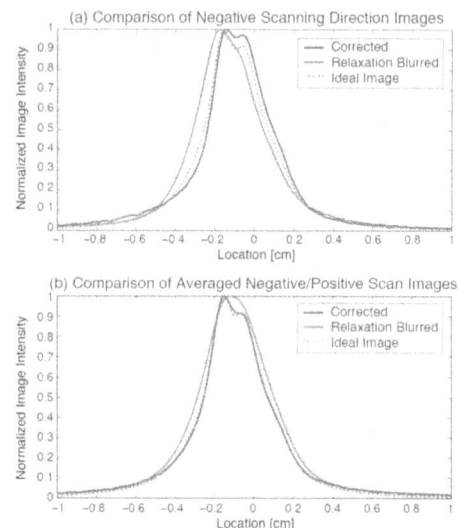

Figure 2: 1-D cross-sections of the 2D results from Fig. 1. Comparison for (a) the images from the negative scanning direction only, and (b) the averages of the images from negative and positive scanning directions. Relaxation corrected image closely matches the ideal image.

ACKNOWLEDGEMENTS This work was supported by the Scientific and Technological Research Council of Turkey through a TUBITAK 3501 Grant (114E167), by the European Commission through an FP7 Marie Curie Career Integration Grant (PCIG13-GA-2013-618834), and by the Turkish Academy of Sciences through TUBA-GEBIP 2015 program.

REFERENCES
[1] P.W. Goodwill and S.M. Conolly. *IEEE Trans Med Imaging*, 30(9):1581-90, 2011. doi: 10.1109/TMI.2001.2125982.
[2] E. U. Saritas *et al.*, *J Magn Reson*, 229:116-126, 2013. doi: 10.1016/j.jmr.2012.11.029.
[3] L.R. Croft *et al.*, *IEEE Transactions on Medical Imaging*, 31(12), 2335–2342, 2012. doi: 10.1109/TMI.2012.2217979.
[4] G. Onuker *et al.*, *Proc of International Workshop on Magnetic Particle Imaging.* doi: 10.1109/IWMPI.2015.7107042
[5] K. Lu *et al.*, *IEEE Transactions on Medical Imaging*, 32(9):1565-1575, 2013. doi: 10.1109/TMI.2013.2257177.

Enhancing the sensitivity in Magnetic Particle Imaging by Background Subtraction

K. Them[1,3,*], M. G. Kaul[2], C. Jung[2], M. Hofmann[1,3], T. Mummert[2], F. Werner[1,3], T. Knopp[1,3]

[1] Section for Biomedical Imaging, University Medical Center Hamburg-Eppendorf, Hamburg, Germany
[2] Department of Diagnostic and Interventional Radiology, University Medical Center Hamburg-Eppendorf, Hamburg, Germany
[3] Institute for Biomedical Imaging, Hamburg University of Technology, Hamburg, Germany
[*] Corresponding author, e-mail: k.them@uke.de

INTRODUCTION Biomedical applications such as cell tracking and angiography require the detection of low concentrations of sperparamagnetic iron oxide nanoparticles (SPIOs) for imaging purposes. Magnetic particle imaging (MPI) enables a quantitative and time-resolved localization of SPIO distributions. The purpose of this work is to investigate a method that can improve the sensitivity of MPI measurements, by subtracting the background signal that has been acquired using an empty scanner tube (in the absence of SPIOs). Additionally, we have also applied a frequency filter that is capable of removing non-static portions of the background signal. It should be noted that a BG subtraction and a frequency filter has been utilized previously [1], [2]. However, the influence of BG subtraction in combination with a frequency cutoff on the image quality and sensitivity has not been investigated so far, and we aim to address this gap in the literature with our results.

MATERIAL AND METHODS The experiments were performed on a preclincal MPI scanner (Philips preclinical MPI package with a Bruker preclinical MPI system). In MPI the SPIO distribution c is obtained by solving the linear system of equations

$$Sc = u$$

where S is the system matrix and u is the measurement signal. In this work it is proposed to apply a BG subtraction and solve

$$\left(\tilde{S} - \tilde{S}^{BG}\right)c = \tilde{u} - \tilde{u}^{BG}$$

where the index BG denotes a background measurement and the index ~ denotes a frequency selection.

RESULTS It has been demonstrated that a BG subtraction in combination with a frequency cutoff enhances the sensitivity of the considered MPI scanner up to a factor of 10. MPI images of in-vivo mouse measurements showed that a BG subtraction can also improve the detection limit in-vivo. The bolus entering the FOV through the inferior vena cava was identified earlier when a BG subtraction was applied.

CONCLUSION In-vivo mouse experiments show that for early time points from when the tracer enters the vena cava a reconstructed image of sufficient quality can only be obtained when a background subtraction is performed. A background subtraction in combination with a frequency cutoff lowers the detection limit for SPIOs in MPI which closes the gap to future medical applications.

ACKNOWLEDGEMENTS We gratefully acknowledge funding and support of the German Research Foundation (DFG, grant number AD 125/5-1).

REFERENCES
[1] J. Weizenecker, B. et.al., "Three-dimensional real-time in vivo magnetic particle imaging," *Physics in Medicine and Biology*, vol. 54, no. 5, p. L1, 2009. [Online]. Available: http://stacks.iop.org/0031-9155/54/i=5/a=L01
[2] T. Knopp and T. M. Buzug. Springer, Berlin/Heidelberg, 2012. doi: 10.1007/978-3-642-04199-0.

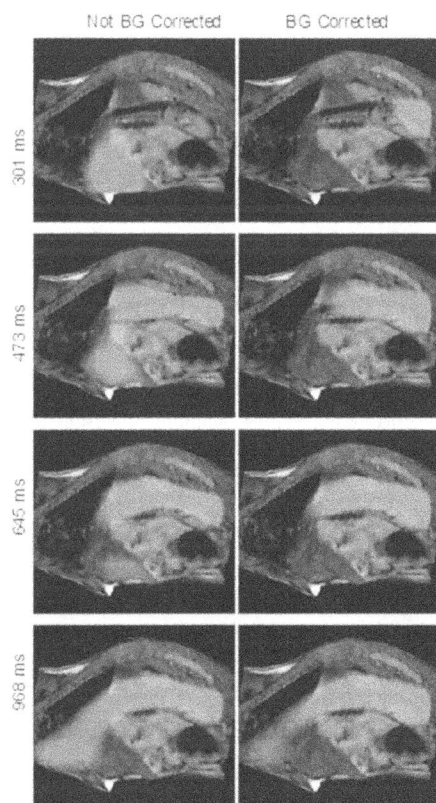

Figure 1: MRI/MPI overlay

Correction of Blurring due to a Difference in Scanning Direction of Field-Free Line in Projection-Based Magnetic Particle Imaging

Kenya Murase[*], Kazuki Shimada, Natsuo Banura

Department of Medical Physics and Engineering, Graduate School of Medicine, Osaka University, Osaka, Japan
[*] Corresponding author, email: murase@sahs.med.osaka-u.ac.jp

INTRODUCTION Magnetic particle imaging (MPI) is a recently introduced imaging method [1] that allows imaging of the spatial distribution of magnetic nanoparticles with high sensitivity, high spatial resolution, and high imaging speed. More recently, we developed a system for projection-based MPI with a field-free-line (FFL) encoding scheme [2]. In such a system, projection data are usually acquired by moving the FFL in a zigzag in order to make the acquisition time as short as possible and a difference in the projection data occurs depending on the scanning direction of the FFL, resulting in blurring in the reconstructed images. To enhance the reliability of projection-based MPI, it is necessary to correct for the blur due to a difference in the scanning direction of FFL. The purpose of this study was to develop a method for correcting for blurring due to a difference in the FFL scanning direction, and to investigate the validity and usefulness of this method in phantom experiments.

MATERIAL AND METHODS First, we assumed that the MPI signal at position x ($S(x)$) is given by

$$S(x) = S_{adiab}(x) \otimes \frac{1}{\xi} e^{-x/\xi} \qquad (1)$$

where $S_{adiab}(x)$ is the adiabatic MPI signal, i.e., the signal without delay at position x and ξ denotes the signal-delay constant. Note that ξ has a unit of mm. Second, the correction of the blur was performed by deconvolving Eq. (1) based on singular value decomposition [3] to obtain $S_{adiab}(x)$ from $S(x)$. The ξ value for correction (ξ_c) was obtained from

$$\xi_c = \arg\min_{\xi} \left\| S_{adiab}^P(x) - S_{adiab}^N(x) \right\|_2 \qquad (2)$$

where $S^P_{adiab}(x)$ and $S^N_{adiab}(x)$ denote the deconvolved MPI signal at position x in the positive and negative directions, respectively, and $\|\cdots\|_2$ represents the 2-norm.

To validate the above method, we performed phantom experiments with a line and A-shaped phantoms filled with Resovist® [4] using the Osaka MPI scanner II, which is an extended version of our previous scanner [2]. To investigate the effect of the velocity of FFL (v), we varied v as 0.62, 1.25, 1.67, and 2.0 mm/s and analyzed the correlation between ξ_c and v. The profiles of the line phantom (2 mm in diameter) along the line passing through the center of the MPI image were obtained and the full width at half maximum (FWHM) was calculated by fitting the profiles by Gaussian function. We also validated our method using the A-shaped phantom, which consisted of silicon tubes 1.5 mm in diameter and filled with 500 mM Fe Resovist®.

RESULTS Figures 1 shows the sinograms of the line phantom before (left) and after correction of the blur (right) for $v = 2$ mm/s. In Fig. 1, the vertical and horizontal axes represent each projection angle and the distance along the projection direction, respectively. The shift of projection data due to a difference in the scanning direction was clearly observed (left), while this shift

disappeared after correction of the signal delay using our method, in which ξ_c was taken as 0.95 mm.

There was a significant linear correlation between the ξ_c value and v (r=0.997) (plot not shown).

Figure 2 shows the FWHM values of the line phantom before (closed circles) and after correction of the blur (open circles) as a function of v. The FWHM value increased linearly with v and it significantly ($P<0.05$) decreased after correction of the blur at v = 1.67 mm/s and 2.0 mm/s.

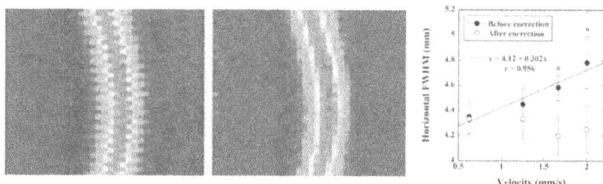

Figure 1: Sinograms of a line phantom before (left) and after correction of signal delay (right).

Figure 2: Relationship between FWHM and v before (●) and after correction of blur (○).

Figure 3 shows a comparison between the MPI images of the A-shaped phantom before (left) and after correction of the blur (right) for $v = 2.0$ mm/s. In this case, ξ_c was taken as 0.95 mm. The reduction of the blur was clearly demonstrated.

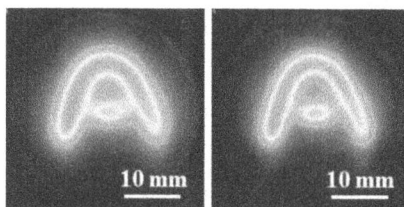

Figure 3: MPI images of an A-shaped phantom before (left) and after correction of the blur (right).

CONCLUSION We developed a method for correcting for blurring caused by a difference in the scanning direction of FFL. Our results suggest that our method will be useful for improving and enhancing the reliability of projection-based MPI.

ACKNOWLEDGEMENTS This work was supported by a Grant-in-Aid for Scientific Research (Grant No. 25282131) from the Japan Society for the Promotion of Science (JSPS).

REFERENCES
[1] B. Gleich and J. Weizenecker. *Nature*, 435(7046):1217–1217, 2005. doi: 10.1038/nature03808.
[2] K. Murase, S. Hiratsuka, R. Song, and T. Takeuchi. *Jpn. J. Appl. Phys.*, 53(6):067001, 2014. doi: 10.7567/jjap.53.067001.
[3] K. Murase, M. Shinohara, and Y. Yamazaki. *Phys. Med. Biol.*, 46(12):3147–3159, 2001. doi: 10.1088/0031-9155/46/12/306.
[4] K. Murase, M. Aoki, N. Banura, K. Nishimoto, A. Mimura, et al. *Open J. Med. Imaging*, 5(2):85–99, 2015. doi: 10.4236/ojmi.2015.52013.

Sensitivity enhancement for stem cell monitoring in Magnetic Particle Imaging

Kolja Them[1,3], J. Salamon[2], M. G. Kaul[2], Claudia Lange[4], H. Ittrich[2], Tobias Knopp[1,3]

[1]Section for Biomedical Imaging, University Medical Center Hamburg-Eppendorf, Hamburg, Germany
[2]Department of Diagnostic and Interventional Radiology, University Medical Center Hamburg-Eppendorf, Hamburg, Germany
[3]Institute for Biomedical Imaging, Hamburg University of Technology, Hamburg, Germany

INTRODUCTION Stem cell therapies to accelerate tissue repair, magnetic fluid hyperthermia for cancer therapy, and targeted drug delivery based on SPIONs require fast and sensitive visualization of the SPION distribution. Magnetic particle imaging (MPI) allows direct imaging of the SPION distribution with positive contrast as well as high temporal and spatial resolution. First MPI images of labeled stem cells were shown in [1], where a detection limit of about 10^4 cells was reached. A SPION labeled stem cell study in a rat was reported in [2]. However, localization and identification of SPION concentrations in MPI low enough for medical applications is still challenging due to artifacts. There are different calibration methods available in MPI which can be used for image reconstruction. In measurement based calibrations, a small sample of SPIONs ("δ-sample") is shifted through a certain region in the scanner tube and measured at certain positions via external magnetic fields to obtain the system characteristics. The degree of freedom of mechanical rotation of the SPIONs is restricted depending on the binding of the SPION shell to the environment. Therefore, in different biological environments SPIONs respond differently to oscillating external magnetic fields, which has been confirmed by magnetic particle spectroscopy. One issue which has not been investigated so far is the influence of different SPION environments, used for calibration, on the occurrence of artifacts in stem cell images.

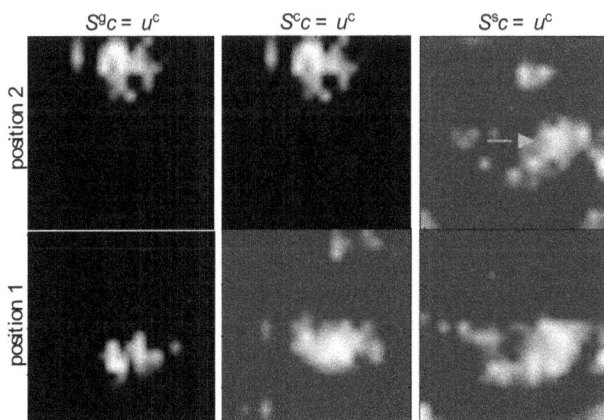

FIGURE 1: An MSC phantom is located at two different positions. The images are reconstructed with a SPION-in-gel SF (left side), an MSC SF (middle) and a watery SF (right side). The identification of the MSC phantom is only possible when the MPI scanner is calibrated with immobilized SPIONs (left and middle cases). A large artifact (marked by the red arrow) occurs when using the watery SF.

MATERIAL AND METHODS The measurement based system matrices containing SPIONs in gel or water (both 250~mmol~(Fe)/l) were obtained using a cubic delta-sample of size 3 x 3 x 3 mm. The SPIONs loaded MSC system function were obtained using a 250μl Eppendorf tube as delta-sample of similar size and shape. The drive field FOV of 37 x 37 x 19 mm was generated using a selection field with a gradient of 1.5 T/m and a drive field amplitude of 14 mT. 30 averages were chosen.

RESULTS In fig. 1, two different positions (1 (down) and 2 (up)) of a SPION labeled stem cell phantom with approximately 8.8 μg Fe are reconstructed using the same three system functions as described above. Using either the gel or the MSC system function for calibration, the MSC phantom can be identified correctly at the two different positions. Using the SPION-in- saline system function for calibration, where large artifacts (marked by a red arrow) occur, led to incorrect SPION-MSC localization in case 1. For the SPION in saline system function, no reconstruction parameters were found for which the MSC phantom could be clearly identified at both positions.

CONCLUSION In conclusion, it was shown that calibrating using rotationally immobilized SPIONs provides superior image quality for SPION-labeled stem cells with reduced artifacts compared to images obtained using rotationally mobile SPIONs. The reduction in artifacts is also valid for small and large amounts of magnetically labeled MSC. The increase in image quality allows identification of a significantly smaller number of SPION-MSC in MPI images (fig. 1) and prevents false localization of SPION-MSC. The corresponding enhancement in sensitivity reduces the detection limit of low concentrated SPION distributions required for future clinical applications.

ACKNOWLEDGEMENTS WE GRATEFULLY ACKNOWLEDGE FUNDING AND SUPPORT OF THE GERMAN RESEARCH FOUNDATION (DFG, GRANT NUMBER AD 125/5-1).

REFERENCES

[1] E. Saritas and et.al, "Magnetic particle imaging (mpi) for nmr and mri researchers." *J Magn Reson.*, vol. 229, pp. 116–26, 2013.
[2] B.Zheng,T.Vazin,P.Goodwill,A.Conway,A.Verma,E. Saritas, D. Schaffer, and S. Conolly, "Magnetic particle imaging tracks the long-term fate of in vivo neural cell implants with high image contrast," *Scientific Reports*, vol. 5, p. 14055, 2015.

Towards the Characterization of Distortion Artifacts in Elongated Trajectory MPI

Annika Hänsch[*], Christian Kaethner, Aileen Cordes, Thorsten M. Buzug[*]

Institute of Medical Engineering, Universität zu Lübeck, Lübeck
[*] Corresponding author, email: {haensch,buzug}@imt.uni-luebeck.de

INTRODUCTION Magnetic particle imaging (MPI) is a quantitative imaging technique that allows for the detection of superparamagnetic nanoparticles [1]. Due to safety reasons, the size of the field of view (FOV) is limited in clinical application scenarios [2]. An approach to increase the FOV size in axial direction within the safety constraints combines a 2D movement of the field free point (FFP) on a Lissajous trajectory with a continuous shift in axial direction, thus yielding an elongated trajectory [3]. The determination of the elongation length, i.e. the distance covered in axial direction, is a trade-off between acquisition speed and subsampling of the imaging region. With increasing elongation length, distortion artifacts can occur in the reconstructed volume. In this contribution, we investigate the artifacts that arise when increasing the elongation length and thus the subsampling of the FOV.

MATERIAL AND METHODS In order to examine the occurring distortion artifacts, we compare the reconstructions to a distance map that we compute for each elongation length individually. The map contains the minimal Euclidean distances to the elongated Lissajous trajectory for all points on a cell-centered grid that corresponds to the discretized FOV. These distances seem suitable since the signal contributed by the nanoparticles also depends on their distance to the FFP [1, 4]. Motivated by the elongation in axial direction, a possible alternative is a distance map containing axial distances to the trajectory at Lissajous node points [3, 5].

We perform simulation experiments on ideal magnetic fields using a FOV with a size of 2 cm $\times$ 2 cm $\times$ 8 $X_s G^{-1}$ discretized into 80 $\times$ 80 $\times$ 5 volume elements. Here, G is the selection field gradient strength in the axial direction and X_s corresponds to the region around an FFP where a significantly high particle signal can be achieved [4]. The applied selection field gradient strength is 0.75 $Tm^{-1}\mu_0^{-1}$ in x- and y-directions and 1.5 $Tm^{-1}\mu_0^{-1}$ in z-direction. The 2D Lissajous trajectory of the FFP is generated in the xy-plane with drive field frequencies $f_x = 26.881$ kHz and $f_y = 26.041$ kHz. The elongation in z-direction is simulated using a linear focus field as described in [3]. Based on the consideration that 2 $X_s G^{-1}$ should be an upper limit to the elongation length to avoid loss of information [3], we choose the elongation lengths 2 $X_s G^{-1}$ and 8 $X_s G^{-1}$ for our simulations. The simulated phantom consists of pipes arranged on a Cartesian grid with a diameter of 1.5 mm, filled with undiluted Resovist® (Bayer Schering Pharma AG, Berlin, Germany) with a particle diameter of 30 nm and $X_s = 1.1$ $mT\mu_0^{-1}$ [4].

RESULTS The used phantom and the resulting reconstructions are shown in Fig. 1. For an elongation length of 2 $X_s G^{-1}$, the upper limit to avoid information loss, we obtain satisfying reconstruction results and observe slight distortion artifacts at the edges of the FOV. When we increase the elongation length in order to exceed the limit of 2 $X_s G^{-1}$, we notice increasing distortion artifacts as seen for 8 $X_s G^{-1}$ in Fig. 1. In the respective

distance maps, also shown in Fig. 1, long distances to the trajectory are represented by bright voxels. A visual comparison of the reconstructions and the distance maps yields a match between highly distorted areas and areas with long distances to the elongated FFP trajectory.

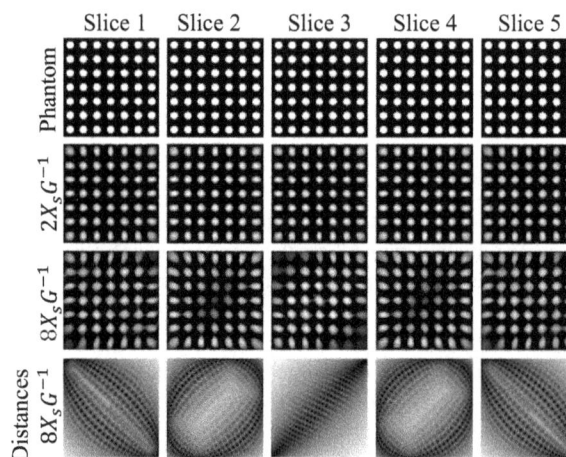

Figure 1: Phantom and reconstructions for elongation lengths 2 $X_s G^{-1}$ and 8 $X_s G^{-1}$. The distance map is shown for 8 $X_s G^{-1}$, where bright voxels correspond to long distances.

CONCLUSION It has been shown that reconstructions based on an elongated trajectory within a limit of 2 $X_s G^{-1}$ result in a good match to the phantom. Exceeding this limit causes distortion artifacts. From our observations, we deduce a link between distortions in the reconstruction and the minimal distances of the concerned areas to the elongated FFP trajectory.

ACKNOWLEDGEMENTS The authors gratefully acknowledge the financial support of the German Research Foundation (DFG, grant number BU 1436/9-1).

REFERENCES
[1] B. Gleich and J. Weizenecker. *Nature*, 435(7046):1214—1217, 2005. doi: 10.1038/nature03808.
[2] E. U. Saritas, P. W. Goodwill, G. Z. Zhang and S. M. Conolly. *IEEE Trans. Med. Imag.*, 32(9):1600-1610, 2013. doi: 10.1109/TMI.2013.2260764.
[3] C. Kaethner, M. Ahlborg, G. Bringout, M. Weber and T. M. Buzug. *IEEE Trans. Med. Imag.*, 34(2):381-387, 2015. doi: 10.1109/TMI.2014.2357077.
[4] B. Gleich. Springer Vieweg, Wiesbaden, 2014. doi: 10.1007/978-3-658-01961-7.
[5] W. Erb, C. Kaethner, P. Dencker and M. Ahlborg. *Dolomites research notes on approximation*, 8:23-36, 2015. doi: 10.14658/pupj-drna-2015-Special_Issue-4.

Fiducial-Based Geometry Planning and Image Registration for Magnetic Particle Imaging

F. Werner[a,b*], C. Jung[c], M. Hofmann[a,b], R. Werner[d], J. Salamon[c], D. Säring[d], M. G. Kaul[c], K. Them[a,b], O. M. Weber[e], T. Mummert[c], G. Adam[c], H. Ittrich[c], T. Knopp[a,b]

[a] Section for Biomedical Imaging, University Medical Center Hamburg-Eppendorf, Hamburg, Germany
[b] Institute for Biomedical Imaging, University of Technology Hamburg-Harburg, Hamburg, Germany
[c] Department of Diagnostic and Interventional Radiology, University Medical Center Hamburg-Eppendorf, Hamburg
[d] Institute for Computational Neuroscience, University Medical Center Hamburg-Eppendorf, Hamburg
[e] Philips Medical Systems DMC GmbH, Hamburg, Germany
* Corresponding author, email: f.werner@uke.de

INTRODUCTION Magnetic Particle Imaging (MPI) is a quantitative imaging modality that allows to visualize distributions of superparamagnetic iron oxide nanoparticles (SPIOs) [1]. Naturally occurring magnetic moments within biological tissue cannot be visualized, which explains the lack of anatomical information in MPI [2]. Most imaging modalities require a precise geometry planning in order to place the region of interest (ROI) within the field of view (FOV). The missing morphological information in MPI complicates geometry planning and the interpretation of MP images. For this reason, MPI data should be fused with morphological data acquired via e.g. Magnetic Resonance Imaging (MRI). To solve the positioning problem and to facilitate fusion with MR data we developed bimodal fiducial markers that can be visualized by MRI as well as by MPI.

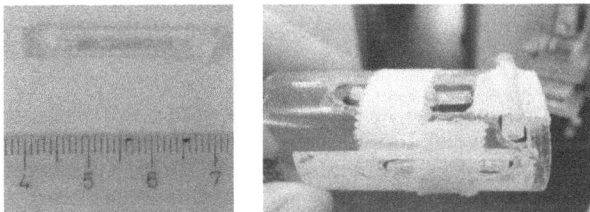

Figure 1: Left: Bimodal fiducial marker: The inner glass capillary is filled with SPIO tracer, the outer flexible tube contains MR contrast agent. Right: Spatial arrangement of the three markers.

MATERIAL AND METHODS The developed bimodal fiducial markers (Figure 1, left) each consist of two tubes arranged coaxially to each other. The inner glass capillary is filled with SPIO tracer (10 mmol(Fe)L^{-1}), while the outer PVC tube contains MR contrast agents (0.75 mmol(Gd-DTPA)L^{-1}). The spatial arrangement of three markers (Figure 1, right) should guarantee a unique determination of the orientation of 3D objects assuming the marker phantom being placed on top of the object. Furthermore, an algorithm for automated marker extraction in both, MR and MP images, and rigid registration is established. Experiments are carried out using a preclinical 7 T MRI system (Bruker) and a preclinical MPI system (Bruker/Philips). A first experiment demonstrates the ROI alignment within the FOV using the marker phantom. The phantom is placed inside the MPI scanner. By using an online reconstruction tool its position is adjusted until the ROI lies within the FOV. Next, MPI and MRI datasets of the same object are acquired [3] to test the automated registration algorithm. The test object is a kiwi fruit with an additional SPIO filled glass capillary placed inside the kiwi. The test object with the marker phantom is sequentially measured with the MRI and the MPI scanner.

RESULTS In the experiments the markers enabled a guided and in turn fast and precise placement of the ROI within the FOV. By means of the marker-based registration, the MR data could be successfully overlaid with the MPI data as can be seen in Figure 2. Both, the rod inside the kiwi fruit and the fiducial markers are correctly aligned.

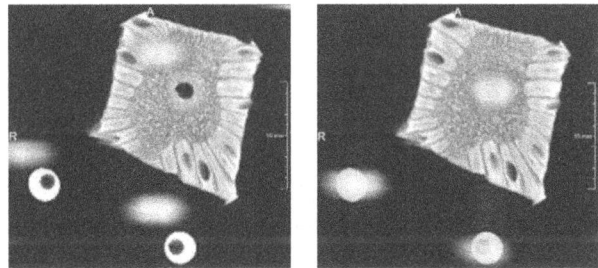

Figure 2: Transversal MRI and MPI slices of the kiwi fruit with the marker phantom on top, before (left) and after registration (right).

CONCLUSION The use of fiducial marker is a promising tool for image acquisition in MPI. They enable exact positioning of the object to be imaged. Furthermore, MP-MR images can be easily registered by using the markers as reference points. A reliable interpretation of MP images is thus possible.

ACKNOWLEDGEMENTS We gratefully acknowledge funding and support of the German Research Foundation (DFG, grant number AD 125 / 5 - 1).

REFERENCES
[1] B. Gleich and J. Weizenecker. *Nature*, 435(7046):1217—1217, 2005. doi: 10.1038/nature03808.
[2] T. Knopp and T. M. Buzug. Springer, Berlin/Heidelberg, 2012. doi: 10.1007/978-3-642-04199-0.
[3] M. G. Kaul, O. Weber, U. Heinen, A. Reitmeier, T. Mummert, C. Jung, N. Raabe, T. Knopp, H. Ittrich, and G. Adam. Fortschr Röntgenstraße, 187(05):347-352,2015.347.

Predicting 2D MPI imaging performance using a conventionally acquired or a hybrid 2D system function

Hanne Medimagh[a,*], Thorsten M. Buzug[a,*]

[a] Institute of Medical Engineering, University of Luebeck, Ratzeburger Allee 160, 23562 Luebeck, Germany
[*] Corresponding authors, email: {medimagh,buzug}@imt.uni-luebeck.de

INTRODUCTION Magnetic Particle Imaging (MPI) is an emerging medical imaging modality capable of quantifying the concentration of superparamagnetic iron oxide nanoparticles. Concerning scanner and sequence design as well as particle development in MPI with system function (SF) based image reconstruction, there is immense interest in the prediction of MPI imaging performance. For particle characterization, the current state-of-the-art is performing (zero-dimensional) Magnetic Particle Spectroscopy (MPS) and selecting particles with a strong signal. Experience has shown, however, that the imaging performance of particles which exhibit a promising spectrum is often worse than expected. Even when the SF is studied, which contains information about the field sequence experienced by the particles, the imaging performance does not always correspond to expectations derived from the signal level. We believe that this may be due to the fact that the inverse nature of the imaging problem has not been accounted for.

Figure 1: From left to right: Lissajous, Cartesian, radial, spiral and Archimedean trajectory. From top to bottom: FFP trajectory; reconstructed MPI image; signal level (common colormap) and energy level (common colormap) for a conventionally acquired SF (values practically identical for a hybrid SF; not shown); local conditioning for a conventionally acquired SF: $\log 10(\mathrm{cond}(S_{\mathrm{local}}))$ (colorbar from 9 to 14.2, from white to black); comparison of the local conditioning for a conventionally acquired and a hybrid SF: $(\mathrm{cond}(S_{\mathrm{local}}))_{\mathrm{norm}} - (\mathrm{cond}(S_{\mathrm{local}}^{\mathrm{hybrid}}))_{\mathrm{norm}}$ (colorbar from -10^{-4} to 10^4; values multiplied by 0.1 for the Cartesian trajectory).

A recently published approach for the characterization of the imaging performance of magnetic tracers uses a 1D hybrid MPI spectrum, but still involves an image reconstruction step with regularization and the analysis of the resulting images, which is indirect, intricate and subjective [1]. By contrast, in this contribution we present considerations toward the prediction of MPI imaging performance based on the direct analysis of the corresponding system function (SF) alone, taking into account the inverse nature of the imaging problem. The concept was evaluated using the example of different field free point (FFP) trajectories, which extends previous studies on the topic of trajectory analysis [2, 3].

MATERIAL AND METHODS Using proprietary software, five FFP trajectories with deliberately low trajectory density were compared regarding their imaging performance in a simulated imaging process using a SF based image reconstruction approach. In the first part, the SF was acquired conventionally (i.e. by measuring a delta sample at different positions within the MPI scanner). In the second part, the acquisition of a 2D hybrid SF (i.e. a SF measured using a 2D MPS) was simulated [4]. Signal level, energy level and local conditioning of the SFs were determined.

RESULTS Both a high signal level and a good conditioning are required for a high imaging performance (see Fig. 1). Results indicate that 2D hybrid SFs can be exploited for the prediction of imaging performance instead of conventionally acquired SFs.

CONCLUSION When studying the interplay of scanner, sequence and particles (e.g. see [5]) in MPI with SF based image reconstruction, it is important to keep in mind that a good conditioning of the SF is essential, next to a high signal level. The proposed concept for the prediction of 2D MPI imaging performance may be a valuable tool in scanner and sequence design and optimization as well as particle characterization. Utilizing hybrid SFs instead of conventionally acquired SFs will substantially facilitate and accelerate the process in practice.

ACKNOWLEDGEMENTS The authors gratefully acknowledge the financial support of the German Federal Ministry of Education and Research (BMBF) under grant number 13N11090 and of the European Union and the State Schleswig-Holstein (Programme for the Future – Economy) under grant number 122-10-004.

REFERENCES
[1] D. Schmidt et al. Biomedical Engineering/Biomedizinische Technik, 60 (s1), 2015. doi: 10.1515/bmt-2015-5010.
[2] T. Knopp et al. *Physics in Medicine and Biology*, 54: 385–397, 2009. doi:10.1088/0031-9155/54/2/014.
[3] H. Wojtczyk et al. *IWMPI 2013*, 2013. doi: 10.1109/IWMPI.2013.6528351.
[4] M. Graeser et al. *Book of Abstracts of the IWMPI 2015*, 96, 2015. doi: 10.1109/IWMPI.2015.7107078.
[5] M. Graeser et al. submitted.

Experimental Results on 3D Real-Time Magnetic Particle Imaging of Large Fields-of-View

Jürgen Rahmer[a,*], Bernhard Gleich[a], Claas Bontus[a], Ingo Schmale[a], Joachim Schmidt[a], Oliver Woywode[b], and Jörn Borgert[a]

[a] Philips Technologie GmbH Innovative Technologies, Research Laboratories, Röntgenstraße 24-26, 22335 Hamburg, Germany
[b] Philips Medical Systems DMC GmbH, Röntgenstraße 24-26, 22335 Hamburg, Germany
* Corresponding author, email: juergen.rahmer@philips.com

INTRODUCTION In 3D real-time MPI, the size of the volume covered by the drive fields is limited due to nerve stimulation and patient heating issues [1,2]. To achieve larger spatial coverage, focus fields have been introduced to shift the volume encoded by drive fields in space. Using different stationary offsets, a larger imaging volume can be addressed [3]. However, to reduce acquisition time, data acquisition has to be performed also in non-stationary situations, while the focus fields are changing rapidly [4]. As the dynamic focus field variation for a given sequence also depends on the eddy current response of the system, the actual field variation has to be determined experimentally using Hall sensors. Based on the measured field values, correct shift-compensated system functions can be generated for reconstruction [5]. Whereas the proof of principle of this approach was delivered using 2D focus field shifts, the present contribution focuses on experimental results obtained with different fast 1D, 2D, and 3D focus field sequences on a preclinical MPI demonstrator system.

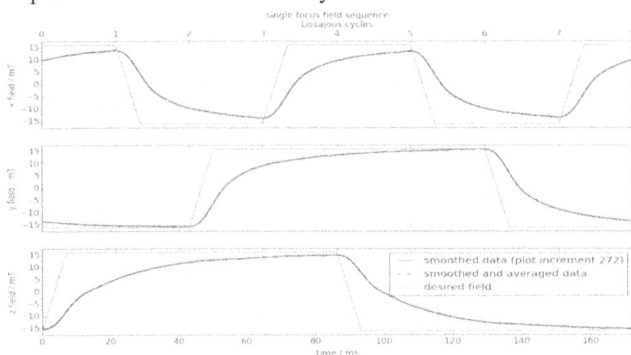

Figure 1: Measured (blue) and desired (green) focus fields for a sequence of 2 × 2 × 2 stations.

MATERIAL AND METHODS The pre-clinical demonstrator was operated using 3D drive field excitation at frequencies 24.5, 26.0, and 25.3 kHz with an amplitude of 16 mT. The applied selection field gradient was 2.5 T/m in z direction, so that a volume of 32 × 32 × 16 mm³ was covered by the drive fields. Focus fields were applied to shift the volume to different stations with a small overlap. One sequence consisted of 2 × 2 × 2 stations, which were addressed sequentially. Time per station varied between experiments, ranging from 1 to 9 Lissajous cycles (duration of one cycle $T_L = 21.5$ ms). Fig. 1 displays the field variation measured using three orthogonal Hall sensors during a sequence using 2 cycles per station (total sequence duration 172 ms). For each cycle, an image was reconstructed using a shift-compensated system function derived from a measured static

system function [5]. All images from one focus field sequence were then combined into a single image.

RESULTS The field measurements displayed in Fig. 1 exhibit large discrepancies between desired (green) and actual (blue) field values which are caused by slew rate limitations of the amplifiers and eddy currents inside the field generator. Especially in the direction with the most frequent field changes (x), a stationary field value is not reached. As the trajectory density is high enough for good spatial encoding, an image can be reconstructed from the acquired data using compensated system functions, as shown in Fig. 2. Nonetheless, artifacts arise from field-compensating the system functions. In addition, at this point no effort has been put into artifact-free stitching of sub-volumes. Further work is required to improve image quality.

CONCLUSION The implemented approach enables 3D imaging over large volumes at acquisition times short enough for real-time imaging.

ACKNOWLEDGEMENTS This work was supported by the German Federal Ministry of Education and Research (BMBF grants FKZ 13N9079 and 13N11086).

REFERENCES

[1] I. Schmale et al., "MPI Safety in the View of MRI Safety Standards." IEEE Transactions on Magnetics 51, no. 2 (2015): 1–4. [2] E. Saritas et al., "Magnetostimulation Limits in MPI." IEEE Transactions on Medical Imaging 32, no. 9 (2013): 1600–1610.
[3] B. Gleich et al., "Fast MPI Demonstrator with Enlarged Field of View." In Proc. ISMRM, 18:218 (2010).
[4] J. Rahmer et al., "Fast Continuous Motion of the Field of View in Magnetic Particle Imaging" Proc. IWMPI, IEEE, (2013).
[5] J. Rahmer et al., "Automated Derivation of Sub-Volume System Functions for 3D MPI", Proc. IWMPI (2014).

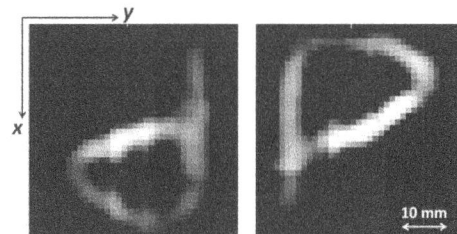

Figure 2: Parallel slices of a 3D volume reconstructed from the 2 × 2 × 2 focus field sequence displayed in Fig. 1. The volume extends over 60 × 63 × 34 mm³ and has been acquired in 172 ms. The phantom consisted of two P-shaped tubes of inner diameter 0.8 mm filled with diluted Resovist (Bayer Schering Pharma, Germany) at a concentration of 6 mmol(Fe)/l. Vertical (z axis) center to center distance of the two Ps was 15 mm.

Rotational Drift Spectroscopy (RDS): Measuring Fast Relaxing Magnetic Nanoparticle Ensembles

M.A. Rückert[a,*], A. Vilter[a], P. Vogel[a], V.C. Behr[a]

[a] Department of Experimental Physics 5 (Biophysics), University of Würzburg, Würzburg
* Corresponding author, email: Martin.Rueckert @physik.uni-wuerzburg.de

INTRODUCTION Rotational drift occurs in weak rotating magnetic fields that are not strong enough to rotate the magnetic particle synchronously. The resulting average rotational drift strongly depends on the magnetic field strength, the properties of the particles and their interaction with the environment. This has been used in [1] for building a highly sensitive sensor for single bacteria detection based on directly measuring the rotation of a single microparticle. Rotational Drift Spectroscopy (RDS) [2,3] is a novel modality for measuring the rotational drift of magnetic particles ensembles, e.g., ferrofluids or magnetic tracer material. The RDS signal is short lived and the 3 ms dead time after the initial alignment of the particles was the major limitation of the first setup presented in [2]. In [3] simulated results of a different measurement sequence were presented in order to reduce the dead time by orders of magnitude. The presented work shows a different approach for overcoming the limitation caused by a long dead time between initial alignment and measurement. It allows the measurement of fast relaxing magnetic particle suspensions that could not be measured with the previous approach in [2].

MATERIAL AND METHODS The setup consists of an orthogonal coil pair used in [2]. The initial pulse used in [2] was replaced by applying an offset field oriented within the rotating plane of the rotating magnetic field and along the receiver coil using a permanent magnet. The offset field was about 2 mT and the rotating field was between 10 mT and 20 mT. The two coil pairs where driven by f_x=49.3 kHz and f_y=50.0 kHz for the x- and y-direction respectively. This results in a rotating magnetic field that periodically changes its rotating direction with a repetition frequency of f_{rep}=f_x-f_y=700 Hz. In [2] this was used in order to create a signal echo train. The presented work uses the fact that the presence of an offset field results in a unidirectional magnetization build up between the rotating direction changes, where the external magnetic field is a linear oscillating field vector. Simulations based on the Langevin equation suggest that this effect is as efficient as the unipolar pulse simulated in [4]. The advantage of this sequence is that it contains only two frequencies and is free of any field switching, which allows to suppress the induction of the rotating magnetic field using only a low-pass filter without the pulse suppression circuit in [2]. The sample size was 40 µl particle suspension in 5 mm glass tubes.

RESULTS Figure 1 shows the signal of fluidMAG-UC/C 200 nm (chemicell, Germany, dashed line) and LS-008 (Lodespin Lab, USA, full line) for two different rotating magnetic field amplitudes. Figure 2 shows the signal of fluidMAG-UC/C 200 nm mixed with different amounts of sugar (dry).

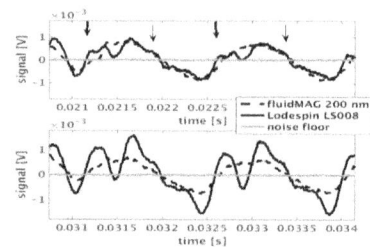

Figure 1: Comparison of two signals for two different magnetic particle suspensions. **Above:** Rotating magnetic field is 20 mT. **Below:** Rotationg magnetic field is 10 mT. The arrows mark the time points where the rotating direction changes. The time between the thick or thin arrows is the repetition time (1.4 ms).

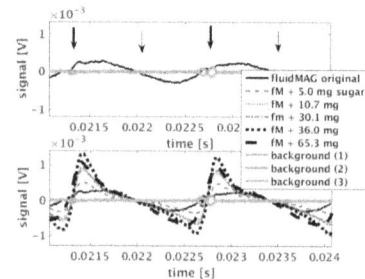

Figure 2: Above: Signal of original suspension of fluidMAG-UC/C 200 nm (40 µl, black line) and background signal (gray). **Below:** signal after adding different amounts of sugar.

CONCLUSION The measurements demonstrate the original limitations of [2] due to the dead time can be overcome. The signal shows a strong dependency on particle type, rotating field strength and changes of the particle suspension. This makes RDS a promising spectroscopic technique for highly sensitive biosensing applications and potentially high resolution imaging.

ACKNOWLEDGEMENTS This work is funded by the DFG (BE 5293/1-1). The authors thank Lodespin Lab for providing us samples of their particles.

REFERENCES
[1] McNaughton et. al, „Single bacterial cell detection with nonlinear rotational frequency shifts of driven magnetic microspheres", Applied physics letters, 2007.
[2] Rückert et. al., "Rotational Drift Spectroscopy for Magnetic Particle Ensembles", IEEE Proc. on IWMPI Berlin, 2014.
[3] Rückert et. al., "Zero dead time rotational drift spectroscopy for magnetic particle imaging", IWMPI Istanbul, 2015.
[4] Rückert et. al., "Simulating the Signal Generation of Rotational Drift Spectroscopy", IEEE Proc. on IWMPI Berlin, 2014

Dependence of Brownian and Néel Time Constants on Magnetic Field

Frank Ludwig [a,*], Jan Dieckhoff [a], Dietmar Eberbeck [b]

[a] Institut für Elektrische Messtechnik und Grundlagen der Elektrotechnik, TU Braunschweig, Germany
[b] Physikalisch-Technische Bundesanstalt, Berlin, Germany
* Corresponding author, email: f.ludwig@tu-bs.de

INTRODUCTION In Magnetic Particle Imaging (MPI), magnetic nanoparticles (MNP) are exposed to large ac and dc magnetic fields. Therefore, for a model-based reconstruction as well as for the estimation whether their dynamics are dominated by the Brownian or the Néel mechanism, the magnetic field dependence of the MNP time constant is of importance. Especially, the realization of mobility MPI as proposed by Wawrzik et al. [1] relies on the fact that at least part of MNP follows the sinusoidal drive field via the Brownian mechanism. In this contribution, Brownian and Néel time constants are experimentally studied as a function of ac magnetic fields with amplitudes up to 9 mT applying ac susceptibility and compared with theoretical expressions.

MATERIAL AND METHODS AC susceptibility measurements were performed with the rotating magnetic field (RMF) setup described in detail in [2] utilizing two crossed Helmholtz coils for single and combined ac and dc field excitations. AC and dc field amplitudes up to 9 mT can be realized, the accessible frequency ranges from 2 Hz to 9 kHz. Measurements were carried out on single-core magnetite nanoparticles with 20 nm (SHP20), 25 nm (SHP25) and 30 nm (SHP30) core diameter from Ocean Nanotech. In order to determine Brownian time constants in the accessible frequency window, SHP25 and SHP30 MNP were dispersed in water-glycerol mixtures (30:70 volume ratio) with a nominal viscosity value of 26.9 mPa·s at 25°C. The Néel time constant was measured on freeze-dried SHP20 and SHP25 samples.

The measured Brownian time constants in dependence of ac field amplitude were fitted with

$$\tau_{B,H} = \frac{\tau_B}{\sqrt{1 + 0.126\xi^{1.72}}} \qquad (1)$$

as derived by Yoshida and Enpuku [3] from solving the Fokker-Planck equation. Here τ_B is the Brownian time constant in zero field, $\xi = mB/(k_BT)$ with ac flux density amplitude B, magnetic moment m and thermal energy k_BT. For fitting the measured Néel time constants, the expression originally derived by Brown [4] for MNPs with their easy axes parallel to a dc magnetic field was used [5]:

$$\tau_{N,H} = \frac{\sqrt{\pi}\tau_{N0}}{\sigma^{1/2}\left(1 - h^2\right)} \times$$

$$\times \left[(1+h)\exp\left(-\sigma(1+h)^2\right) + (1-h)\exp\left(-\sigma(1-h)^2\right)\right]^{-1} \qquad (2)$$

Here τ_{N0} is assumed to be 10^{-9} s [6], $\sigma = KV_c/(k_BT)$ and $h = B/B_k$ with $B_k = 2K/M_s$. K is the effective anisotropy constant, M_s the saturation magnetization and V_c the core volume. Since (2) holds only for easy axes parallel to the field and for h < 0.4 [5], the measured Néel time constants were alternatively fitted with a phenomenological equation resembling (1):

$$\tau_{N,H} = \frac{\tau_N}{\sqrt{1 + A\xi^C}} \qquad (3)$$

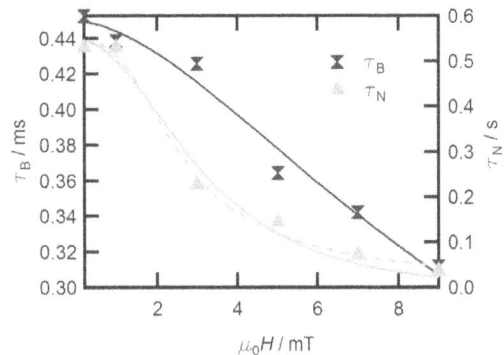

Figure 1: Dependence of Brownian τ_B and Néel time constants τ_N on ac field amplitude. Lines show fits with (1) for τ_B and with (2) and (3) (full and dashed line, resp.) for τ_N.

RESULTS Fig. 1 depicts the Brownian and Néel time constants measured on SHP25 in dependence of ac field amplitude along with the best fits. In small fields, the Brownian time constant of the 25 nm MNP is well below the Néel time constant even in a medium with a viscosity of about 25 mPa·s. Although the Néel time constant decays much faster with ac field amplitude, its value at 9 mT amplitude is still well above the corresponding Brownian time constant. Extrapolating equations (1), (2) and (3) to B = 25 mT, values of 159 µs, 56 µs and 9.4 ms are obtained, respectively. Since MPS measurements with 25 mT excitation field amplitude on a suspended and freeze-dried SHP25 sample show pronounced differences (especially at low frequencies), it must be concluded that at least part of the MNP follows the excitation via the Brownian mechanism. Consequently, equation (2) does not correctly describe the dependence of τ_N on ac field amplitude for samples with randomly oriented easy axes.

CONCLUSION The dependence of Brownian and Néel time constants on ac field amplitude was experimentally studied and compared to theoretical models. Whereas there is a proper model for the Brownian time constant, a proper model for the field dependence of the Néel time constant of a sample with random distribution of easy axes is still lacking.

ACKNOWLEDGEMENTS Financial support by the European Commission Framework Programme 7 under NANOMAG Project (NMP-LA-2013-604448) and DFG priority program SPP1681 (TR408/8-1) are acknowledged.

REFERENCES
[1] T. Wawrzik, F. Ludwig, and M. Schilling, Springer Proceedings in Physics, Volume 140:16—23, 2012.
[2] J. Dieckhoff, M. Schilling, and F. Ludwig, *Appl. Phys. Lett.*, 99:112501, 2011.
[3] T. Yoshida and K. Enpuku, *Japan. J. Appl. Phys.*, 48:127002, 2009.
[4] W. F. Brown, Jr., *Phys. Rev.*, 130(5):1677—1686, 1963.
[5] W. T. Coffey, P. J. Cregg, and Yu. P. Kalmykov, *Adv. Chem. Phys.* 83:263—464, 1993.
[6] P. C. Fannin and S. W. Charles, *J. Phys. D: Appl. Phys.* 27:185—188, 1994.

Harmonic phases of the nanoparticle magnetization and their variation with temperature.

Eneko Garaio[a,*], Juan-Mari Collantes[a], Jose Angel Garcia[b,c], Fernando Plazaola[a], Irati Rodrigo[a] and Olivier Sandre[4,*]

[a] Elektrizitatea eta Elektronika Saila, UPV/EHU, 644.48080 Bilbao. Spain
[b] Fisika Aplikatua II Saila, UPV/EHU, 644.48080 Bilbao. Spain
[c] BC Materials (Basque Center for Materials, Application and Nanostructures) 48040 Leioa. Spain
[d] Laboratoire de Chimie des Polymères Organiques, UMR 5629 CNRS/Université de Bordeaux, Bordeaux. France
*Corresponding authors, email: eneko.garayo@ehu.es, olivier.sandre@enscbp.fr

INTRODUCTION Magnetic fluid hyperthermia is a promising cancer therapy in which magnetic nanoparticles act as heat sources activated by an external AC magnetic field. The nanoparticles, located near or inside the tumour, absorb energy from the magnetic field and then heat up the cancerous tissues. In addition, the spatial distribution of the nanoparticles can be acquired by magnetic particle imaging (MPI), a novel technique based on the harmonics of the nanoparticle magnetization [1].

During the hyperthermia treatment, it is crucial to control the temperature of different tissues: too high temperature can cause undesired damage in healthy tissues through an uncontrolled necrosis. However, the current thermometry in magnetic hyperthermia presents some important technical problems. The widely used optical fibre thermometers only provide the temperature in a discrete set of spatial points. Moreover, surgery is required to locate these probes in the correct place. In this scope, we propose here a non-invasive method to measure the temperature of a magnetic sample. The approach is based on monitoring the thermal dependence of the high order harmonic phases of the nanoparticle dynamic magnetization. The method is non-invasive and it does not need any additional probe or sensor attached to the magnetic nanoparticles. Moreover, this method has the potential to be used together with the magnetic particle imaging technique to map the spatial distribution of the temperature, possibly concomitant with magnetic hyperthermia.

MATERIAL AND METHODS A water dispersed maghemite nanoparticle sample was prepared by the alkaline co-precipitation route. Afterwards, a size-grading process based on successive phase separation by an added electrolyte was applied in order to isolate a fraction with diameters, around 15-20 nm.

In order to obtain the spectrum of the MNP sample magnetization, first the AC hysteresis loops or dynamic magnetization, were measured by the lab-made AC magnetometer [2]. Afterward, the Fast Fourier Transform (FFT) of the magnetization signal was performed in order to obtain the complex harmonics with their corresponding magnitudes and phases (φ_n). To measure the thermal dependence of the harmonics, the temperature of the sample was varied in the biological range between 30 °C and 50 °C and the first 7 harmonics were measured continuously. The actual temperature of the sample was measured by an optical fibre thermometer with the probe in direct contact with the colloidal suspension. The process was repeated at different AC magnetic field frequencies (75, 302, 676 and 1030 kHz) and intensities.

RESULTS As shown in figure 1, the high order harmonic phases (3rd, 5th and 7th) of maghemite nanoparticle magnetization increase with temperature in an almost linear way for all the measured frequencies. The temperature can be obtained *via* the previously measured temperature dependence of harmonic phases [3]. Figure 2 shows the temperature evolution of the sample obtained from the phase of the 3rd harmonic with a previous calibration.

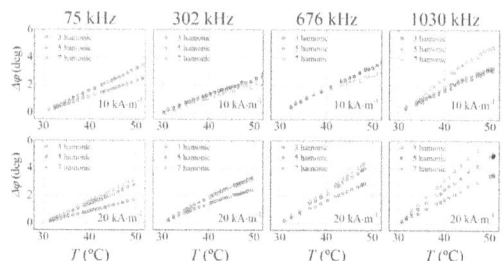

Figure 1: Variations of harmonic phases (relative to the phase at 30 °C) with temperature.

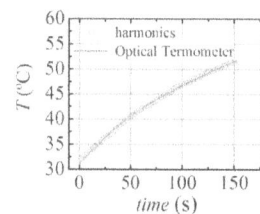

Figure 2: Temperature evolution obtained from the harmonic phases compared with the actual temperature measured by a fibre optic probe

CONCLUSION We propose here the measurement of high order harmonic phase as a non-invasive method to obtain the temperature of a magnetic nanoparticle sample. The method can possibly be combined with MPI to obtain 3D thermal mapping.

ACKNOWLEDGEMENTS The authors thank the European COST action TD1402 "Multifunctional Nanoparticles for Magnetic Hyperthermia and Indirect Radiation Therapy" (RADIOMAG). This work has been financed by the Basque Government under grant IT-443-10.

REFERENCES

[1] B. Gleich and J. Weizenecker. *Nature*, 435(7046):1217—1217, 2005.

[2] E. Garaio, J. Collantes, F. Plazaola, J. Garcia, and I. Castellanos-Rubio, *Measurement Science and Technology*, vol. 25, p. 115702, 2014.

[3] E. Garaio, J-M. Collantes, J. A. Garcia, F. Plazaola, and O. Sandre, "Harmonic phases of the nanoparticle magnetization: an intrinsic temperature probe", *Applied Physics Letters* 107: 123103, 2015

Heat Transfer Simulation for Optimization and Treatment Planning of Magnetic Hyperthermia Using Magnetic Particle Imaging

Natsuo Banura, Atsushi Mimura, Kohei Nishimoto, Kenya Murase[*]

Department of Medical Physics and Engineering, Graduate School of Medicine, Osaka University, Osaka, Japan
[*] Corresponding author, email: murase@sahs.med.osaka-u.ac.jp

INTRODUCTION Magnetic hyperthermia (MHT) is a strategy for cancer treatment using the temperature rise of magnetic nanoparticles (MNPs) under an alternating magnetic field (AMF). An accurate knowledge of the spatial distribution and amount of MNPs in tumors is crucial for designing an optimal treatment planning of MHT that can prevent insufficient heating in the targeted region and overheating in the healthy tissue. The purpose of this study was to develop a system for heat transfer simulation for optimization and treatment planning of MHT using magnetic particle imaging (MPI) [1].

MATERIAL AND METHODS First, we performed phantom experiments to investigate the relationship between the pixel value of the MPI image and the temperature rise of MNPs under an AMF. Samples filled with various iron concentrations of MNPs (Resovist®) were prepared and were imaged using our MPI scanner [2]. These samples were also heated using the AMF. The specific absorption rate (SAR) was calculated from the initial slope of the temperature rise and the regression equation between the MPI pixel value and SAR was obtained.

Second, we imaged tumor-bearing mice injected with Resovist® using our MPI scanner [2] and X-ray CT scanner. To generate the geometry of a tumor-bearing mouse, four regions (soft tissue, bone, spine, and sinus) were extracted from the X-ray CT images using K-means clustering, and the tumor region was segmented by setting a region of interest (ROI) manually. After these segmentations, the contours of these regions were extracted by edge detection method, and these edge images were converted into drawing exchange format (DXF) and were imported into COMSOL® (software based on finite element method) as the geometry. We also generated the ROI images from the MPI images by taking the threshold value for extracting the contour as 40% of the maximum MPI pixel value within the ROI. After the ROI images thus obtained by MPI were co-registered to the tumor regions, they were imported into COMSOL®. Finally, Pennes' bioheat equation [3] was solved using COMSOL® to simulate the heat transfer in the geometry of a tumor-bearing mouse. Pennes' bioheat equation is given by

$$\rho c \frac{\partial T}{\partial t} = \kappa \nabla^2 T + Q + SAR + \rho_b c_b \omega (T_b - T) \quad (1)$$

where k, ρ, c, Q, T, and ω denote the thermal conductivity, density, specific heat, metabolic heat generation rate, temperature, and perfusion rate for tumor or healthy tissue, respectively. ρ_b, c_b, and T_b denote the density, specific heat, and temperature of blood, respectively. In this study, we used the thermophysical data in the literature [4]. The SAR in (1) was calculated from the pixel values in the ROI images using the regression equation obtained above. The initial temperatures of the tumor and healthy tissue were set at 36 °C. The surfaces of the tumor and healthy tissue were subject to a natural convection boundary condition (T=28 °C and Q=3.7 W/m²).

RESULTS There was an excellent correlation between the MPI pixel value (x) and the SAR value (y, W/cm³) (r=0.956) with a regression equation of y = 0.0542x + 0.0527, from which the MPI pixel value was converted to the SAR value in (1).

Figures 1(a) and 1(b) show the fusion image between the MPI and X-ray CT images of a tumor-bearing mouse injected with Resovist® (250 mM) and the temperature distribution at 20 min after the start of MHT obtained by COMSOL®, respectively.

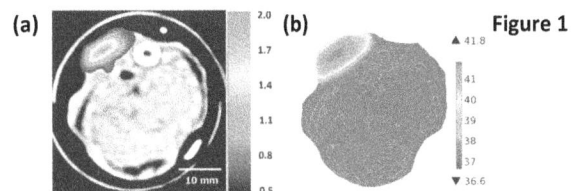

Figure 1

Figure 2(a) shows the comparison between the time course of the temperature in the tumor during MHT obtained by simulation (○) and that obtained experimentally (●). A good agreement was observed between them. Figure 2(b) shows the simulation results of the time course of the temperature in the tumor for various blood perfusion rates in the tumor (ω_t), demonstrating that the temperature largely depended on ω_t.

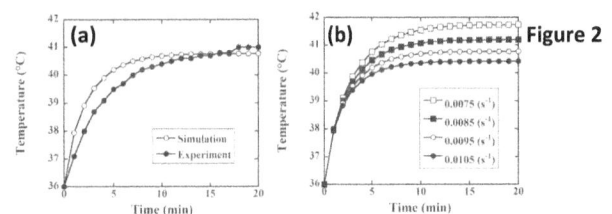

Figure 2

CONCLUSION Our results suggest that our system for heat transfer simulation using MPI will be useful for optimization and treatment planning of MHT.

ACKNOWLEDGEMENTS This work was supported by a Grant-in-Aid for Scientific Research from the Japan Society for the Promotion of Science (JSPS) and the Japan Science and Technology Agency (JST).

REFERENCES
[1] B. Gleich and J. Weizenecker. *Nature*, 435(7046):1217–1217, 2005. doi: 10.1038/nature03808.
[2] K. Murase, M. Aoki, N. Banura, K. Nishimoto, A. Mimura, et al. *Open J. Med. Imaging*, 5(2):85–99, 2015. doi: 10.4236/ojmi.2015.52013.
[3] H. H. Pennes. *J. Appl. Physiol.*, 1(2):93-122, 1948.
[4] A. Attaluri, S. K. Kandala, M. Wabler, H. Zhou, C. Cornejo, et al. *Int. J. Hyperthermia*, 31(4):359-374, 2015. doi: 10.3109/02656736.2015.1005178.

Magnetic Nanoparticle Temperature Estimation Using Dual-Frequency Magnetic Filed

Kai Wei [a], Shiqiang Pi [a], Wenzhong Liu [a,*]

[a] School of Automation, Huazhong University of Science and Technology, Wuhan, China
* Corresponding author, email: lwz7410@hust.edu.cn

INTRODUCTION Magnetic nanothermometer using magnetic nanoparticle (MNP) has been intensively studied due to its potential for precise and non-invasive temperature probing. For instance, the magnetization or susceptibility of MNPs induced in an AC magnetic field was recently used as the thermometric properties for temperature probing [1, 2]. However, compared with single-frequency magnetic field, the harmonic of MNPs magnetization induced in dual-frequency magnetic field are more abundant. Therefore, we can more easily construct mathematic model for assessing the temperature. In this paper, a novel method using MNPs under dual-frequency magnetic field was proposed for noninvasive, robust, and precise temperature probing.

MATERIAL AND METHODS Magnetization of the MNPs under low frequency ac magnetic field follows the first-order Langevin function. When the amplitude of the applied magnetic field is considerably small, the Taylor expression form of magnetization of the Langevin function can be written as

$$M = \Phi m_s \left[\frac{Hm_s}{3kT} - \frac{1}{45}\left(\frac{Hm_s}{kT}\right)^3 + \frac{2}{945}\left(\frac{Hm_s}{kT}\right)^5 \right.$$
$$\left. - \frac{1}{4725}\left(\frac{Hm_s}{kT}\right)^7 + \frac{2}{93555}\left(\frac{Hm_s}{kT}\right)^9 \right]$$

(1)

where, m_s is the effective magnetic moment of the MNP, Φ is the concentration of the MNP, k is the Boltzmann constant, T is the absolute temperature, and H is the excitation field. Let

$$H = H_0[\sin(\omega_1 t) + \sin(\omega_2 t)], x = \frac{H_0 m_s}{kT} \text{ and } y = \Phi m_s$$

(2)

where H_0 denotes the amplitude of applied dual-frequency magnetic field, whereas ω_1 and ω_2 represents angular frequency. Substituting Eq. (2) into Eq. (1), we can obtain:

$$M = \sum_{j=1}^{5} A_{2j-1} sin[(2j-1)\omega_1 t] + \sum_{j=1}^{5} C_{2j-1} sin[(2j-1)\omega_2 t]$$
$$+ \sum_{\substack{i=1 \text{ or } j=1 \\ and\ i+j=odd}}^{5} B_{i+j} sin[(i\omega_1 \pm j\omega_2)t]$$

(3)

where, A_{2j-1}, C_{2j-1}, and B_{i+j} are the amplitudes of the harmonics which are the function of tow variables, x and y. Then, the temperature estimation model can be described by the amplitudes of the harmonics. From the methods, we can find that we can choose different harmonics to construct temperature estimation model. Therefore, temperature measurement is converted into harmonics measurement of the MNP magnetization. In order to ensure the temperature measurement precision, we can choose the

maximum amplitudes of the MNP, which have high signal to noise ratio (SNR), to construct temperature estimation model. Simulation was performed to analysis the feasibility of the novel method introduced herein. The following parameters were used in the simulation: $m_s = 2 \times 10^{-19}$ emu, $\omega_1 = 2\pi \times 96$ rad/s, $\omega_2 = 2\pi \times 384$ rad/s, $H_0 = 50$ Gs, SNR = 80 dB.

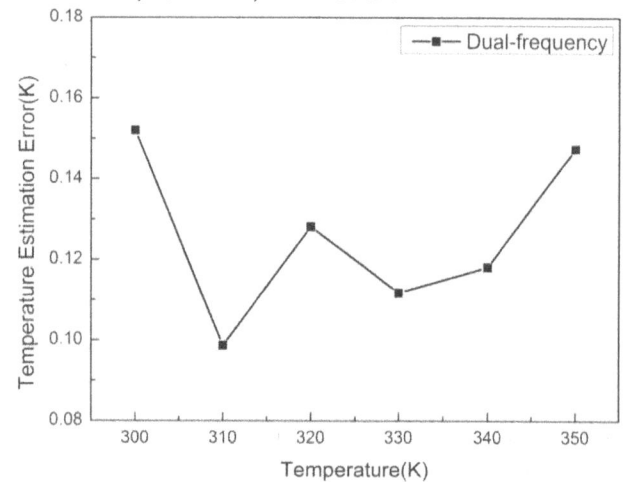

Figure 1: Simulation of temperature estimation errors using the harmonics of MNPs magnetization induced in dual-frequency magnetic field.

RESULTS Fig. 1 shows the simulation result of temperature estimation using dual-frequency magnetic field. From the simulation result, we found that the novel method presented herein allows precise temperature estimation. The maximum temperature estimation errors is less than 0.2K.

CONCLUSION The purpose of this paper is to discuss a novel temperature measurement method based on the magnetic response of the MNPs under dual-frequency magnetic field. The simulation result shows that the harmonic of MNPs magnetization induced in dual-frequency magnetic field has the potential for temperature probing. From the Eq. (3), it is easy to find that we can choose different harmonics to construct temperature estimation model. Moreover, the measurement accuracy of harmonic amplitudes plays an important role in the temperature estimation.

ACKNOWLEDGEMENTS This work was supported by 61571199 (NSFC) and Hubei Provincial project of 2014AEA048.

REFERENCES
[1] J. B. Weaver, A. M. Rauwerdink and E. W. Hansen. *Med.Phys*, 2009. doi: http://dx.doi.org/10.1118/1.3106342.
[2] J. Zhong, W. Z. Liu, et al. *Nanotechnology*, 2012. doi: http://dx.doi.org/10.1088/0957-4484/23/7/07570

3D-GUI Simulation Environment for MPI

P. Vogel [a,*], M.A. Rückert [a], V.C. Behr [a]

[a] Department of Experimental Physics 5 (Biophysics), University of Würzburg, Würzburg
* Corresponding author, email: Patrick.Vogel@physik.uni-wuerzburg.de

INTRODUCTION Simulations are useful tools for testing and understanding novel ideas and approaches before their have to be built. For Magnetic Particle Imaging (MPI) several simulation frameworks has been presented [1][2].

Describing a full MPI experiment requires several different simulation environments: static and dynamic magnetic fields, inductive signals generated by emulating receive chains, the dynamics of superparamagnetic iron-oxide nanoparticles (SPIONs), spinsystems, determination of trajectories, calculation of system matrices, and many more.

However, for each issue a simulation tool can be found, but it is difficult to find a tool, which combines all these environments. In this abstract a simulation tool is introduced, which contains all the mentioned environments and embeds those in an intuitive 3D-graphical user interface (GUI).

MATERIAL AND METHODS For the simulation of static and dynamic magnetic fields several commercial and open-source software is offered, whereby different types of solvers are used to depict the magnetic fields. For the simulation of quasistatic and dynamic magnetic fields with frequencies below 10 MHz, which is sufficient for common MPI experiments, the law of Biot-Savart is valid [3].

Faraday's Law can be employed for simulating the induced signal generated by a receive chain.

Coils and conductors can be set by using a scripting toolbox, which offers the possibility of creating arbitrarily geometries. The induced signal is calculated by using volume based sensitivities areas.

The behavior of a particle system can be described basically by the Langevintheory [4] or more sophisticated theories like a second order modified mean-field theory [5], which also takes particle-particle interaction into account.

Additional information about the sample can be obtained by using a dynamic Blochsolver. This solver can describe an isochromat in arbitrarily magnetic fields, which can be useful to understand a MRI signal in different environments [6].

Additional features are live 3D displaying and manipulation of the simulated coil system, realtime preview of magnetic field distributions and phase maps, realtime calculation and displaying of trajectories for MPI, pre-calculation of systemmatrices for the reconstruction, live preview of induced signals, and many more.

RESULTS Fig. 1. (a) displays the 3D model of a TWMPI scanner consisting the excitation coils (i) and a receive coil (ii). The geometries of the coils can be easily manipulated. In Fig. 1. (b) and (c) a realtime preview of the magnetic field distribution and the phase-map are displayed, which allow to determine the hardware parameters of the device. By tracking the field free point (FFP) the trajectory can be described and the data can be used for the reconstruction process (Fig. 1. (d)). An easy-to-use editor gives the possibility of generating different particle distributions for simulating signal performance.

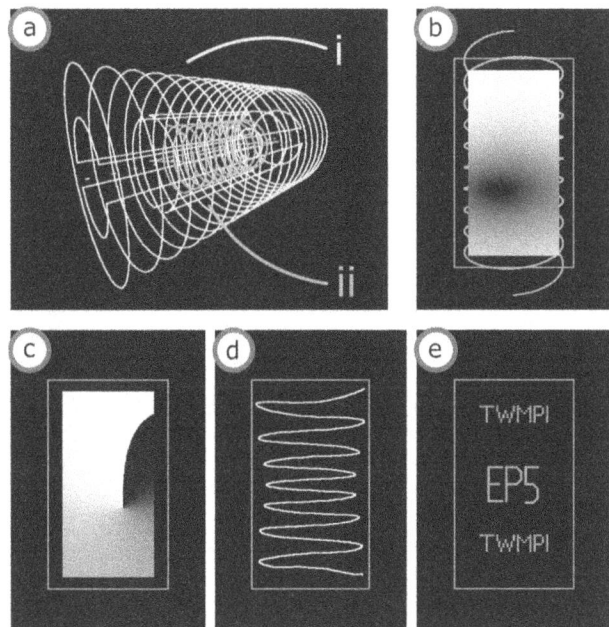

Figure 1: (a) 3D preview of the TWMPI system: (i) are the coils, which are used for excitation, (ii) represents the receive coil. Top view on the TWMPI system: the receive coil can be seen overlapped by the magnetic field distribution (b) and the phase map (c). The image (d) shows an example of the trajectory inside a TWMPI scanner. (e) Example for an arbitrarily distribution of superparamagnetic material.

CONCLUSION The presented simulation environment is a helpful tool for hardware development in MPI and provides a better understanding of signal generation influenced by several physical parameters. The intuitive 3D-GUI as well as additional feature for the information workup makes this simulation software to a powerful tool for MPI scientists.

ACKNOWLEDGEMENTS This work was supported by the BMBF (FKZ 1745X08), the EU (IDEA project 279288) and the DFG (BE-5293/1-1).

REFERENCES

[1] W.T. Smolik, et al., *Proc. on IWMPI Istanbul*, p27, 2014.
[2] M. Straub, et al., *IEEE Trans. Magn.*, 51(2):6501204, 2015. doi: 10.1109/TMAG.2014.2329733.
[3] C. Volkmar, et al., *ISSE*, 36:210-215, 2013. doi: 10.1109/ISSE.2013.6648244.
[4] M.A. Rückert, et al., Biomed. Tech., 58(6):593-600, 2013. doi: 10.1515/BMT.2013.0015.
[5] J.-P. Gercke, et al., *Proc. on IWMPI Lübeck*, p73f, 2010.
[6] P. Vogel, et al., *Proc on ICMRM Munich*, L062, 2015.

Elevator speeches 2:

Tracer Materials

Biocompatible Magnetite Nanoparticles as Tracer Material for Magnetic Particle Imaging

Corinna Stegelmeier[*], Ankit Malhotra, Kerstin Lüdtke-Buzug

Institute of Medical Engineering, University of Lübeck
[*] Corresponding author, email: stegelmeier@imt.uni-luebeck.de

INTRODUCTION Nowadays, nanoparticles are used in many interesting fields, such as environmental technology or biomedicine.

Magnetic particle imaging is a very promising method for medical diagnostics and is making use of the magnetic behavior of magnetite nanocrystals. These consist of only a single domain and show superparamagnetic behavior when exposed to magnetic fields and can therefore be used as tracers for magnetic particle imaging.

Compared to other imaging techniques MPI offers advantages, such as its high temporal and spatial resolution as well as its high sensitivity and the lack of ionizing radiation.

Conventional synthesis methods in water include precipitation processes from supersaturated alkaline solutions in the presence of a biocompatible and water-soluble stabilizers like dextran. This method yields particles suitable for MPI in one step but with broad size distribution and without control over the particle shape.

Alternatively, high temperature decomposition methods in high boiling organic solvents provide an excellent opportunity to precisely tailor particle shape and size and thus their magnetic properties. [1]

Since biocompatibility is needed for medical applications, a phase transfer into aqueous systems is required which is the topic of this work. For this purpose water-soluble homopolymer ligands, such as poly(ethylene-oxide) (PEO), pose well suited systems. Once functional groups that effectively bind to the particle surface are introduced into the ligand, a "brush-like" way of binding is enabled that creates a hydrophilic shell of PEO around the nanoparticles. [2,3] The experimental strategy is illustrated in Fig.1.

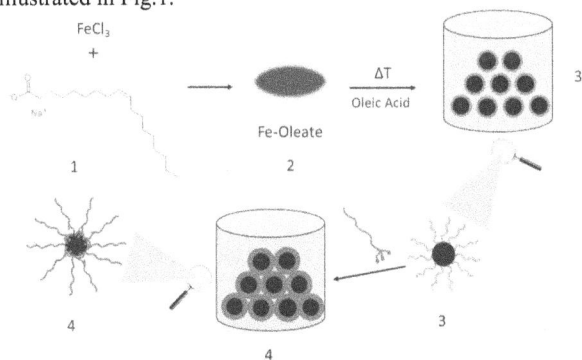

Figure 1: Schematic description of the experimental strategy. The primary materials include $FeCl_3$ and Na-Oleate (1) which are reacted to Fe-Oleate (2). This complex is decomposed to yield Fe_3O_4 nanoparticles with a layer of oleic acid for stabilization (3). This layer is replaced by a larger biocompatible homopolymer (PEO) with a functional anchor group.

MATERIAL AND METHODS Fe_3O_4-nanoparticles were prepared via the thermal decomposition of iron oleate.

TEM images were prepared using the Zeiss LEO EM922 Omega and the Jeol 1011, respectively. The hydrodynamic diameters were determined by photon cross correlation spectroscopy (PCCS) in chloroform using the Nanophox system by Sympatec.

RESULTS After particle synthesis the resulting dark brown material is well soluble in organic solvents due to its hydrophobic shell. After ligand exchange stable particles were obtained with higher hydrodynamic diameters due to the larger polymer ligand. The results are summarized in Fig.2.

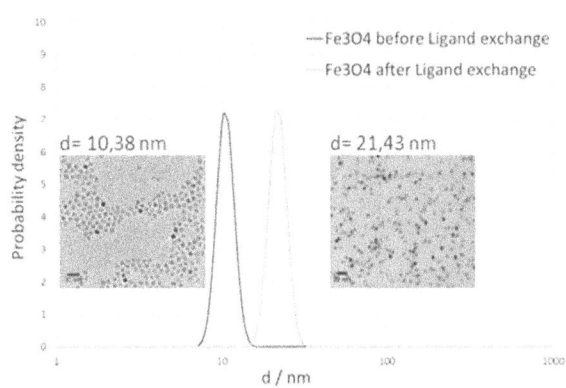

Figure 2: Particle size distributions of the iron oxide nanoparticles before (black) and after ligand exchange (gray). The two inlets show the TEM images with 20 nm scale bars.

CONCLUSION A phase transfer method for monodisperse iron oxide nanoparticles from organic to aqueous systems was described. The particles where characterized with TEM and PCCS revealing a growth of the hydrodynamic diameter due to the binding of the larger polymer ligand. By adjusting the core diameter, the magnetic properties of biocompatible magnetite crystals can selectively be tailored which is disired for applications in MPI.

ACKNOWLEDGEMENTS This work has been supported by the German Federal Ministry of Education and Research under the grant numbers 13N11090 and 13GW0069A and the European Union and the state Schleswig-Holstein (Program for the Future Economy) under the grant number 122-10-004.

REFERENCES

[1] J. Park, J. An, Y. Hwang, J.-G. Park, H.-J. Noh, J.-Y. Kim, J.-H. Park, N.-M. Hwang, T. Hyeon. *Nature Materials*, 3:891—895, 2004. doi: 10.1038/nmat1251.
[2] S. Ehlert, S. M. Taheri, D. Pirner, M. Drechsler, H.-W. Schmidt, S. Förster. *ACS Nano*, 8(6):6114-22, 2014. doi: 10.1021/nn5014512.
[3] M. S. Nikolic, M. Krack, V. Aleksandrovic, A. Kornowski, S. Förster, H. Weller. Angew. Chem. Int. Ed., 45:6577–65, 2006. doi: 10.1002/anie.20060220.

Continuous Synthesis of Single-Core Iron Oxide Nanoparticles for Biomedical Applications

Abdulkader Baki[a], Regina Bleul[a,*], Christoph Bantz[a], Raphael Thiermann[a], Michael Maskos[a,b*]

[a] Nanoparticle Technologies Department, Fraunhofer ICT-IMM, Carl-Zeiss-Straße 18–20, 55129 Mainz, Germany
[b] Institut für Physikalische Chemie, Johannes Gutenberg-Universität Mainz, Jakob-Welder-Weg 11, 55128 Mainz, Germany
* Corresponding author, email: regina.bleul@imm.fraunhofer.de, michael.maskos@imm.fraunhofer.de

INTRODUCTION Magnetic Particle Imaging (MPI) is a powerful tomographic imaging technique, which provides a high sensitivity and spatial resolution [1]. It is based on the non-linear magnetization behavior of magnetic nanoparticles as tracer material. To ensure the further development of the performance of this promising technique, new efficient tracers are required. Resovist®, a contrast agent for Magnetic Resonance Imaging (MRI) approved in 2002, was a promising candidate for this application, but it is not available on the market anymore. However, it is still commonly used as gold standard for comparison of new tracers for magnetic particle imaging [2]. According to theoretical calculations, the optimum size for MPI tracer materials is about 30 – 60 nm in diameter [3,4]. However, the synthesis of single-core magnetite nanoparticles in this size range is not easily accessible by common synthesis routes. Besides the core size, it is also challenging to realize long-term colloidal stability in aqueous media. To address these issues, we developed a continuous synthesis based on micromixing technology.

MATERIAL AND METHODS The microfluidic mixing set-up consists of HPLC pumps and a micromixer. The particle formation can be described as a two-stage process. The first step is nucleation, followed by particle growth. Depending on temperature, mixing ratio, total flow rate and residence time, size and morphology of the magnetic nanoparticles can be adjusted.

Figure 1: Different Micro mixers and their mixing profiles: Left: Slit Interdigital Micro Mixer, Right: Caterpillar Micro Mixer.

RESULTS Our method allows an easy, fast and highly reproducible manufacturing of single-core iron oxide nanoparticles in aqueous dispersion. We are able to adjust the particle size from 15 nm in diameter to almost 100 nm and influence their morphology, for example disk-like or spherical. First measurements with magnetic particle spectroscopy have already shown promising results concerning the performance as MPI tracers. Further investigations are in progress. It is also planned to evaluate the use of our magnetic nanoparticles for other (bio)medical applications.

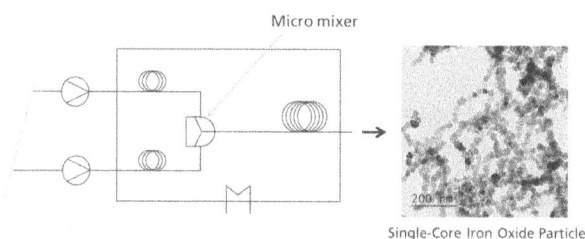

Figure 2: Schematic representation of continuous manufacturing of magnetic single-core iron oxide nanoparticles.

CONCLUSION Magnetic single-core nanoparticles are promising candidates not only as tracers for magnetic particle imaging but also for other (bio)medical applications as for hyperthermia, magnetic targeting and magnetically triggered drug release. To guarantee an optimum performance, different applications demand magnetic particles of different sizes. Thus, a versatile method, which allows the reliable production of different nanoparticle sizes, is beneficial. The presented manufacturing technique allows the synthesis of stable-dispersed magnetic single-core iron oxide particles with good control over size and morphology. In summary, our continuous synthesis method combines good scalability with high reproducibility of particle characteristics.

ACKNOWLEDGEMENTS We thank Norbert Löwa, Dietmar Eberbeck and Dr. Lutz Trahms from Physikalisch-Technische Bundesanstalt (PTB) for their kind assistance with magnetic particle spectrometry measurements and their expert advice. We gratefully acknowledge financial support from the European Regional Development Fund (EFRE). R.B. thanks the German Academic Exchange Service (DAAD) for a PhD fellowship DAAD-PKZ:D/11/43560 and the Fraunhofer-Gesellschaft for the support within the Fraunhofer TALENTA program.

REFERENCES
[1] B. Gleich and J. Weizenecker. *Nature*, 435(7046):1217–1217, 2005. doi: 10.1038/nature03808.
[2] Taupitz, M., Schmitz, S. & Hamm, B. *Rofo-Fortschritte Auf Dem Gebiet Der Rontgenstrahlen Und Der Bildgebenden Verfahren*, 175, 752 – 765, 2003.
[3] Eberbeck, D., Wiekhorst, F., Wagner, S. & Trahms, L. *Applied Physics Letters* 98, 182502, 2011. doi: 10.1063/1.3586776
[4] Gleich, B. Method and apparatus for improved determination of spatial non-agglomerated magnetic particle distribution in an area of examination. Patent, WO2004091398 A2 (2004).

Diffusion-Controlled Synthesis of Magnetic Nanoparticles

David Heinke[1,*], Nicole Gehrke[1], Daniel Schmidt[2], Uwe Steinhoff[2], Thilo Viereck[3], Hilke Remmer[3], Frank Ludwig[3], Andreas Briel[1]

[1] nanoPET Pharma GmbH, Berlin, Germany
[2] Physikalisch-Technische Bundesanstalt, Berlin, Germany
[3] Institue of Electrical Measurement and Fundamental Electrical Engineering, Technische Universität Braunschweig, Germany
[*] Corresponding author, email: david.heinke@nanopet.de

INTRODUCTION Apart from the technological advancements of the Magnetic Particle Imaging (MPI) scanner and image reconstruction, the future of MPI crucially relies on the development of high performing tracers. An interesting material, not only for MPI, are biogenic iron oxide nanoparticles due to their superior magnetic properties [1,2]. It is a fact, however, that the production of such particles is extremely challenging. Here, we present a route to synthesize magnetic iron oxide nanoparticles mimicking the controlled conditions of biomineralization by reducing the diffusion rate of the reactants.

MATERIAL AND METHODS The synthesis of nanoparticles was conducted via controlled coprecipitation in an agarose-gel network. The increased viscosity resulted in a reduction of the diffusion rates of the reactants. Subsequently, the trapped particles (NPIO-105) were released by hydrogen peroxide treatment and electrostatically stabilized by acid treatment. The particles were characterized by dynamic light scattering, transmission electron microscopy (TEM), high-resolution TEM (HR-TEM) and electrophoretic mobility measurements. The magnetic particle spectrum (MPS) was recorded at a drive field with an amplitude of $10\ mT/\mu_0$ and a frequency, f_0 of 25 kHz. The effective magnetic core sizes were estimated via static magnetization measurements.

RESULTS The particles synthesized here exhibit a mean hydrodynamic diameter of about 55 nm and a zetapotential of +47 mV.

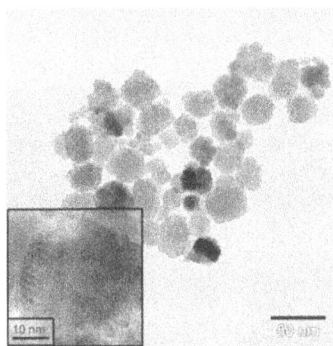

Figure 1: TEM and HR-TEM micrograph (inset) of NPIO-105.

Via TEM micrographs (example in Fig. 1) a mean crystallite size of 24 nm and a standard deviation of 10 nm was determined. The continuous crystal lattices visible on HR-TEM micrographs (example in Fig. 1 inset) imply that most of the particles are single crystals, although their rough appearance suggests that they are composed of several small crystallites.
The fit of the static magnetization curve to a bimodal size distribution (not shown) shows a large mode of particles having effective magnetic core sizes in the theoretically predicted ideal size range for MPI of 30 nm [3].

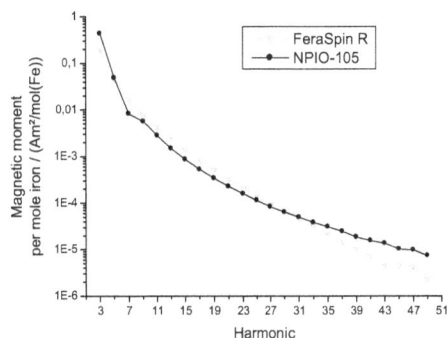

Figure 2: MPS measured at $10\ mT/\mu_0$ and $f_0 = 25$ kHz.

The third harmonic of the MPS exceeds that of our benchmark FeraSpin™ R by a factor of 2.3 (see Fig. 2). However, the steep decay causes less intense harmonics to be observed as from harmonic number 7. This may be attributed to interparticular magnetic interactions, considering the absence of a protective polymeric coating keeping the particles at distance, as well as to the large particle size and consequently to a disability of their magnetic moments to follow the sinusoidal excitation field of 25 kHz. Surprisingly, at higher harmonics (from harmonic number 29 onwards), the decay becomes less steep causing similar and even higher signal intensities compared to FeraSpin R. AC susceptibility as well as fluxgate magnetorelaxometry measurements are being conducted in order to explain these effects.

CONCLUSION In this work, magnetic nanoparticles having high MPS amplitudes in both the low and high harmonics range were synthesized using a controlled coprecipitation process via gel diffusion. Future work will involve the further characterization of the synthesis pathway and its biomineralization-mimicking mechanisms to optimize the MPI properties of the particles.

ACKNOWLEDGEMENTS This work was supported by the European Commission Framework Programme 7 under the NanoMag project [grant agreement no 604448] and the DFG priority program SPP1681 (LU800/4-1). Furthermore, the authors thank Mihály Pósfai for the HR-TEM micrographs.

REFERENCES
[1] D. Faivre and D. Schüler, Magnetotactic bacteria and magnetosomes. Chem. Rev., 108(11): 4875—4898, 2008.
[2] A. Kraupner, D. Heinke, R. Uebe, D. Eberbeck, N. Gehrke, D. Schüler and A. Briel. *IWMPI 2014 Book of Abstracts*, 141—142, 2014.
[3] B. Gleich and J. Weizenecker. *Nature*, 435(7046):1217—1217, 2005. doi: 10.1038/nature03808.

Development and Physicochemical Characterization of Continuously Manufactured Single-Core Iron Oxide Nanoparticles

Christoph Bantz [a,*], Regina Bleul [a], Abdulkader Baki [a], Raphael Thiermann [a], Norbert Löwa [b], Dietmar Eberbeck [b], Lutz Trahms [b], Michael Maskos [a,c*]

[a] Nanoparticle Technologies Department, Fraunhofer ICT-IMM, Carl-Zeiss-Straße 18–20, 55129 Mainz, Germany
[b] Physikalisch-Technische Bundesanstalt, Abbestraße 2–12, 10587 Berlin, Germany
[c] Institut für Physikalische Chemie, Johannes Gutenberg-Universität Mainz, Jakob-Welder-Weg 11, 55128 Mainz, Germany
[*] Corresponding author, email: christoph.bantz@imm.fraunhofer.de, michael.maskos@imm.fraunhofer.de

INTRODUCTION Superparamagnetic iron oxide nanoparticles have been utilized for medical applications already since the 1990s. For instance, Endorem® and Resovist® are trade names for contrast agents for Magnetic Resonance Imaging (MRI) approved in 1994 and 2002 [1]. Another application for superparamagnetic iron oxide nanoparticles is Magnetic Particle Imaging (MPI), a powerful tomographic imaging technique. MPI is based on the non-linear magnetization behavior of magnetic nanoparticles as tracer material and provides a high sensitivity and spatial resolution [2]. For the further development of MPI performance, new efficient tracers are required. Regarding the synthesis of new tracer materials, two main challenges have to be overcome: (i) The synthesis of single-core magnetite nanoparticles in the optimum size range (about 30 – 60 nm [3]) is not easily accessible by common synthesis routes. (ii) For a good colloidal stability even in complex aqueous media, an aqueous synthesis route is preferable; otherwise, any agglomeration during the transfer from organic solvent to water needs to be prevented. Furthermore, an efficient (preferably sterically) stabilizing agent is required to protect the tracer material from undesired aggregation.

We developed a continuous synthesis of single-core iron oxide nanoparticles based on micromixing technology. By a comprehensive characterization of the nanoparticle samples, we evaluate particle size and agglomeration state [4]. This is necessary to be able to fine-tune the particles' properties by selecting the appropriate process parameters.

MATERIAL AND METHODS A comprehensive characterization of the physicochemical properties of a colloidal sample requires the application of different, complementary characterization techniques. Therefore, we applied Transmission Electron Microscopy for imaging (TEM, both after dry preparation as well as under cryogenic conditions) and Analytical Centrifugation (Differential Centrifugal Sedimentation, DCS) and Dynamic Light Scattering (DLS) as ensemble methods.

RESULTS The analysis reveals that the continuously manufactured particles exhibit different sizes (ranging from 15 – 100 nm) and different shapes (e.g. spherical or disk-like), examples are shown in Fig. 1. These properties are adjustable by different process parameters such as mixing ratio, total flow rate, residence time and temperature. The effect of reaction temperature is presented in Fig. 2: With increasing temperature, the average particle diameter increases. This selected example shows the importance of single reaction parameters: As a few nanometers in core diameter can have a significant effect on the particles' performance as MPI tracers, the selection of the optimum reaction conditions is of great importance.

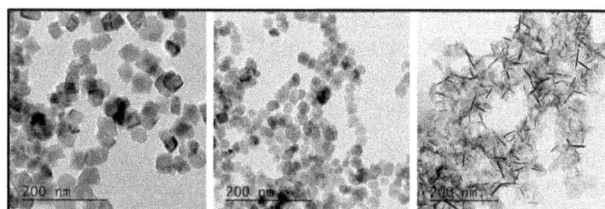

Figure 1: TEM micrographs of magnetic iron oxide nanoparticles of different size and shape.

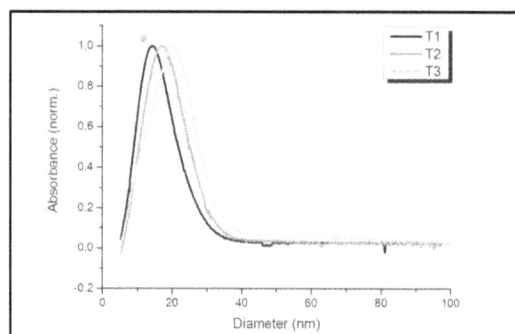

Figure 2: Analytical Centrifugation: Influence of reaction temperature (T_1-T_3) on the average particle size.

CONCLUSION Only by a comprehensive characterization of the physicochemical properties of the synthesized single-core iron oxide nanoparticles, we are able to gain a deep understanding of the influence of the single reaction parameters. As the developed continuous synthesis technique allows a good control over the reaction parameters, we are, in summary, able to accurately adjust the particles' specific physicochemical characteristics. Thus, the production of magnetic nanoparticles of a distinct average size and of certain shapes is feasible to use them for Magnetic Particle Imaging, but also for other biomedical applications, e.g. hyperthermia or magnetic drug targeting.

ACKNOWLEDGEMENTS We gratefully acknowledge financial support from the European Regional Development Fund (EFRE).

REFERENCES
[1] Taupitz, M., Schmitz, S. & Hamm, B. *Rofo-Fortschritte auf dem Gebiet der Röntgenstrahlen und der Bildgebenden Verfahren*, 175, 752–765, 2003.
[2] Gleich, B., Weizenecker, J. *Nature*, 435(7046), 1217–1217, 2005. doi: 10.1038/nature03808.
[3] Eberbeck, D., Wiekhorst, F., Wagner, S. & Trahms, L. *Applied Physics Letters* 98, 182502, 2011. doi: 10.1063/1.3586776
[4] Bantz, C., Koshkina, O., Lang, T., …, Maskos, M. *Beilstein Journal of. Nanotechnology* 5, 1774-1786, 2014. doi:10.3762/bjnano.5.188

Formation of a Protein Corona on Magnetic Nanoparticles Affects Nanoparticle-Cell Interactions

A. Weidner[a], C. Gräfe[b], M. v.d. Lühe[c,d], C. Bergemann[e], J.H. Clement[b,d], F.H. Schacher[c,d], S. Dutz[a,*]

[a] Institut für Biomedizinische Technik und Informatik (BMTI), Technische Universität Ilmenau, Germany
[b] Klinik für Innere Medizin II, Abteilung Hämatologie und Internistische Onkologie, Universitätsklinikum Jena, Germany
[c] Institut für Organische Chemie und Markomolekulare Chemie (IOMC), Friedrich-Schiller-University Jena, Germany
[d] Jena Center for Soft Matter (JCSM), Friedrich-Schiller-University Jena, Germany
[e] Chemicell GmbH, Berlin, Germany
* Corresponding author: silvio.dutz@tu-ilmenau.de

INTRODUCTION When magnetic nanoparticles (MNP) are exposed to the blood circulation, a protein corona consisting of various components is formed immediately. The composition of the corona as well as their amount bound to the particle surface is dependent on different factors [1]. The current composition of the formed protein corona might be of major importance for cellular uptake of magnetic nanoparticles. The aim of our study is to analyze the formation of the protein corona during *in vitro* serum incubation in dependence of composition of incubation medium as well as to investigate the particle-cell interactions of serum coated MNP.

MATERIAL AND METHODS Cytotoxic polyethylenimine (PEI) coated nanoparticles were incubated in fetal calf serum (FCS) for defined times and temperatures within a water bath to form the protein corona on the surface of the particles. To control amount of proteins binding on MNP surface, the FCS concentration of incubation medium (FCS + cell medium) was varied. Before and after the incubation the physical properties of the particles were determined by a variety of methods (zeta-potential, VSM – vibrating sample magnetometry, TGA - thermogravimetric analysis, and TEM – transmission electron microscopy). Additionally, the incubated nanoparticles were applied to SDS polyacrylamide gel electrophoresis (SDS-PAGE). Protein bands were visualized by silver staining. The effect of incubated particles on cell viability was tested for human brain microvascular endothelial cells (HBMEC) by the CellTiter Glo™ Cell Viability Assay and for long-term viability up to 96 h by real time cell analysis (RTCA). Interactions of coated MNP and serum incubated MNP with HBMEC were investigated by means of flow cytometry of fluorochrome-labelled particles.

RESULTS Immediately after the contact of MNP and FCS, a protein corona is formed on the surface of the coated MNP. Since the zeta potential of FCS incubated MNP varies as a function of FCS concentration in incubation media, it is clearly demonstrated, that FCS concentration has an influence on the formation of the protein corona [2]. This effect was confirmed by means of SDS-PAGE where a dependency of corona amount and protein size distribution within corona on FCS concentration was found (Figure 1).
Cell toxicity assays revealed no cytotoxic effect of PEI coated MNP, which were covered by a protein corona. Long-term viability assays (RTCA) showed that the protein corona might mask the cytotoxic effect of polyethylenimine (PEI). Flow cytometry indicate that FCS coating reduces the particle-cell interaction of cytotoxic PEI coated MNP.

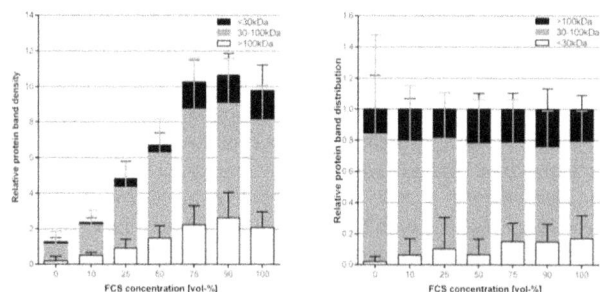

Figure 1: Amount and size distribution of the corona proteins after incubation with FCS, determined by SDS-PAGE.

CONCLUSION The influence of FCS concentration in incubation media on protein corona formation was confirmed. The protein corona shows no cell irritating effects and may mask cytotoxic effects of core/shell particles. Due to the possibility of heating by magnetic losses (additionally to the external heating) ferromagnetic multicore nanoparticles are very interesting model particles for ongoing investigations on corona formation kinetics as well as the investigation of particle-cell interactions and biological fate of serum incubated MNP after cellular uptake by means of *in vitro* and *in vivo* experiments.

ACKNOWLEDGEMENTS This work was supported by Deutsche Forschungsgemeinschaft (DFG) via priority program SPP 1681 (FKZ: DU 1293/4-1, SCHA1640/7-1, CL202/3).

REFERENCES
[1] A. Weidner, C. Gräfe, M. v.d. Lühe, H. Remmer, J.H. Clement, D. Eberbeck, F. Ludwig, R. Müller, F.H. Schacher, S. Dutz. Preparation of Core-Shell Hybrid Materials by Producing a Protein Corona Around Magnetic Nanoparticles. Nanoscale Research Letters. 2015;10(1):992.
[2] C. Gräfe, A. Weidner, M. v.d. Lühe, C. Bergemann, F.H. Schacher, J.H. Clement, S. Dutz. Intentional formation of a protein corona on nanoparticles - Serum concentration affects protein corona mass, surface charge, and nanoparticle-cell interaction. Int J Biochem Cell Biol.

Development of Magnetic Nanocarriers Based on Thermosensitive Liposomes and Their Visualization Using Magnetic Particle Imaging

Shuki Maruyama, Kohei Enmeiji, Kazuki Shimada, Kenya Murase[*]

Department of Medical Physics and Engineering, Graduate School of Medicine, Osaka University, Osaka, Japan
[*] Corresponding author, email: murase@sahs.med.osaka-u.ac.jp

INTRODUCTION The use of nanoparticles in diagnosis and therapy for cancer is progressively growing among the most important developments in drug delivery systems (DDS). Thermosensitive liposomes (TSLs) are one of the most promising tools for cancer therapy when used in combination with local hyperthermia. Recently, a new imaging method called magnetic particle imaging (MPI) has been introduced, which allows imaging of the spatial distribution of magnetic nanoparticles (MNPs) [1]. The TSLs encapsulating both MNPs and anticancer drugs have a potential to be applied to diagnosis as contrast agents in MPI and to chemotherapy using drug release induced by magnetic hyperthermia (MH) under an alternating magnetic field (AMF). The purpose of this study was to develop magnetic TSLs by encapsulating MNPs into the inner cavity of TSLs and to investigate the feasibility of visualizing them using MPI and their usefulness as nanocarriers in DDS.

MATERIAL AND METHODS First, the lipid mixture composing of 1,2-dipalmitoyl-sn-glycero-3-phosphocholine (DPPC) and polyoxyethylene (20) stearyl ether (Brij78) at a molar ratio of 96 : 4 was used to prepare the TSLs loaded fluorescent dye (calcein) by thin film hydration method. The temperature-dependent release of calcein from the TSLs was measured using the self-quenching phenomenon. Briefly, the samples of the TSL solution containing calcein were incubated for 20 min at each temperature level starting from 27 °C and increasing to 34, 36, 38, 40, 42, 44, and 50 °C. The fluorescence intensity of each sample was measured using a plate reader.

Second, we encapsulated both MNPs and calcein into the inner cavity of TSLs and performed phantom experiments. In this study, M-300 (magnetite, Fe_3O_4) [2] was used as a source of MNPs at various iron concentrations ranging from 0.75 to 1.2 M Fe. Untrapped MNPs were removed by washing repeatedly with phosphate buffered saline and by filtration using centrifugation.

Finally, the TSL solution and aqueous solution outside the TSLs were put into cylindrical polyethylene tubes (6 mm in diameter and 5 mm in length, 100 μL) separately (Fig. 1(a)) and were imaged using our MPI scanner [3, 4] (Fig. 1(b)). We then investigated the correlation between the average MPI value and the initial iron concentration. In this study, the MPI value was defined as the pixel value of the transverse image reconstructed from the third-harmonic signals [4], and the average MPI value was obtained by drawing a circular region of interest (ROI) with the same area as that of the above tube (115 pixels) and calculating the mean of the pixel values within the ROI. After phantom experiments, the net iron concentration in the TSLs was determined using potassium thiocyanate method. We also heated the magnetic TSLs using our system for MH [5, 6] at an AMF frequency of 600 kHz and a peak amplitude of 3.5 kA/m and measured the time course of the temperature rise with a fluorescence optical fiber thermometer. We then investigated the correlation between the average MPI value and the temperature rise. We also investigated the correlation between the average MPI value and calcein release from the magnetic TSLs.

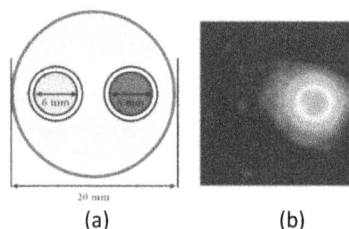

Figure 1: Ilustration (a) and MPI image (b) of a phantom having two tubes 6 mm in diameter. The left tube contains aqueous solution after filtration of magnetic TSLs, while the right tube contains only magnetic TSLs.

RESULTS The calcein release from the TSLs significantly increased at about 38-40 °C, while it was stable at lower temperature (27-36 °C). The magnetic TSLs were visualized using our MPI scanner [3, 4]. Figure 1(b) shows a typical example of the MPI image of the magnetic TSLs encapsulating both MNPs (1.2 M Fe) and calcein.

There were significant correlations between the average MPI value and the initial iron concentration (r=0.801) and between average MPI value and the temperature rise (r=0.908). In addition, calcein was more rapidly released with an increase of the average MPI value, suggesting that the average MPI value can be used for estimating the drug release from the magnetic TSLs induced by MH.

CONCLUSION Our results suggest that MPI is useful for enhancing the effectiveness of chemotherapy using DDS with magnetic TSLs, because MPI can visualize magnetic TSLs and estimate the drug release from the TSLs induced by MH.

ACKNOWLEDGEMENTS This work was supported by a Grant-in-Aid for Scientific Research (Grant No. 25282131) from the Japan Society for the Promotion of Science (JSPS).

REFERENCES
[1] B. Gleich and J. Weizenecker. *Nature*, 435(7046):1217–1217, 2005. doi: 10.1038/nature03808.
[2] K. Murase, A. Mimura, N. Banura, K. Nishimoto, and H. Takata. *Open J. Med. Imaging*, 5(2):56–65, 2015. doi: 10.4236/ojmi.2015.52009.
[3] K. Murase, S. Hiratsuka, R. Song, and T. Takeuchi. *Jpn. J. Appl. Phys.*, 53(6):067001, 2014. doi: 10.7567/jjap.53.067001.
[4] K. Murase, R. Song, and S. Hiratsuka. *Appl. Phys. Lett.*, 104(25): 252409, 2014. doi: 10.1063/1.4885146.
[5] K. Murase, J. Oonoki, H. Takata, R. Song, A. Angraini, et al. *Radiol. Phys. Technol.*, 4(2):194-202, 2011. doi: 10.1007/s12194-011-0123-4.
[6] K. Murase, M. Aoki, N. Banura, K. Nishimoto, A. Mimura, et al. *Open J. Med. Imaging*, 5(2):85–99, 2015. doi: 10.4236/ojmi.2015.52013.

Quantitative biodistribution studies of optimized MPI tracers radiolabeled for multimodal SPECT/CT imaging

Hamed Arami[a], Kathayoun Saatchi[b], Eric Teeman[a], Alyssa Troksa[a], Haydin Bradshaw[a], Urs O. Häfeli[b], and Kannan M. Krishnan[a]

[a] Department of Materials Sciences & Engineering, University of Washington, Seattle, Washington, USA
[b] Faculty of Pharmaceutical Sciences, University of British Columbia, Vancouver, Canada
* Corresponding author, email: kannanmk@uw.edu

INTRODUCTION Building on the robust PMAO-PEG co-polymer coating platform [1] for functionalizing our optimized MPI tracers [2], we radiolabeled them for SPECT/CT contrast to enable accurate determination of their biodistribution and pharmacokinetics. SPECT/CT imaging provide high tracer mass sensitivity, enabling accurate, quantitative estimation of NPs concentration in the main clearance organs (*i.e.*, liver, spleen and kidneys). This complements our earlier NIRF imaging of similar MPI tracers to determine their biodistribution with anatomical sensitivity [3].

MATERIAL AND METHODS Iron oxide NPs with median core size of 27 nm (Fig1 A,B), optimal for MPI (Fig 1C), were synthesized and coated with PMAO in chloroform and transferred to water, showing a negative zeta potential ($\sim$-30 to -50 mV) due to presence of carboxyl groups on their surface. A branched polyethylene imine (PEI, 10 kDa, Polyscience) was added to these NPs (PEI:NPs=10 mg:1 mg), sonicated for $\sim$1 h, and purified, showing a positive zeta potential ($\sim$+30 to +60mV) due to presence of cationic PEI on their surface. For radiolabeling and SPECT imaging, the PEI coated particles were modified with the chelator *p*-SCN-bz-NOTA (Macrocyclics, USA) and then radiolabeled with Gallium-67 (^{67}Ga, $T_{1/2}$ = 78.3 h). For this purpose, the particles were diluted with NaHCO$_3$ (0.1 N) and incubated with *p*-SCN-bz-NOTA solution at 19 °C overnight. The particles were concentrated using an Amicon vial (30 kDa MWCO), washed twice and re-suspended in PBS. NOTA were bound to the particles through a thiourea bond. Particles were radiolabeled by adding ^{67}GaCl$_3$ to the suspension and incubating at room temperature with mixing, followed by Amicon concentration and washing twice with PBS (92% labeling efficiency). The radiolabeled particles were dispersed in PBS for the SPECT/CT biodistribution studies. Three female C57Bl/6 mice (Charles River Laboratories) were injected via tail vein with the tracer (120 µl, radiolabelled PEI-MNPs). The animals were scanned individually in a SPECT/CT immediately and at 4 h post injection and sacrificed. Following each SPECT acquisition, a whole body CT scan was performed to obtain anatomical information and both images were registered. For quantitative analysis, SPECT data were reconstructed with ordered subset expectation maximization algorithm (OS-EM) using 6 iterations of 16 subsets and 0.4 mm^3 voxel size. All organs were then counted for radioactivity.

RESULTS MPS analyses of the tracers showed an excellent dm/dH response with a narrow FWHM (Fig 1C), suitable for high resolutions MPI. The branched PEI used in conjugation is a cationic polymer with abundant number of available amine groups, making it a suitable platform for radiolabeling with ^{67}Ga-NOTA. SPECT/CT imaging (Fig 1D) and biodistribution studies confirmed that PEI-coated NPs were only accumulated in liver and spleen 4 h after injection, with almost no traces of NPs in

kidneys, lungs and heart (Fig 1E). Quantitative measurements of the radioactivity from all organs indicated the biodistribution to be predominantly in the liver (92±10% of the input dose) and spleen (5±1%) with trace distributions, within experimental error, in the brain and kidneys.

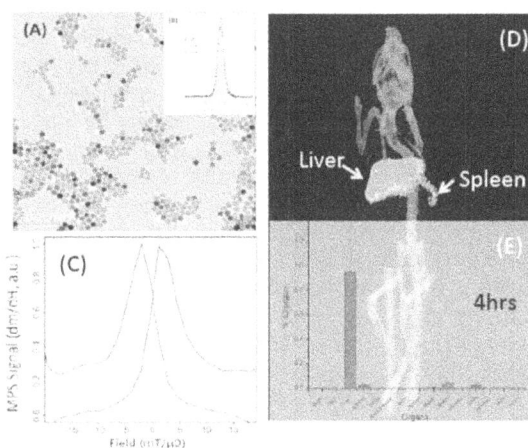

Figure 1: TEM image (A), size distribution histogram (B, inset) and (C) MPS (dm/dH) plot of the optimized tracers. (D) Typical SPECT image of a rodent model showing the biodistribution of the 67Ga-NOTA-labeled MPI tracers in mice, 4 h after tail vein injection. (E) NPs were only accumulated in liver and spleen with no trace of NPs observed in kidneys or other organs.

CONCLUSION Amine functionalization of the PMAO coated MPI tracers with PEI, enables facile radiolabeling of ^{67}Ga-NOTA, using a thiourea bond, suitable for quantitative SPECT/CT imaging. Absence of radioactive tracers in the kidneys, 4 h post tail-vein injection, confirmed by the high tracer mass sensitivity of SPECT/CT imaging, is promising for the safe applications of these tracers for in vivo MPI imaging. The flexible PMAO-PEG platform further provides opportunities for various conjugation strategies, ideal for multimodal MPI/CT/NIRF/MRI/CT/SPECT imaging, and each one with distinct advantages. Details of the development of a single nanoparticle tracer, with a broad functionalizing platform, providing complementary imaging capabilities, will also be presented.

ACKNOWLEDGEMENTS This research was supported by NIH grant 2R42EB013520-02A1. SPECT/CT imaging protocol was approved by the UBC –ACC, according to CCAC guidelines.

REFERENCES
[1] A.P. Khandhar *et al*, *IEEE Trans. Mag.* **51** (2015), 5300304.
[2] R.M. Ferguson *et al*, *IEEE Trans. Med. Imag.* **34** (2015), 1077.
[3] H. Arami *et al*, *Biomaterials*, **52** (2015) 251-261.

Magnetic Separation to Extract Suitable Cells for MPI Cell Tracking

Angela Ariza de Schellenberger [a,*], Norbert Löwa [b], Jörg Schnorr [a], Harald Kratz [a], Matthias Taupitz [a], Frank Wiekhorst [b]

[a] Department of Experimental Radiology, Charité – Universitätsmedizin Berlin, Germany
[b] Physikalisch-Technische Bundesanstalt, Abbestr. 2-12, 10587 Berlin, Germany
* email: Angela.Ariza@charite.de

INTRODUCTION Magnetic Particle Imaging (MPI) used as a tool for sensitive and quantitative cell tracking could become one of the most relevant applications of this emerging technique [1]. This requires the preparation of cell samples loaded with magnetic nanoparticles (MNP) suitable for MPI. Therefore, the designation of suitable MNP maintaining their superior MPI performance after internalization, biological effects of the internalized MNP or the optimization of MNP loading [2] is presently in the focus of investigation. Studies revealed that MNP loading of in vitro labeled cells is uneven due to cellular division and different internalization rates caused by different cell cycle phases [3]. This complicates the systematic design and evaluation of experiments and further reduces the recovery rate in a cell tracking experiment. To address this challenging task we investigated the feasibility to magnetically separate cells of high MNP loading using a commercial column separator which we extended here to separate at different (smaller) magnetic fields. To evaluate the separation efficacy we used Magnetic Particle Spectroscopy (MPS) as this technique constitutes a fast and accurate tool for cell preserving quantification of MNP internalized by cells [3]. Since MPS is based on the same physical principle as MPI, MPS signals reflect the MPI performance of MNP loaded cells. In combination with a standard cell counting procedure the separated fractions were analyzed to find the cell fraction of highest MNP loading.

MATERIAL AND METHODS Cells: Non phagocytic, murine bone marrow derived mesenchymal stem cells (MSC) are labeled in vitro with Resovist®, (RES) (AG Schering) and multicore iron oxide nanoparticles (MCP) (Charité) designed for MPI. Intracellular MNP uptake for 24 h was implemented by means of an optimized protocol to obtain efficient (> 10 pg(Fe)/cell) intracellular labeling. For magnetic cell separation different procedures were evaluated and optimized according to applicability, amount of harvested cells, reproducibility and non-specific absorption of cells in the separation column.
Magnetic separation: Magnetic separation was performed using a commercially available separation column (LS column, Miltenyi Biotec, GER). Typically, the separation is performed using the corresponding separator magnet (MidiMACS™, Miltenyi Biotec, GER) which provides a fixed field strength of 0.5 T. Additionally, smaller magnetic fields between 40 mT and 0.5 mT were generated by a customized 4 layer copper coil (108 mm height, 60 mm diameter). MPS-quantification: To quantify the MNP amount in cells the MPS signal (B_{excit}=25 mT, f_{excit}=25 kHz) of a sample was compared with a reference of known MNP content. Fluorescence-activated cell sorting (FACS) was used to quantify the cell amount.

RESULTS We found, that RES and MCP loaded MSC could be successfully separated at magnetic field strengths below 5 mT (Fig. 1a). From FACS followed that nearly 100% of the cells were captured if the standard setup (B=500 mT) was used. Even at a 100-times lower magnetic field B=5 mT most of the cells were still trapped by the column. Significant separation of cells took place between 1 mT and 2.5 mT. The upper graph of Fig. 1 shows that cells separated at 1 mT had on average almost twice the iron amount per cell compared to the initial state (for MCP) as quantified by MPS. Interestingly, the iron amount per cell decreased for lower separation fields or rather less cells have been trapped. We attribute this to unspecific absorption of cells which is independent of the cellular iron amount. This was confirmed by separations without using a magnetic field (B=0 mT) where ~15% of the cells were captured in the column. Furthermore, we found that separation of cells loaded with RES or MCP did not affect their MPS performance (iron normalized amplitude and shape of the spectrum).

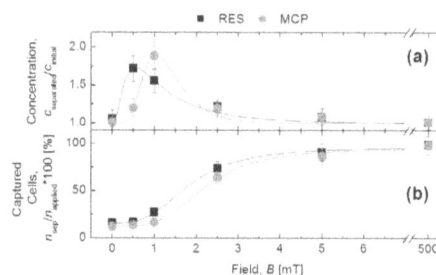

Figure 1: Separation of RES and MCP loaded MSC using separation fields between 500 mT and 0.5 mT. (b) The cell amount, as quantified by FACS, is shown as the ratio between separated and applied cells. (a) The ratio between the averaged cellular MNP loading after and before separation was calculated to estimate the effectiveness of the separation process.

CONCLUSION Magnetic cell separation was successfully carried out using a commercial column separator adapted to provide smaller magnetic fields. We found that the optimum field range for enrichment of highly loaded cells differs for different MNP types. Furthermore, MPS in combination with FACS has proven to be ideally suitable to accompany the development of an effective cell labeling procedure and to determine the cell fraction of highest MNP loading after separation.

ACKNOWLEDGEMENTS The research was supported by German Research Foundation, through DFG Research Unit FOR917 (Nanoguide) and German Ministry of Education and Research under Grant FKZ 13N11092 (MAPIT).

REFERENCES

[1] J. Borgert, J. D. Schmidt, I. Schmale, J. Rahmer, C. Bontus, B. Gleich, B. David, R. Eckart, O. Woywode, J. Weizenecker, J. Schnorr, M. Taupitz, J. Haegele, F. M. Vogt, and J. Barkhausen. *J. Cardiovasc. Comput. Tomogr.*, 6(3): 149—153, 2012. doi: 10.1016/j.jcct.2012.04.007.
[2] W.C. Poller, N. Löwa, F. Wiekhorst, M. Taupitz, S. Wagner, K. Möller, G. Baumann, V. Stangl, L. Trahms, and A. Ludwig. *J. Biomed. Nanotech.*, 12(2): 337—346, 2016. doi: 10.1166/jbn.2016.2204.
[3] J. A. Kim, C. Åberg, A. Salvati, and K. A. Dawson. *Nature Nanotech.*, 7(1): 62—68, 2011. doi: 10.1038/nnano.2011.191.

Evaluation of harmonic signal from blood-pooling magnetic nanoparticles for magnetic particle imaging

Satoshi Ota*, Ryuji Takeda, Tsutomu Yamada, Yasushi Takemura

a Department of Electrical and Computer Engineering, Yokohama National University, Japan
* Corresponding author, email: ota-satoshi-gw@ynu.jp

INTRODUCTION Magnetic nanoparticles (MNPs) which show high harmonic contribute to high resolution of magnetic particle imaging (MPI) [1]. In this study, the harmonic signals of iron oxide nanoparticles were estimated by the measurement of ac magnetization curves. MPI signal is generated by applying an ac magnetic field to MNPs accumulated in the affected parts. In order to optimize MNPs for MPI, the dependence of MPI signal intensity on the core diameter of MNP has been researched [2]. In addition, it has been shown that the third harmonic intensity is influenced on the distribution of core diameter of MNP [3].

MATERIAL AND METHODS Carboxymethyl-diethylaminoethyl dextran coated maghemite nanoparticles were measured under the ac magnetic fields in 4 and 8 kA/m of amplitude and 1–100 kHz of frequency range. The core diameters of these MNPs were 4, 5 and 8 nm. These MNPs were dispersed in water and fixed with agar, which were prepared as the liquid and fixed samples, respectively. It has been shown that the blood-pooling time of these MNPs are longer than the conventional ones because of less recognizability by phagocytes due to the albumin binding in blood [4]. In addition, these MNPs showed superparamagnetism because coercivities of dc hysteresis loops were marginal.

RESULTS Figure 1 (a) shows the dependence of the third harmonic intensity on frequency in the fixed samples under 8 kA/m of magnetic field. It is shown that the third harmonic intensity of 8-nm-diameter MNP was larger than that of 4- and 5-diameter MNPs. The third harmonic intensity of 4-nm-diameter MNP was higher than that of 5-nm-diameter one despite the smaller diameter of 4-nm-diameter MNP than 5-nm-diameter one. It is indicated that the independence of the third harmonic intensities on the core diameters of MNPs was influenced by the core

diameter distributions [3]. Figure 1 (b) shows the M_3/M_1 ratio in the fixed samples. M_1 and M_3 indicates the magnetization of fundamental and third harmonics, respectively. The M_3/M_1 ratio shows the third harmonic intensity normalized by the fundamental one, which indicates the process of magnetization reversal. With respect to the M_3/M_1 ratio, the difference between 4- and 8-nm-diameter MNPs was marginal in contrast to the significant difference between 5- and 8-nm-diameter ones despite the lower intensities of the third harmonic in 4- and 5-nm-diameter MNPs than that in 8-nm-diameter one (Fig. 1). Moreover, both the third harmonic intensity and M_3/M_1 ratio decreased with the increase of frequency in Fig. 1. It is confirmed that magnetization response in the high frequency was more linear than that in the low frequency.

CONCLUSION The third harmonic intensity was independent on the core diameters of MNPs. In addition, it is indicated that the evaluation of both the third harmonic intensity and M_3/M_1 ratio contribute for the optimization of MNPs for MPI.

ACKNOWLEDGEMENTS Magnetic particles used in this study were supplied from Meito Sangyo Co., Ltd.. This work was partially supported by the JSPS KAKENHI Grant Number 15H05764 and 26289124.

REFERENCES

[1] B. Gleich, J. Weizenecker, *Nature*, **435**, 1214 (2005).

[2] A. Tomitaka, R. M. Ferguson, A. P. Khandhar, S. J. Kemp, S. Ota, K. Nakamura, Y. Takemura, K. M. Krishnan, *IEEE Trans. Magn.*, **51**, 6100504 (2015).

[3] T. Yoshida, N. B. Othman, K. Enpuku, *J. Appl. Phys.*, **114**, 173908 (2013).

[4] N. Nitta, K. Tsuchiya, A. Sonoda, S. Ota, N. Ushio, M. Takahashi, K. Murata, S. Nohara, *Jpn. J. Radiol.*, **30**, 832 (2012).

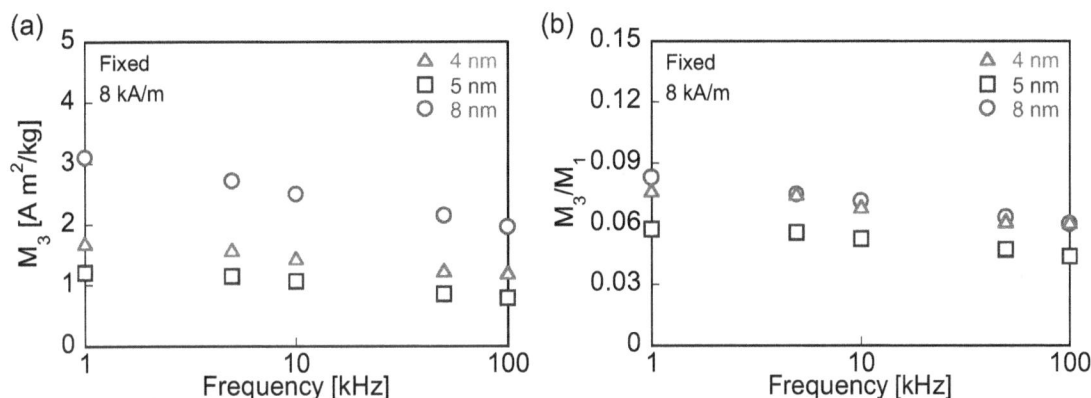

Fig. 1 Frequency dependences of (a) the third harmonic intensity and (b) the ratio of the third and fundamental harmonics.

Does a highly concentrated sample generate a better system function?

Olaf Kosch [a,*], Norbert Löwa [a], Frank Wiekhorst [a], Lutz Trahms [a]

[a] Department 8.2 Biosignals, Physikalisch-Technische Bundesanstalt, Berlin, Germany
* Corresponding author, email: olaf.kosch@ptb.de

INTRODUCTION Lissajous-based Magnetic Particle Imaging (MPI) enables a highly sensitive detection of magnetic nanoparticles (MNP) with high temporal and spatial resolution [1]. Since an exact analytical model of the nonlinear dynamic magnetization behavior of MNP is not yet available, image reconstruction requires a system function (SF) relating the spatial distribution of MNP to the generated signal in the scanner [2, 3]. To this end, MPI SF acquisition uses a point-like MNP reference sample of high concentration to obtain best signal-to-noise-ratio (SNR). This presupposes that dynamic magnetization behavior of the reference sample is concentration independent. We investigated the SF of Resovist® (Bayer HealthCare, GER) and its precursor Ferucarbotran (Meito Sangyo, JPN) at different concentrations to verify this assumption.

MATERIAL AND METHODS We have investigated Resovist® (c(Fe) = 100, 200 and 500 mmol/l) and it precursor Ferucarbotran (c(Fe) = 1 mol/l) by means of Magnetic Particle Spectroscopy (MPS) using a sample volume of 30 µl, an averaging time of 10 s and a flux density of 12 mT. A key objective was to find differences in the spectral magnetic response depending on iron concentration. Furthermore, we measured and compared the MPI SFs at all four concentrations. The SFs were recorded using a gradient field of 2.5 T/m, a drive field of 12 mT in x-, y- and z- direction and 100 averages. We employed different recorded SFs to reconstruct MPI images of a hosepipe phantom housing MNP at different concentrations c(Fe) = 50, 100 and 500 mmol/l. We applied the Kaczmarz algorithm with 20 iterations, a regularization = 10^{-5} and an average of 8 frames for reconstruction.

RESULTS The MPS harmonics exhibited a shallower decay for lower concentrated Resovist®. In the MPI SF patterns, we observed different spatial distributions depending on concentration. Fig.1 depict the SF patterns of the mode = 2 x f_x + 2 x f_y + 1 x f_z at z = 1 mm. The spatial pattern created with lower concentrated Resovist® is closer to the ideal particle SF corresponding to a rich harmonic spectrum resulting in a more checkerboard-like arrangement [3] whereas highly concentrated particles shows little deviations from a circular pattern. The reconstruction of low concentrated MPI tracer distributions (50 and 100 mmol/l) applying the SF derived from a high concentrated sample results in distributions dominated by noise. In opposite the SF recorded with a sample at 100 mmol/l works in reconstruction even with the five times lower SNR.

CONCLUSION The concentration dependence in the SF can perturb the reconstruction results of magnetic particle imaging of a lower concentrated particle distribution as shown for Resovist®. Therefore, using a SF recorded with a sample of highest concentration can be worse the image reconstruction of a distribution at lower concentration even if the SNR of the SF is higher. We refer this to an increased magnetic interaction between MNP of Resovist® at higher concentrations as shown in

[4]. Thus, we suggest to use a Resovist® concentration below 200 mmol/l for SF recording, especially for *in vivo* experiments where low MNP concentrations are expected.

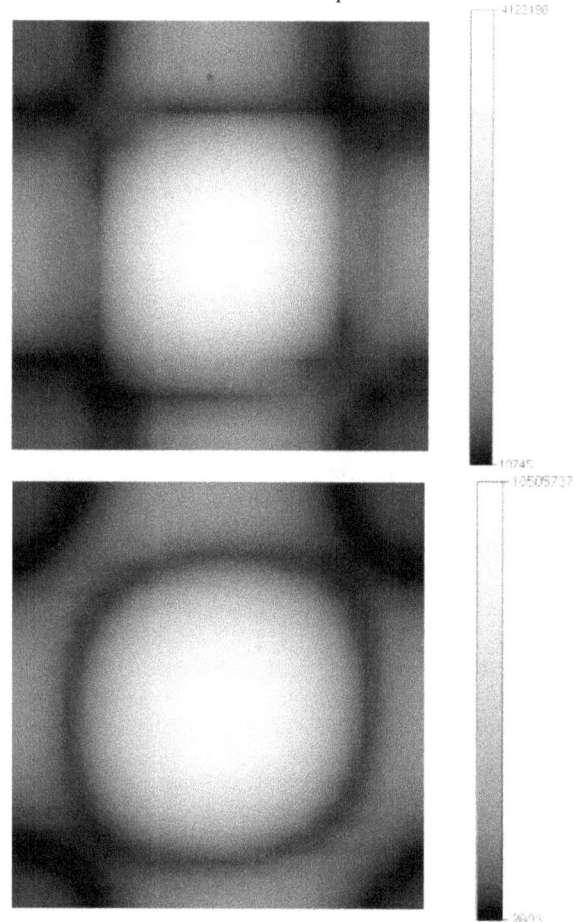

Figure 1: System function (absolute values) of Resovist®, sample volume: 8µL, in xy-plane at z = 1 mm for mode = 2,2,1 at c(Fe) = 100 mmol/l (above) and c(Fe) = 500 mmol/l (below).

ACKNOWLEDGEMENTS The research was supported by the German Ministry for Education and Research (Magnetic Particle Imaging Technology, Grant No. FKZ 13N11092).

REFERENCES
[1] J. Weizenecker, J. Borgert and B. Gleich. *Phys. Med Biol*, 52 6363–6374, 2007. doi:10.1088/0031-9155/52/21/001.
[2] J. Rahmer, J. Weizenecker, B. Gleich and J. Borgert. *BMC Med Imaging* 9:4, 2009. doi: 10.1186/1471-2342-9-4.
[3] J. Rahmer, J. Weizenecker, B. Gleich and J. Borgert. *IEEE Trans. Med Imaging*, 31:6, 2012. doi: 10.1109/TMI.2012.2188639.
[4] N. Löwa, P. Radon, O. Kosch, F. Wiekhorst. *Int. J. of MPI*, 1:1, 2016. [in press]

In vivo measurement und comparison of two Magnetic Particle Imaging tracer: LS-008 and Resovist

Michael G. Kaul[a,*], Caroline Jung[a], Johannes Salamon[a], Tobias Mummert[a], Martin Hofmann [a,b], Scott J. Kemp[c], R. Matthew Ferguson[c], Amit P. Khandhar[c], Kannan M. Krishnan[c,d], Harald Ittrich[a], Gerhard Adam[a], Tobias Knopp[a,b]

[a] Department of Diagnstic and Interventional Radiology, University Medial Center Hamburg, Germany
[b] Hamburg University of Technology, Hamburg, Germany
[c] LodeSpin Labs, Seattle, Washington, USA
[d] Materials Science and Engineering Department, University of Washington, Seattle, USA
[*] Corresponding author, email: mkaul@uke.uni-hamburg.de

INTRODUCTION The development of Magnetic Particle Imaging (MPI) relies on the improvement of three aspects: hardware, software which is mainly image reconstruction, and tracers. While the second and partly the first are independent of the biological and medical experiments, tracers are the interface; tracers interact with biological tissue and at the same time they are the source of signal generation [1]. It is known that Resovist (Bayer-Schering), consisting of a widely spread particle size distribution, is a reliable MPI tracer, even though only a small fraction of its volume generates the MPI signal. It is therefore straight foreword to substitute it with an improved tracer. The goal of this study was to perform *in vivo* MPI measurements with a new monodisperse and size-optimized MPI tracer, LS-008 (LodeSpin), and to compare it with the performance of Resovist.

MATERIAL AND METHODS *In vivo* measurements, approved by a local animal care committee, were carried out in six healthy mice divided in two equally sized groups. The first group received 30mL LS-008 (89mM) and the second received diluted Resovist of the same concentration and volume. Tracer injections were performed with a syringe pump during a dynamic MPI scan. For anatomic referencing MRI was applied before the MPI measurements [2]. We used fiducial makers to position heart, lung and liver in the central part of the field of view. Reconstruction and post processing was performed offline with in house written framework (Julia language and ImageJ). First, we reconstructed the frames of the first pass of the tracer on a single frame base to depict the major vessels. To check the accumulation in the liver we reconstructed a later phase after six minutes with averaged frames.

RESULTS *In vivo* measurements were performed successfully. While both tracers can visualize the propagation of the bolus through the inferior vena cava there was a clear quality difference visible (fig.1). MPI with LS-008 did show less temporal fluctuation artifacts. Focusing on time frames, when the bolus leaves the heart a pronounced difference becomes visible. With LS-008 the aorta can be nicely distinguished form the caval vein (fig.2). Furthermore, a vessel passing the liver and a vessel structure leading in the direction of the head could be observed (fig.3). This is missing using Resovist.

CONCLUSION MPI can be significantly improved by applying more effective tracers. LS-008 shows a clear improvement with respect to the delineation and resolving a larger number of vessels in comparison to Resovist. Therefore, in aspects of quality and quantity LS-008 is clearly favorable. While Resovist has a fast accumulation in liver, LS-008 possesses a slower blood clearance and larger circulation time. However, this result is not presented here, even though it is of major interest for long lasting experiments where a stationary phase is of importance.

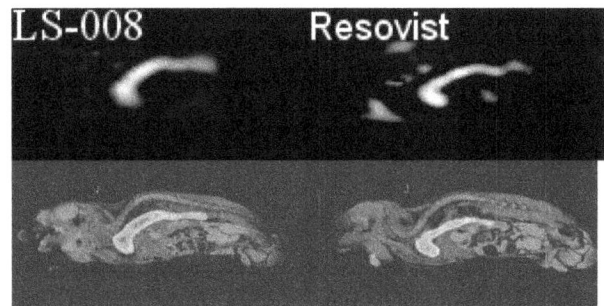

Figure 1: Bolus inflow, through the inferior vena cava to the heart. of (left) LS-008 and (right) Resovist showing (top) MPI signal only and (bottom) co-registered with MRI.

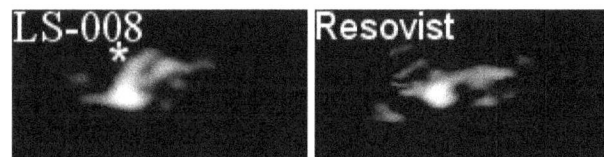

Figure 2: In a later phase the aorta descendens (*) leaving the heart becomes visible (left) with LS-008 while (right) with Resovist is not. The amount of artifacts increases with Resovist

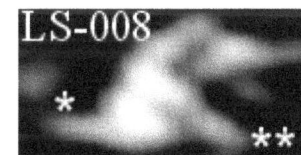

Figure 3: Five seconds after arrival in the heart with LS-008 a vessel (**) crossing the liver becomes visible and later on it is more pronounced. Another vessel structure (*) leaving in direction of the head, as well as parts of aorta caval vein, are visible as well.

ACKNOWLEDGEMENTS We gratefully acknowledge funding and support of the German Research Foundation. (DFG, grant number AD 125/5-1). Work at UW/LSL was supported by NIH grants 1R41EB013520-01 and 2R42EB013520-02A1

REFERENCES
[1] R. M. Ferguson, A. P. Khandhar, S. J. Kemp, et al., "Magnetic particle imaging with tailored iron oxide nanoparticle tracers.," IEEE transactions on medical imaging, 34(5), 1077–1084, 2015.
[2] Michael G Kaul, Oliver Weber, Ulrich Heinen, Aline Reitmeier, Tobias Mummert, Caroline Jung, Nina Raabe, Tobias Knopp, Harald Ittrich, Gerhard.Adam. Combined Preclinical Magnetic Particle Imaging and Magnetic Resonance Imaging: Initial Results in Mice. *Fortschr Röntgenstr* 187: 347–352, 2015

Correlation of MPS with Colorimetric Iron Content Measurements

Lisa Wendt, Kerstin Lüdtke-Buzug[*]

Institute of Medical Engineering, Universität zu Lübeck, Lübeck
[*] Corresponding author, email: luedtke-buzug@imt.uni-luebeck.de

INTRODUCTION For most of today's common imaging modalities, contrast agents or tracer materials may improve image quality or contrast. However, even tracer materials based on iron show toxic properties in higher concentrations. For magnetic particle imaging (MPI), for instance, it is therefore necessary in preclinical and, in particular, for clinical applications to determine the optimal amount of the tracer material. Currently, in MPI the standard tracer material is based on superparamagnetic iron oxide nanoparticles.

Iron oxide can occur in various forms, e.g. as magnetite, a mixture of iron (II)-oxide and iron (III)-oxide, which is a black, heat-resistant, ferrimagnetic substance, having spinel structure, which is insoluble in water. For signal generation in MPI, ideally, the tracer material should have superparamagnetic properties. This can be realized by using single-domain sized nanoparticles of the material. One common way to prepare SPIONs suitable for MPI applications is the coprecipitation of magnetite in alkaline solution.

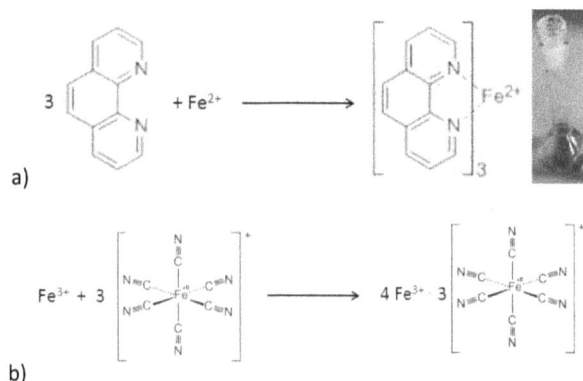

Figure 1: a) Phenanthrolin iron complex, and b) Prussian blue complex

MATERIAL AND METHODS In this project, two different photometric methods are used for the determination of the iron content, the Prussian blue method and the Phenanthroline method.

The concentration is determined by measuring the absorption of the colored metal complexes formed with potassium ferrocyanide or phenanthroline solved in water or buffer (see Fig. 1). To obtain precise results of the iron concentrations of different solutions a commercial iron standard for atomic absorption spectroscopy (Sigma-Aldrich) was used as reference.

The colorimetric analysis is based on the change in the intensity of the color of a solution with variations in the concentration. The optimal wavelength for analyzing the Prussian blue complex is 690 nm, for the phenathroline compelex the wavelength has to be changed to 525 nm.

For the sensitivity measurements with magnetic particle spectroscopy (MPS) different dilution series were prepared. The SPIONs have been diluted with demin. water or buffer (MPS measurement conditions: temperature 25 °C, field strength

25 mT, averaging periods 10, sample volume 10 µL, repetitions 12500, frequency 25 kHz).

In addition, dry powder nanoparticles (<50 nm particle size, Sigma-Aldrich) were used as reference for the MPS measurements. The particles were used as a powder without any further preparation and in a biopolymer matrix, respectively.

RESULTS The MPS measurements and the results of colorimetric measurements of the tracer material's iron content show an excellent correlation. In Fig. 2 it can be seen for five different concentrations in a dilution series that consistent results over all relevant harmonics can be obtained by MPS.

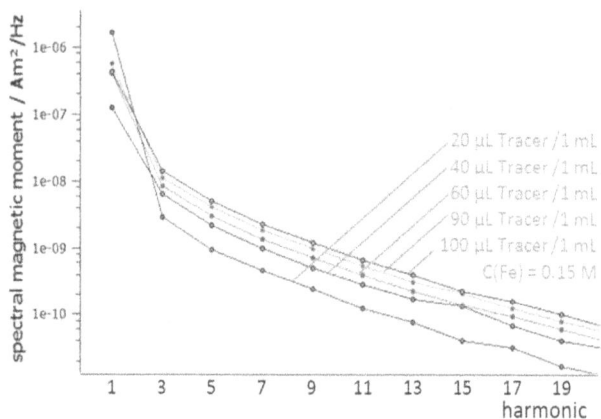

Figure 2: MPS for a dilution series of SPION samples

CONCLUSION For preclinical and clinical applications of MPI it is necessary to know the optimal iron concentration of the tracer material due to toxic effects of iron. Here, optimal means the minimal iron content at acceptable MPI signal strength. It has been demonstrated that MPS measurements can be correlated to the iron content of the SPIONs.

ACKNOWLEDGEMENTS The research leading to these results has received funding from the European Union Seventh Framework Programme (FP7/2007-2013) under grant agreement no 604448.

REFERENCES

[1] B. Gleich and J. Weizenecker. *Nature*, 435(7046):1217—1217, 2005. doi: 10.1038/nature03808.

[2] T. Knopp and T. M. Buzug. Springer, Berlin/Heidelberg, 2012. doi: 10.1007/978-3-642-04199-0.

[3] G. Jander, E. Blasius, J. Strähle, *Lehrbuch der analytischen und präparativen anorganischen Chemie*, Ausg.16, Verlag Hirzel, 2006, ISBN 9783777613888

[4] F. Jančik, Zur o-Phenanthrolin-Methode der Eisenbestimmung im Blutserum, *Fresenius Zeitschrift für Analytische Chemie 11*/1967; 232(6):441-442. DOI:10.1007/BF00532393

[5] S. Biederer et al.. *J. Phys. D: Appl. Phys.*, 42(20): 205007, 2009. doi: 10.1088/0022-3727/42/20/205007.

Magnetic Particle Spectrometry of LS-008 driven at 153 kHz, 15 mT/ μ_0

R. Matthew Ferguson[a,*], Amit P. Khandhar[a], Scott J. Kemp[a], and Kannan M Krishnan[a,b]

[a]Lodespin Labs, PO Box 95632, Seattle WA 98145, [b]Department of Materials Science & Enginerering, University of Washington, USA
* Corresponding author, email: matt@lodespin.com

INTRODUCTION Ideal MPI tracers are close to the superparamagnetic-ferrimagnetic transition, where thermal energy and coherent rotation both contribute to magnetic moment reversal [1]. Varying the drive field amplitude, H_0, or frequency, ω, (collectively the slew rate, ωH_0) alters the balance of contributions from thermal energy and coherent rotation, affecting the tracer's magnetic response [2]. To accommodate safety limits, Philips clinical prototype scanner is reportedly being designed to operate around 150 kHz drive field; a higher frequency than has been standard for MPI. In this contribution, we analyze the MPS behavior of single-core tracers at 153 kHz, including LS-008, a long-circulating high-resolution MPI tracer.

MATERIAL AND METHODS Iron oxide nanoparticles of varying magnetic core size, including LS-008, were synthesized using an iron oleate thermolysis method [3] and coated with poly(maleic anhydride-alt-1-octadecene) grafted with 20 kDa methoxy-PEG-amine. Sample iron concentration was determined by ICP-OES. MPI performance was analyzed with a custom magnetic particle spectrometer operating at 25 kHz and 153 kHz, with amplitude variable from 0-50 mT/ μ_0 (25 kHz) and 0-15mT/μ_0 (153 kHz). Samples were measured in triplicate after first subtracting the background signal generated by an empty volume. Differential susceptibility, $\chi_d = dm/dH$, and harmonic spectra were determined according to established methods.

RESULTS Magnetic behavior was influenced by the MPS drive frequency. At 153 kHz, for each of the samples studied $\chi_d max$ (the maximum χ_d) was reduced by 10-20% compared to 25 kHz. In figure 1a, $\chi_d(H)$ is presented for LS-008. $\chi_d max$ was 13% less at 153 kHz (15 mT/μ_0,) than at 25 kHz. A shift in the position of $\chi_d max$ and a broadening of the peak full width at half maximum were also observed. This is consistent with previous findings that greater slew rate caused χ_d to stretch toward greater fields, likely due to reduced role of thermal fluctuations in the magnetic reversal [2]. At 7.5 mT/μ_0, the morphologically different d_C=33nm sample did not saturate, so $\chi_d max$ was not determined.

The effect of frequency, though real, was modest, as reflected in the similar profile of the odd harmonics for the two frequencies at 15 mT/μ_0. We note some artifacts due to system noise in the measurements at 153 kHz that appear to introduce secondary peaks (fig1a) and a bump in the 5th harmonic intensity. Table 1 summarizes the MPS results using $\chi_d max$,. The reduction in intensity when increasing the frequency to 153 kHz with 15 mT/μ_0 amplitude varied from 10-20%.

CONCLUSION We have presented MPS results measured at 153 kHz and confirmed that slew rate impacts MPI tracer behavior, including reducing $\chi_d max$ and increasing the $\chi_d(H)$ peak width. MPI tracers typically provide the most ideal $\chi_d(H)$ response for slow-changing fields, but the relatively small difference (~15%) between 25 kHz and 153 kHz, makes LS-008 promising for MPI over a wide range of frequencies.

ACKNOWLEDGEMENTS Work at LSL and UW supported by NIH/NIBIB R42EB013520.

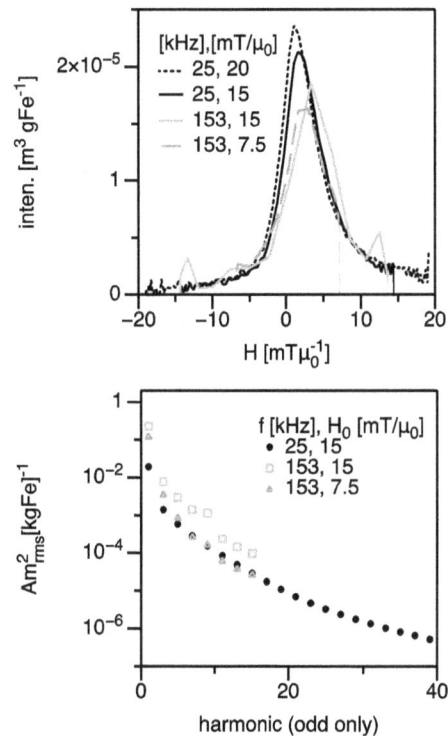

Figure 1: LS-008 MPS at 25 kHz and 153 kHz drive field and different amplitudes. a) dynamic susceptibility, χ_d, and b) harmonic spectrum.

Table 1: Table of results comparing performance at 25 kHz and 153 kHz drive frequency, and 15 and 7.5 mT/μ_0 drive amplitude.

| | | $\chi_d max$, (x10⁻⁵) [m³/kgFe] | | |
| | | 25 kHz | 153 kHz | |
	d_C [nm]	15 mT/μ_0	15 mT/μ_0	7.5 mT/μ_0
A	21.0	1.50	1.20	1.17
LS-008	24.9	2.13	1.86	1.63
B	28.1	2.33	2.10	1.93
C	33.1	2.09	1.85	n/a

REFERENCES
[1] S. A. Shah, D. B. Reeves, R. M. Ferguson, J. B. Weaver, and K. M. Krishnan, *Physical Review B*, vol. 92, no. 9, p. 094438, Sep. 2015.

[2] S. A. Shah, R. M. Ferguson, and K. M. Krishnan, *J. Appl. Phys.*, vol. 116, no. 16, p. 163910, Oct. 2014.

[3] S. J. Kemp, R. M. Ferguson, A. P. Khandhar, and K. M. Krishnan, presented at the Magnetic Particle Imaging (IWMPI), 2015 5th International Workshop on, 2015

MPS study on new MPI tracer material

Christina Debbeler [a,*], Cathrine Frandsen [b], Nicole Gehrke [c], Cordula Grüttner [d], David Heinke [c], Christer Johansson [e], Anja Johl [d], María del Puerto Morales [f], Miriam Varón [b], Kerstin Lüdtke-Buzug[a,*]

[a] Institute of Medical Engineering, Universität zu Lübeck, Germany
[b] Department of Physics, Technical University of Denmark, Kongens Lyngby, Denmark
[c] nanoPET Pharma GmbH, Berlin, Germany
[d] Micromod Partikeltechnologie GmbH, Rostock, Germany
[e] Acreo Swedish ICT AB, Göteborg, Sweden
[f] Instituto de Ciencia de Materiales de Madrid, Spain
* Corresponding author, email: {debbeler,luedtke-buzug}@imt.uni-luebeck.de

INTRODUCTION Since the introduction of the new imaging modality Magnetic Particle Imaging (MPI) in 2005 [1], the construction of scanners as well as the improvement of scanner topologies and simulation and reconstruction approaches has been a rapid process of worldwide interest. With a rising number of working MPI systems and Resovist® (Bayer Schering Pharma, Berlin, Germany) as a gold standard tracer that is no longer commercially available, the research effort on MPI suited tracer material has increased as well. The most common way to synthesize tracers (superparamagnetic iron oxide nanoparticles, SPION) for biomedical applications is coprecipitation, although numerous other approaches are becoming more prevalent as the synthesis of SPIONs with uniform core diameters becomes more desirable. Dextran is often used to provide colloid stability and biocompatibility, but there are also numerous experiments using for example dimercaptosuccinic acid or silica as coating material. The produced magnetite nanoparticles should be of monodisperse size and roughly spherical shape [2]. The size of the particles is a tradeoff between the steepness of the magnetization curve, the anisotropy and the relaxation of the particles [3]. Within the scope of the EU financed NanoMag project [4], which objectives are the standardization, improve-ment and redefinition of manufacturing technologies as well as analyzing methods, newly synthesized particles were investigated regarding their suitability as MPI tracers using Magnetic Particle Spectroscopy (MPS) [5].

MATERIAL AND METHODS The investigated particles were synthesized within the NanoMag project and show varying characteristics. Synthesis methods vary from coprecipitation and oxidative precipitation to polyol approaches using different core materials like maghemite or maghemite/magnetite mixtures. For colloidal stabilization, various coating materials as dextran, carboxydextran, polyacrylic acid, polyacrylic acid sodium salt or citrate were used. The iron concentration of the undiluted samples ranges from 1.0 mg/ml to 7.3 mg/ml. Additionally, the core diameter of the particles ranges from roughly 5 nm to 33 nm, while the hydrodynamic diameter ranges from about 39 nm to 176 nm.

For measurements, the MPS setup of Biederer et al. [5] was used. For each sample measurement, 10 µl of the undiluted particle suspensions were prepared in a micro test tube. Measurements took place with an excitation frequency of 25 kHz, a field strength of 20 mT and an acquisition time of 5 s.

RESULTS The results of the MPS measurements are shown in figure 1. Due to varying iron concentrations of the samples, the spectral magnetic moments of all measurements were normalized to the iron concentration. Especially samples FeraSpin™ L, NPG3310, MM-08-01 and DTU-01 show a slightly better performance than Resovist®. The decay of the spectral magnetic

moment is less steep than for the other particles. This results in a higher number of detected harmonics, subsequently improving the resolution of the MPI system.

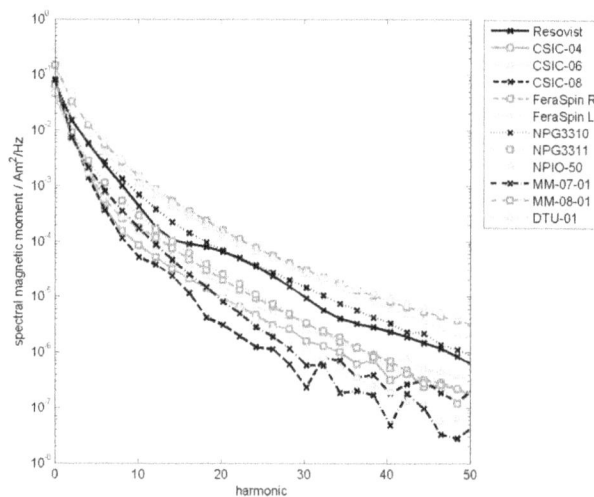

Figure 1: Amplitude spectra of different particles obtained from MPS measurements. Shown are only the odd harmonics, which were normalized to the iron concentration.

CONCLUSION First results show that particles with varying characteristic parameters show an acceptable performance in MPS measurements. The detected magnetization signal is comparable to or even better than the performance of Resovist®. Since MPS can only give a first impression of the performance of the particles as an MPI tracer, the next step should be the performance of measurements in existing MPI scanner setups to determine the suitability of the particles as tracer for MPI.

ACKNOWLEDGEMENTS The research leading to these results has received funding from the European Union Seventh Framework Programme (FP7/2007-2013) under grant agreement no 604448 (NanoMag).

REFERENCES
[1] B. Gleich and J. Weizenecker. *Nature*, 435(7046):1217—1217, 2005. doi: 10.1038/nature03808.
[2] K. Lüdtke-Buzug et al.. Springer IFMBE Proceedings, 22, 2343—2346, 2008. doi:10.1007/978-3-540-89208-3_562.
[3] B. D. Cullity and C. D. Graham. John Wiley & Sons, Hoboken USA, 2008. doi: 10.1002/9780470386323.
[4] F. Ludwig et al. *IEEE Transaction on Magnetics*, 50(11):5300204, 2014. doi: 10.1109/TMAG.2014.2321456.
[5] S. Biederer et al. *J. Phys. D: Appl. Phys.*, 42(20): 205007, 2009. doi: 10.1088/0022-3727/42/20/205007.

Imaging Characterization of MPI Tracers Employing Offset Measurements in a two Dimensional Magnetic Particle Spectrometer

Daniel Schmidt [a,*], Matthias Graeser [b], Anselm von Gladiss [b], Thorsten M. Buzug [b], Uwe Steinhoff [a]

[a] Physikalisch-Technische Bundesanstalt, Berlin, Germany
[b] Institute of Medical Engineering, Universität zu Lübeck, Germany
[*] Daniel Schmidt, email: daniel.schmidt@ptb.de

INTRODUCTION Magnetic Particle Spectroscopy (MPS) is a fast and straightforward method for the characterization of potential MPI tracers. High measurement spectra (normalized to the iron content) are interpreted as potentially suitable tracers. As these spectra lack the imaging character of MPI and phantom experiments in a scanner are very time consuming, we proposed a method employing an MPS equipped with an additional magnetic offset field [1], based on the hybrid system function approach [2, 3]. With this setup, we sequentially measured the 1D system function from which we generated software phantoms, thereby obtaining the line resolution of different tracers in dependence on the noise level. In comparison with a real MPI scanner, this setup still lacked the mixed frequencies that arise from the simultaneous excitation with two or three different frequencies as it is the case in 2D or 3D MPI. Here, we present current results of the commercially available MPI tracer FeraSpin™ R, characterized with our method employing a 2D MPS including the mixed frequencies.

MATERIAL AND METHODS We discretized the continuous offset field induced by the gradient coils by increments of $\Delta B = 0.25$ mT (Fig. 1). We then measured the spectrum of each offset combination between 0 mT and 12 mT and mirrored the system function according to [4].

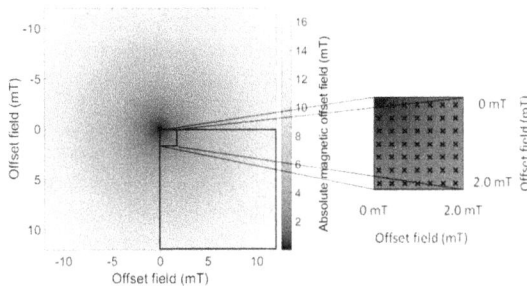

Figure 1: Division of the spatial dependent offset in 0.25 mT increments. Big Frame: Measurement area; Small frame: Magnification of the measurement grid with each cross representing the offset amplitudes of one measurement.

We used these measurements to generate MPI signals of cubic and sinusoidal software phantoms using

$$s_{\text{MPI}} = \sum_{n=1}^{N} m_n \frac{S_n}{A} + W,$$

with s_{MPI} being the artificial MPI signal, m_n being the defined iron mass at location n, S_n/A being the normalized measurement spectrum and W being additional noise. In the reconstruction process, we raised the noise level until we reached the resolution limit of our phantom. At this point we predicted the expected resolution via the spatial frequencies we derived from the harmonics and compared these predictions to the object distance in the phantom.

RESULTS Tab. 1 depicts the predicted resolution for cubic and sinusoidal phantoms in comparison to the actual distance after the noise was raised until the defined objects were barely distinguishable. Note that we defined the resolution as the distance between the centers of the sinusoidal or cubic objects.

Table 1: Predicted resolution in comparison to actual distance between phantom centers

Center Distance	Prediction (Cubic Phantom)	Prediction (Sinusoidal Phantom)
4.8 mm	5.5 mm	4.3 mm
6.4 mm	7.68 mm	6.4 mm
8.0 mm	7.68 mm	7.68 mm

The predictions and actual resolution at the respective noise levels match, especially for sinusoidal phantoms, corresponding to the sinusoidal shape of the spatial frequencies.

CONCLUSION We showed that our approach of offset field supported MPS characterization is also applicable for the 2D case, utilizing a 2D MPS with two excitation directions. This enabled us to perform a 2D imaging characterization in which we could show that the resolution can be predicted by analyzing the spatial harmonics of the MPI signal above noise level. Both this insight and the characterization method is a valuable link between pure spectroscopic characterization and time consuming phantom experiments.

ACKNOWLEDGEMENTS This work was supported by the EU FP7 research program "Nanomag" FP7-NMP-2013-LARGE-7 and the Federal Ministry of Education and Research, Germany (BMBF) under Grant number 13GW0069A. We also thank nanoPET GmbH for providing the MPI tracer FeraSpin™ R.

REFERENCES

[1] D. Schmidt, F. Palmetshofer, D. Heinke, O. Posth and U. Steinhoff. *International Workshop on Magnetic Particle Imaging (IWMPI)*, 2015. doi: 10.1109/iwmpi.2015.7107006

[2] M. Graeser, M. Grüttner, S. Biederer, H. Wojtczyk, W. Tenner, T. Sattel, B. Gleich, J. Borgert, T. Knopp and T. M. Buzug. *44. Jahrestagung der Deutschen Gesellschaft für Biomedizinische Technik im VDE*, 56, 2011.

[3] M. Grüttner, T. Knopp, J. Franke, M. Heidenreich, J. Rahmer, A. Halkola, C. Kaethner, J. Borgert and T. M. Buzug. *Biomedizinische Technik/Biomedical Engineering*, 58(6), 2015. doi: 10.1515/bmt-2012-0063

[4] A. Weber and T. Knopp. *Physics in Medicine and Biology*, 60(10): 4033, 2015. doi: 10.1088/0031-9155/60/10/4033

The Particle Response of Blended Nanoparticles in MPI

Anselm von Gladiss[a,*], Matthias Graeser[a], R. Matthew Ferguson[b], Amit P. Khandhar[b], Scott J. Kemp[b], Kannan M. Krishnan[b,c], Thorsten M. Buzug[a]

[a] Institute of Medical Engineering, University of Luebeck, [b] LodeSpin Labs, USA, [c] Department of Materials Science and Engineering, University of Washington
[*] Corresponding author, email: {gladiss, buzug}@imt.uni-luebeck.de

INTRODUCTION Superparamagnetic iron oxide nanoparticles (SPIONs) are used in Magnetic Particle Imaging (MPI) as tracer material. Gaining knowledge about their behavior is crucial for image reconstruction. In MPI, the spatial particle distribution is often reconstructed using a system matrix. The system matrix encodes the particle signal over every spatial point of the field of view. The acquisition is time-demanding and should happen for every type of SPIONs that is used as the magnetization signature of SPIONs differ. Therefore, it is of great interest to know if particle responses simply add up when blending SPIONs, as then not every blend has to be examined itself.

The response of a single type of SPIONs has been analysed [1]. Recently, blended particle samples are being applied for e.g. simultaneous imaging and magnetic heating (hyperthermia) [2] and color MPI [3] as well. In this contribution, a blended particle sample will be the object of interest. Experiments will be taken out that shall show, if a blended particle sample has the same particle response as the added particle responses of the single components of the blend.

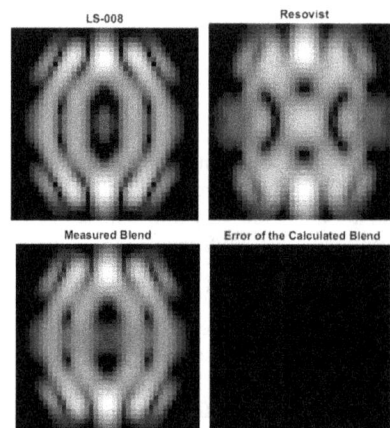

Figure 2: System functions of LS-008, Resovist®, a measured blend of the two and the error of the calculated blend.

RESULTS Fig. 1 shows the relative error in amplitude and the absolute error in phase between a simulated blend created by summing the two measurements and the actual blend for one of the two receive channels. The error in amplitude decreases with increasing frequency and is about 2 % in maximum. The phase error rises with increasing frequency. Most probably, the phase shifts because of thermal effects in the hardware setup.

The amplitude of system functions is shown in Fig. 2 for different samples. The influence of both the LS-008 and Resovist® measurements can be detected when measuring the blend of the two. The error image has been scaled to match the value range of the blended measurement. The error in amplitude is not visible.

CONCLUSION This work experimentally shows that adding up particle responses of single SPION samples is the same as measuring an equivalent blend, provided the blended SPIONs do not interact with each other. Additionally to sensing system matrices compressively [6] and exploiting symmetries of system matrices [5], calculating blends instead of measuring them is another way of reducing the amount of system matrix measurements globally.

ACKNOWLEDGEMENTS The German Research Foundation (DFG) and the Federal Ministry of Education and Research (BMBF) support this project (Grant Numbers BU 1436/10-1 and 13GW0069A). Work at LodeSpin was supported by NIH/NIBIB R42EB013520.

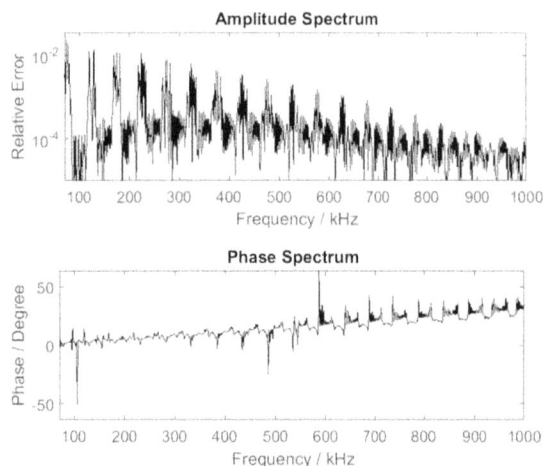

Figure 1: Error in the amplitude and phase spectra from 70 kHz up to 1 MHz. The relative error between the measured and the calculated particle blend is shown in the amplitude. The phase spectrum shows the absolute error.

MATERIAL AND METHODS Both 5 μl of undiluted LS-008 and an 11.5 % Resovist® dilution have been exposed to a two-dimensional sinusoidal excitation field in an Magnetic Particle Spectrometer (MPS) [4]. Due to the dilution of Resovist®, the iron concentration of both samples match approximately. After measuring the two samples, they have been blended and the blend has been measured in the same way.

In a second measurement series, various offset fields have been applied to the excitation field in order to perform a system matrix acquisition. One quadrant of the system matrix has been measured. After that, it has been mirrored as described in [5].

REFERENCES
[1] S. Biederer et al. J. Phys. D: Appl. Phys., 42(20), 2009.
[2] K. Murase et al. OJMI 5(2), 2015. doi: 10.4236/ojmi.2015.52009
[3] J. Rahmer et al. Phys. Med. Biol. 60(5), 2015.
[4] M. Graeser et al. IWMPI 2015. doi: 10.1109/IWMPI.2015.7107078
[5] A. Weber and T. Knopp. PMB, 60(10), 2015.
[6] A. von Gladiss et al. IEEE Trans. Magn. 51(2), 2015. doi: 10.1109/TMAG.2014.2326432

Determining magnetic impurities and nonspecific magnetic nanoparticle adhesion of MPI phantom materials

Patricia Radon*, Norbert Löwa, Felix Ptach, Dirk Gutkelch, Frank Wiekhorst

Physikalisch-Technische Bundesanstalt, Berlin 10587, Germany
*Corresponding author, email: patricia.radon@ptb.de

INTRODUCTION Magnetic Particle Imaging (MPI) is a promising imaging technique to visualize the spatial distribution of magnetic nanoparticles (MNP). At present various research MPI systems exist and a first commercial preclinical MPI scanner is available [1]. To evaluate the performance of a scanner or to facilitate comparison between scanners sophisticated phantoms have to be designed which enable the characterization of sensitivity, resolution, or geometric distortion of stationary as well as moving MNP [2]. The matrix of an MPI phantom hosting the MNP for MPI measurements should be made of inert materials (chemically resistant, impermeable for fluids), should contain no magnetic impurities, exhibit no interaction with MNP and on top be manufacturable with high precision. Additive manufacturing technology (commonly known as 3D printing) allows for the fast and cost-effective development of complex structures and thus is frequently used to fabricate phantoms. However, 3D printing has not yet been full established, partly because printing materials are not sufficiently characterized with respect to MNP interaction or magnetic properties (magnetic impurities). To provide substantiated data for the choice of appropriate MPI phantom materials we investigated different 3D printed test specimen used magnetic particle spectroscopy (MPS), also known as the zero-dimensional MPI. Employing this sensitive technique, we determined the spectral magnetic signal (caused by magnetic impurities) of four commercially available 3D printing materials and quantified the amount of MNP that nonspecifically bind to those materials.

MATERIAL AND METHODS We studied four commercially available polymer materials: E-Shell 200, E-Shell 600, R05 Gray and R05 (envisionTEC GmbH, GER). Of these materials cylinders of 5.3 mm diameter and a 5.3 mm height were 3D printed. The spectral magnetic response of the samples was measured at magnetic excitation fields of 2 mT, 10 mT and 25 mT at 25 kHz using a MPS device (Bruker, Germany). To study the nonspecific binding of MNP a 3D printed container with inner surface of 25 mm^2 was filled with MNP (20 mmol/L). After 5 min the MNP were removed and the container was intensively cleaned with ultra pure water. The measured MPS signal (B_{exc}=25 mT, f_{exc}=25 kHz) of the container was compared with a reference of known MNP content. We used four different commercial MNP types Resovist® (Schering AG, GER), FluidMag (Chemicell, GER), Endorem®, and Feraheme® (AMAG Pharmaceuticals, USA) coated with carboxy dextran, starch, dextran, and polyglucose sorbitol carboxymethylether, respectively. Additionally, we determined the hydrodynamic size distribution and zeta potential (Zetasizer, Malvern, GBR) of the samples.

RESULTS The MPS signals measured at different excitation fields varied significantly for the four different materials. As can be seen in Fig. 1a, the third harmonic amplitude μ_3 normalized to the sample weight of E-Shell 200 was about three times higher than for E-Shell 600 (at B_{exc}=25 mT). For lower excitation fields the differences of the MPS signals between the materials decreased significantly. At B_{exc}=2 mT the MPS signals of all materials were below the spectrometers noise level, so no difference could be detected.

The investigation of nonspecific adhesion of MNP to the 3D printing materials showed that Fluidmag-D (d_{hyd}=50 nm) strongly binds to the tested materials (see Fig. 1b). Compared to the other MNP types the nonspecific binding of Fluidmag-D of about 3.2 ng/mm^2 was determined about eight times higher than for Resovist®. We attribute this to the low zeta potential of this sample (-1.4 mV) compared to all others (between -30 mV and -20 mV).

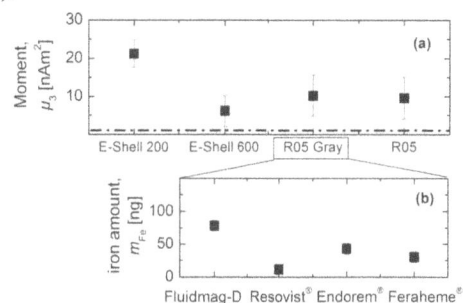

Figure 1: **(a)** MPS measurements on four materials for 3D printing at B_{exc}=25 mT (f_{exc}=25 kHz). **(b)** Nonspecific adhesion of MNP to 3D printing material R05 Gray. Quantification was performed by means of MPS at B_{exc}=25 mT (f_{exc}=25 kHz).

CONCLUSION The characterization of materials to be used to build MPI phantoms is very important as magnetic impurities may vary considerably. Moreover, it is necessary to consider potential nonspecific binding of MNP to a phantom material surface. As this depends on MNP type and phantom material the respective combination should be examined prior to an individual phantom design. Furthermore, MPS has proven to be ideally suitable to assist the development of MPI phantoms as it allows for rapid and sensitive magnetic material testing using small aliquot samples.

ACKNOWLEDGEMENTS The research was supported by German Research Foundation, through DFG Research Unit FOR917 (Nanoguide) and by the European Commission's Framework Programme-7 (project NanoMag, grant number: 604448).

REFERENCES
[1] M.G. Kaul, H. Ittrich, O. Weber, U. Heinen, A. Reitmeier, T. Mummert, C. Jung, N. Raabe, T. Knopp, and G. Adam. *RoFo: Fortschritte auf dem Gebiete der Röntgenstrahlen und der Nuklearmedizin*, 187(5): 347—352, 2015. doi: 10.1055/s-0034-1399344.
[2] U. Heinen, J. Franke, N. Baxan, K. Strobel, H. Lehr, A. Weber, W. Ruhm, A.P. Khandhar, R.M. Ferguson, S. Kemp, K.M. Krishnan, and M. Heidenreich. *Magnetic Particle Imaging (IWMPI), 2015 5th International Workshop on*, 2015. doi: 10.1109/IWMPI.2015.7107033.

Session 2:

Application 1

Keynote:
Potential Clinical Applications of MPI

Dr. med. Harald Ittrich
University Medical Center Hamburg-Eppendorf
(UKE), Germany

Dr. med. Johannes Salamon
University Medical Center Hamburg-Eppendorf
(UKE), Germany

ABSTRACT A new imaging modality using SPIO is magnetic particle imaging (MPI) [1]. This new radiation-free tomographic imaging method provides background-free information about the spatial distribution of SPIO with high temporal and spatial resolution [2]. Feasibility in living organisms was able to be shown in initial preclinical studies [3]. Moreover, with optimization of the SPIO and the equipment hardware, this technology has potential to image nano- and picomolar SPIO concentrations [4], making the application therefore interesting in preclinical and potentially in clinical imaging. Potential MPI application areas include cardiovascular applications (angiographies, cardiac vitality diagnosis, tissue perfusion, endovascular interventions, bleeding source diagnosis) and with optimized SPIO tracers and MPI hardware applications in tumor, molecular, and cellular imaging (passive and active targeting, molecular therapies, cellular labeling and cell monitoring) [5-8].

[1]   Gleich B et al. Nature 2005; 435: 1214-1217
[2]   Borgert J et al. Journal of cardiovascular computed tomography 2012; 6: 149-153
[3]   Weizenecker J et al. Physics in medicine and biology 2009; 54: L1-L10
[4]   Weizenecker J et al. Physics in medicine and biology 2007; 52: 6363-6374
[5]   Rahmer J et al. Physics in medicine and biology 2013; 58: 3965-3977
[6]   Haegele J et al. Fortschr Röntgenstr 2012; 184: 420-426
[7]   Buzug TM. Biomedical engineering 2013; 58: 489-491
[8]   Borgert J et al. Biomedical engineering 2013; 58: 551-556

Color MPI for Cardiovascular Interventions

Julian Haegele[a,*], Sarah Vaalma[a], Nikolaos Panagiotopoulos[a], Jörg Barkhausen[a], Florian M. Vogt[a], Jörn Borgert[b], Jürgen Rahmer[b]

[a] Department of Radiology and Nuclear Medicine, University Hospital Schleswig Holstein, Lübeck, Germany
[b] Philips Technologie GmbH Innovative Technologies, Hamburg, Germany
* Corresponding author, email: julian.haegele@uksh.de

INTRODUCTION Magnetic Particle Imaging (MPI) is a promising future method for cardiovascular imaging and guidance of interventions [1,2]. Recently it was demonstrated that it is possible to modify commercially available devices for cardiovascular interventions, such as a guide wire and a diagnostic catheter, for visualization in MPI by coating them with a dedicated superparamagnetic iron oxide (SPIO) varnish [3]. However, for safe guidance of cardiovascular interventions the devices have to be delineable from each other and the contrasted lumen. This can be achieved by using different concentrations of the same tracer or different coating patterns of the instruments. As this method of differentiation is only based on the intensity of each device in the resulting image, differentiation can still be difficult, especially if the lumen is contrasted by a relatively high concentration of SPIOs.

Color MPI is a method where different colors are assigned to different signal sources to allow for visualization in a single image [4]. In principle, different signal sources can be different tracers or even different aggregation states of a tracer. This could be helpful for the differentiation of devices from each other on the one and from the contrasted vessel lumen on the other side in MPI-guided cardiovascular interventions.

The goal of this work was to use the color MPI approach to differentiate the contrasted lumen of a vessel phantom from a SPIO coated guide wire.

MATERIAL AND METHODS A vessel Phantom (latex rubber) with a diameter of 10 mm was filled with a solution of sodium chloride and Resovist (1:800, I'rom Pharmaceuticals, Tokio, Japan). This setup was placed in the bore of an experimental preclinical MPI-demonstrator (Philips Research, Hamburg). Then a commercially available guide wire (0.035"/0.89 mm diameter, Terumo Radifocus, Standard Type angled, Tokyo, Japan) was modified by a thin coating of a dedicated, newly designed Resovist varnish as described in [3]. This modified guide wire was introduced into the vessel phantom.

The field of view (FOV) was 30.8 x 35.2 x 19.2 mm³ and the voxel size was 1.1 x 1.1 x 0.8 mm³. Image acquisition was conducted with a sample rate of 21.5 ms, corresponding to 46 volumes, i.e. FOVs, per second. For image reconstruction two dedicated system functions were acquired: One system function was measured using a probe of liquid Resovist and one system function was measured with a small scaffold coated with the dried SPIO varnish as described in [3].

RESULTS Using the color-coded image reconstruction it was possible to clearly detect and delineate the very thin guide wire in the contrasted vessel lumen (Figure 1). Because of the high temporal resolution the movement of the guide wire in the vessel phantom could be visualized without motion artifacts. The spatial resolution was around 3 mm in x- and y-direction and 1.5 mm in z-direction. The reconstructed image of the vessel lumen was slightly distorted at the edges, leading to an irregular representation of the normally straight and round vessel phantom.

This was caused by an incomplete signal separation in the color MPI reconstruction process: part of the guide-wire signal was incorrectly assigned to the vessel and leads to artifacts in the vessel image, as both signals were based on Resovist. It has to be noted, that the guide wire was not as homogenously coated as it is desirable, as its design was a proof of principle concept [3].

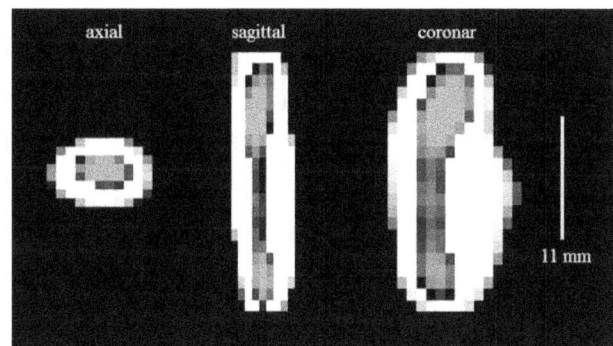

Figure 1: The tip of the catheter is clearly delineable in the lumen of the vessel phantom, notice the curved tip.

CONCLUSION Color MPI allowed a very good differentiation of a Resovist coated guide wire and the lumen of a vessel phantom contrasted with a Resovist solution. This was possible because of the different aggregation states of Resovist and the thus different magnetization responses in MPI. However, as the magnetization responses were not completely different, an absolute differentiation of the different aggregation states and thus the guide wire and the lumen was not possible. This lead to artifacts, i.e. irregular image representation of the vessel borders and the guide wire and thus diminished image quality. In the future, a better delineation could be achieved by using different tracers for the different devices on the one hand and the vessel lumen on the other hand. To differentiate multiple devices from each other, different coating patterns of the same tracers should be a valuable additional method.

ACKNOWLEDGEMENTS JB and JR acknowledge support by the German Federal Ministry of Education and Research (BMBF grants FKZ 13N9079 and 13N11086). The authors thank Dr. Jochen Franke from Bruker Biospin for the fabrication of the coated scaffold.

REFERENCES
[1] P. W. Goodwill, et al. Advanced Materials, vol. 24, pp. 3870-3877, 2012. doi: 10.1002/adma.201200221.
[2] N. Panagiotopoulos, et al. International journal of nanomedicine, vol. 10, pp. 3097-114, 2015. doi: 10.2147/IJN.S70488 ijn-10-3097.
[3] N. Panagiotopoulos, et al. 5th International Workshop on Magnetic Particle Imaging (IWMPI), 2015. doi: 10.1109/iwmpi.2015.7107008.
[4] J. Rahmer, et al. Physics in Medicine and Biology, vol. 60, pp. 1775-1791, 2015. doi: 10.1088/0031-9155/60/5/1775.

The next step towards interventional MPI: Real Time 3D MPI-guided treatment of a vessel stenosis using a blood pool agent and MRI Road Map approach

Johannes Salamon[1,*]; Martin Hofmann[2,3]; Caroline Jung[1]; Michael Gerhard Kaul, Rudolph Reimer[4]; Annika vom Scheidt[5]; Gerhard Adam[1]; Tobias Knopp[1,2,3]; Harald Ittrich[1]

[1]Department of Diagnostic and Interventional Radiology, University Medical Center Hamburg-Eppendorf, Hamburg, Germany
[2]Section for Biomedical Imaging, University Medical Center Hamburg-Eppendorf, Hamburg, Germany
[3]Institute for Biomedical Imaging, Hamburg University of Technology, Hamburg, Germany
[4]Microscopy and Image Analysis, Heinrich Pette Institute, Leibniz Institute for Experimental Virology, Hamburg, Germany
[5]Department of Osteology and Biomechanics, University Medical Center Hamburg Eppendorf, Hamburg, Germany
* Corresponding author, email: j.salamon@uke.de

INTRODUCTION

Magnetic particle imaging (MPI) is a fast, radiation free, sensitive, and quantifiable imaging modality using superparamagnetic iron oxide particles (SPIO) as tracer [1]. Potential MPI utilizations include cardiovascular applications such as angiographies and endovascular interventions [2].

As initial step towards cardiovascular interventions Heagele et. al demonstrated the feasibility of MP-imaging SPIO-labeled endovascular instruments [3] and introduced the first magnetic coating for instrument visualization in MPI with a thickness of about 500 µm [4]. No less important for MPI guided intervention is a real-time visualization for direct feedback [5] using fast reconstruction methods [6].

From a medical point of view one challenge of MPI is the lack of anatomical background information. For compensation a combined workflow for MR and MP imaging [7] or a hybrid MPI/MRI scanner [8] can be used. Both methods combine highly resolved anatomic MR images and dynamic MPI data at high temporal resolution using image fusion techniques.

In this study the feasibility of in-vitro 3D real-time guidewire and balloon-catheter tracking as well as stenosis treatment with a magnetic particle imaging MPI/MRI Road Map approach and an MPI guided approach using a blood pool contrast agent is shown.

MATERIAL AND METHODS

A standard guide wire and balloon-catheter were labeled with a thin layer of magnetic lacquer at the tip of the wire and on both sides of the balloon. A vessel phantom with a stenosis was either filled with saline or super paramagnetic iron oxide particles (MM4) and equipped with fiducial markers for co-registration in preclinical 7T MRI and MPI. In-vitro stenosis treatment was performed inflating the balloon either with MM4 or with saline. MPI data were acquired using a field of view of 37.3x37.3x18.6 mm^3 at a rate of 46 frames/s and real-time reconstructed at a rate of 2 frames/s. For analysis of the magnetic lacquer marks electron microscopy, atomic absorption spectrometry and micro-computed tomography were performed.

RESULTS

The magnetic lacquer consisted of thin (10-20 nm) iron plates with a size of 0.5 to 90µm (10.6 mg Fe/ml). Micro-CT showed a maximum thickness of the lacquer on the guide wire of 100 µm and 120µm on the balloon catheter allowing for coaxial use. In both approaches, the progress of angioplasty was monitored and guided in real-time by MPI. Successful angioplasty was verified by MPI and MRI. Magnetic makers allowed for guidance of interventional devices in the Road Map approach. Fiducial markers enable MRI/MPI image fusion for anatomical orientation.

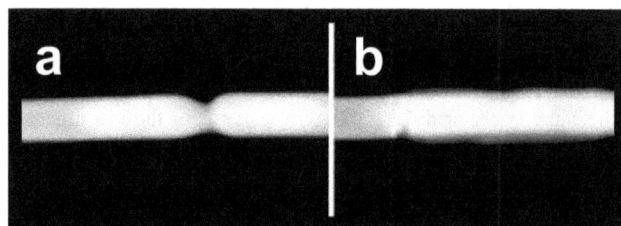

Figure 1: **Road Map Approach,** MRI (grayscale)/MPI (white) fusion image of the vessel phantom with a stenosis before (**a**) and after balloon treatment of the stenosis (**b**).

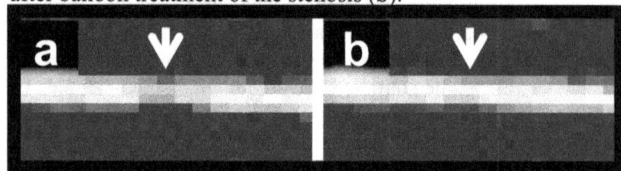

Figure 2: **Blood Pool Agent Approach,** MPI of the vessel phantom with stenosis before (**a**) and after balloon treatment of the stenosis (**b**), white arrows indicating the region of the stenosis.

CONCLUSION

3D real-time tracking of endovascular instruments, MPI guided instrument positioning and PTA is feasible. A combination of MPI/MRI for an anatomical Road Map in addition to interventions with a blood pool agent might emerge as a promising tool for radiation free intervention.

ACKNOWLEDGEMENTS

We thankfully acknowledge funding and support by the German Research Foundation (DFG, grant number AD 125 / 5-1) and the city of Hamburg.

REFERENCES

[1] Gleich B and Weizenecker J. Nature. 2005;435(7046):1214-7. doi: nature03808 [pii]/10.1038/nature03808.
[2] Borgert J et. al. J Cardiovasc Comput Tomogr. 2012;6(3):149-53. doi: 10.1016/j.jcct.2012.04.007.
[3] Haegele J et. al. Radiology. 2012;265(3):933-8. doi: 10.1148/radiol.12120424.
[4] Haegele J et. al. Magn Reson Med. 2013;69(6):1761-7. doi: 10.1002/mrm.24421.
[5] Pablico-Lansigan MH et. al. Nanoscale. 2013;5(10):4040-55. doi: 10.1039/c3nr00544e.
[6] Knopp T et. al. Phys Med Biol. 2010;55(6):1577-89. doi: 10.1088/0031-9155/55/6/003.
[7] Kaul MG et. al. Rofo. 2015;187(5):347-52. doi: 10.1055/s-0034-1399344.
[8] P. Vogel et. al., IEEE Trans Med Imag.,33(10):1954–1959, 2014.

Quantification of Vascular Stenosis Phantoms using Traveling Wave MPI

S. Herz [a,*], P. Vogel [b], V.C. Behr [b], T.A. Bley [a]

[a] Department of Diagnostics and Interventional Radiology, Würzburg University Hospital, Würzburg
[b] Department of Experimental Physics 5 (Biophysics), University of Würzburg, Würzburg
* Corresponding author, email: Herz_S@ukw.de

INTRODUCTION The quantification of vascular stenoses is of high clinical importance for diagnosis and therapy of many cardiovascular diseases such as myocardial infarction and peripheral arterial disease. Magnetic Particle Imaging (MPI) is a promising new tomographic imaging method with a high potential for cardiovascular imaging due to its very high temporal resolution, high sensitivity and a good spatial resolution [1]. MPI can detect the spatial distribution of superparamagnetic iron-oxide nanoparticles (SPIOs) and the signal strength depends directly on the amount of SPIOs, which inherently enables MPI as a quantitative measurement method. The aim of this study was to investigate the ability of traveling wave MPI to quantify stenoses in a phantom model by measuring the signal intensity as well as direct quantification.

MATERIAL AND METHODS The traveling wave MPI (TWMPI) scanner is a MPI scanner which uses a dynamic linear gradient array for generating a moving sinusoidal magnetic field to generate two field free points (FFPs). With two perpendicular saddle-coil pairs the FFPs can be deflected arbitrarily from the symmetry axis to scan a 3D volume. Datasets with a resolution of about 2 mm can be acquired using the TWMPI system (gradient strength 2.7 T/m) in the slice-scanning mode [1][2].
The MPI signal was used for image reconstruction and signal intensity measurements. Custom-made plastic phantoms with a length of 40 mm and an inner diameter of 8 mm served as stenosis models with various degrees of stenosis (0%, 25%, 50%, 75% and 100%). Each stenosis phantom was measured 10 fold. Stenosis phantoms were filled with diluted superparamagnetic iron-oxide nanoparticles (Resovist®, Bayer Pharma AG, Germany; 5μmol (Fe)/ml). In a stenosis the diameter has a quadratic relationship with the signal intensity obtained from the intraluminal SPIOs. Therefore the signal variation profile alongside the stenosis phantoms was used to estimate the degree of stenosis with the following formalism (adapted from NASCET criteria for internal carotid artery stenosis grades) [3]:

$$stenosis = \frac{a-b}{b} \cdot 100\% = \frac{\sqrt{s_a} - \sqrt{s_b}}{\sqrt{s_a}} \cdot 100\%$$

RESULTS With TWMPI we were able to quantify all stenosis grades accurately with a high temporal resolution of 100 ms per image. The 25% stenosis phantom revealed a 32,2%±0,9 stenosis, the 50% stenosis phantom revealed a 49,4%±1,8 stenosis, and the higher grade 75% stenosis phantom revealed a 68,1%±7,7 stenosis. In the 0% stenosis phantom a single maximum signal intensity plateau was obtained whereas in the 100% stenosis the signal intensity minimum dropped to the noise level. The images obtained from the different stenosis phantoms could be qualitatively distinguished, however direct quantification of the stenosis grades was hampered due to geometric distortions in the reconstructed images.

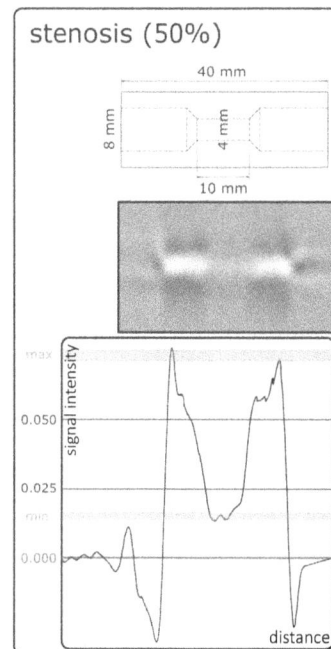

Figure 1: Diagram and TWMPI image of a 50% stenosis phantom filled with diluted SPIOs (Resovist®). The graph below shows the signal intensity profile obtained longitudinally along the reconstructed stenosis image. For stenosis grading the maximum and minimum signal intensity values (red lines) were used.

CONCLUSION With TWMPI accurate quantification of different stenosis grades was feasible using the SPIO signal intensity profile longitudinally along different stenosis in a phantom model. However, direct quantification of the stenosis grades was hampered due to geometric distortions in the reconstructed images. To introduce TWMPI as a competitive vascular imaging method substantial improvements are necessary, especially improvements in scanner hardware and reconstruction procedures.

ACKNOWLEDGEMENTS This work was partially funded by the DFG (BE-5293/1-1).

REFERENCES
[5] J. Borgert, et al., J Cardiovasc Comput Tomogr, 6:149-153, 2012. doi: 10.1016/J.JCCT.2012.04.007.
[6] P. Vogel, et al., *IEEE TMI*, 33(2):400-407, 2013. doi: 10.1109/TMI.2013.2285472.
[7] C. Arning, et al., *Ultraschall in Med*, 31:251-257, 2010.

Session 3:

Methodology 2

Resolution Improvement for X-Space MPI having Low Gradient Field

Hamed Jabbari Asl[a], Jungwon Yoon[a,*]

[a] Robots and Intelligent Systems Lab, School of Mechanical and Aerospace Engineering and ReCAPT, Gyeongsang National University, Republic of Korea
[*] Corresponding author, email: jwyoon@gnu.ac.kr

INTRODUCTION The intrinsic resolution and the image quality, in the x-space magnetic particle imaging (MPI), depend on the native point spread function (PSF) and the gradient field strength. The developed x-space MPI devices use a strong gradient field ($\geq 5\,\text{Tm}^{-1}$) to have high-resolution image [1]. However, this reduces the field of view (FOV), since the field free point can only travel a small area with a limited drive field. To cover the whole workspace, partial field of views are used, and the final image is generated by stitching these images. Although the reported results show satisfactory images, this process increases the imaging time, power consumption and requires an extra effort to connect the images.

On the other hand, the PSF is related to the derivative of the magnetization curve, which is similar to a Gaussian function. Therefore, the obtained image is anisotropic and includes haze due to the long tails of the PSF function. Deconvolution, may improve the x-space MPI image, but it also reduces the signal-to-noise ratio (SNR). Recently, several efforts have been done to overcome the mentioned problem. In [2], two orthogonal line-scan drive fields are combined to achieve the isotropic resolution in a 2D image. The presented results show improved images. However, the new images still have low contrast because of the tails of the 2D PSF. To reduce the image haze, a k-space equalization filter is implemented in [3], which seems to need a proper k-space transfer function to remove the haze with minimal loss of SNR.

Considering the above-mentioned problems, here we propose an image enhancement scheme for x-space imaging which highly reduce the image haze even when a low gradient field is applied.

MATERIAL AND METHODS The proposed approach exploits the property of the Gaussian-like function to reduce the tails of the PSF. The idea is that, if a Gaussian function is multiplied by itself, the new Gaussian function is narrower; i.e., has shorter tails. This property is shown for $\dot{L}(\xi)$ (the time derivative of Langevin function) in Figure 1. In this example, first the second power of $\dot{L}(\xi)$ is computed, then it is normalized to its peak value. This normalization is required to preserve the true contrast of the image. As the figure demonstrates, the new function has shorter tails compared to the original one. Therefore, by this method, the quality of 1D x-space image can be improved.

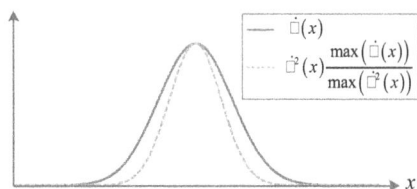

Figure 1: Property of the PSF.

One problem of the proposed approach is that, the normalization, to preserve the contrast of the image, is performed based on the maximum value of the signal. This may change the contrast of the other particles in the FOV of the drive field. This property is not acceptable when a quantitative imaging is of interest. To overcome this problem, we propose to normalize the signal locally instead of normalizing w.r.t. the global maximum. For example, as shown in Figure 2, the FOV is partitioned based on the position of the particles and local maxima of the signal are measured to locally normalize the signal.

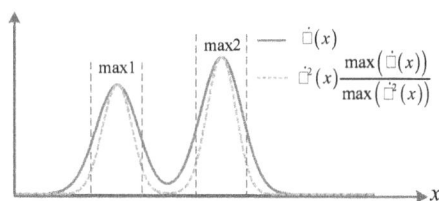

Figure 2: Local normalization of the x-space image.

RESULTS To evaluate the effectiveness of the approach for the multidimensional case, a simulation result for 2D imaging is presented in Figure 3. To perform this simulation, the MPI simulation toolbox, developed in [4], is utilized which is properly modified for the x-space reconstruction.

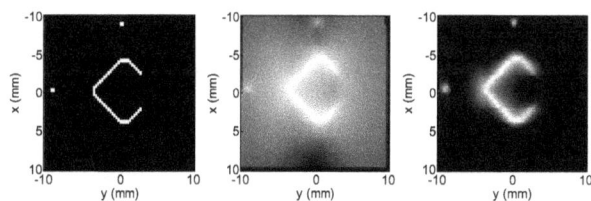

Figure 3: (left): imaging phantom, (middle): original x-space image, (right): x-space image powered by three and locally normalized.

CONCLUSION The proposed approach effectively decreases the haze of the image and the simulation results are comparable to the images obtained by applying a strong gradient field. Therefore, the imaging time and energy consumption of the whole process can be considerably decreased.

ACKNOWLEDGEMENTS This research was supported by the Pioneer Research Centre Program through the National Research Foundation of Korea funded by the Ministry of Education, Science and Technology (NRF 2012-0009524).

REFERENCES
[1] P. W. Goodwill and S. M. Conolly. *IEEE Transactions on Medical Imaging*, 30:1581—1590, 2011. doi: 10.1109/TMI.2011.2125982.
[2] K. Lu, P. Goodwill, and S. Conolly. *2014 International Workshop on Magnetic Particle Imaging (IWMPI)*, 2014.
[3] Kuan Lu and Goodwill, P. and Bo Zheng and Conolly, S. *2015 International Workshop on Magnetic Particle Imaging (IWMPI)*, 2015.
[4] G. Bringout. (2015) MPI Simulation Toolbox. [Online]. Available: https://github.com/gBringout/BasicsMPI.

X-space Deconvolution for Multidimensional Lissajous-based Data-Acquisition Schemes

Aileen Cordes[a,*], Christian Kaethner[a], Mandy Ahlborg[a], Thorsten M. Buzug[a]

[a] Institute of Medical Engineering, Universität zu Lübeck, Lübeck
[*] Corresponding author, email: {cordes,buzug}@imt.uni-luebeck.de

INTRODUCTION In MPI, the reconstruction of the spatial distribution of superparamagnetic iron oxide nanoparticles can be performed by a system matrix inversion. While this method shows promising results, it suffers from the fact that the system matrix acquisition and the image reconstruction can be very time-consuming. An alternative approach is referred to as x-space reconstruction [1,2]. Based on the Langevin theory of paramagnetism, the induced voltage signal can be formulated as a convolution of the nanoparticle distribution and a point spread function. This way, the image can be reconstructed by gridding the voltage signal to the known location of the field free point (FFP). Afterwards, a deconvolution can be applied to increase the spatial resolution. In [3], it was shown that this reconstruction technique leads to good results for MPI systems based on cartesian trajectories. However, the x-space approach is much more challenging for non-linear trajectories. In this contribution, we highlight the difficulties arising in the context of Lissajous trajectories and discuss possible solutions to these problems.

MATERIAL AND METHODS In order to illustrate the challenges arising in the context of Lissajous-based data acquisition schemes, the voltage signal induced by a delta sample located at the center of the field of view (FOV) has been simulated using the Langevin theory and ideal magnetic fields.

RESULTS Fig. 1(b) shows the simulated voltage signals corresponding to the sections of the Lissajous trajectory illustrated in Fig. 1(a). It can be seen that the signal shows a high peak with a small FWHM, if the FFP path directly crosses the position of the nanoparticles. In contrast, an offset of the magnetic field results in a flatter signal curve. Consequently, the signal measured at the intersection point of both FFP paths is not unique [4]. The induced voltage signal does not only depend on the distance between the FFP position and the sample, but also on the direction of the FFP velocity vector with respect to the nanoparticles. This means that no consistent native x-space exists for this data acquisition scheme. Due to this fact, it is not possible to reconstruct the nanoparticle distribution by simply gridding the voltage signal to the known location of the FFP. As presented in [5], only an approximate solution can be obtained by assuming two diagonal reference frames. However, this approach does not take into account that MPI systems that generate curved trajectories are not shift-invariant [6]. Due to the dependency of the measured signal on the orientation of the FFP path, the point spread function (PSF) is different for each spatio-temporal position. Therefore, an exact reconstruction of the nanoparticle distribution requires a time-dependent deconvolution kernel. In order to compute such deconvolution kernels, the induced voltage signal $s_n(t_m) =: s_{mn}$ of a delta sample has to be known for each discrete position $r_n, n = 1, \dots N$ of the sampled FOV and for each discrete time point $t_m, m = 1, \dots, M$. These system responses can be determined by measurements or simulations. Subsequently, the resulting system matrix $S = [s_{mn}]_{n=1,..,N, m=1,...M}$ has to be

inverted. This can for instance be done using the singular value decomposition. The column n of the inverse system matrix S^{-1} contains the deconvolution kernel $d_m(r_n)$ corresponding to the time point t_m. Based on these time-dependent deconvolution kernels the image $f(r_n)$ can finally be calculated as follows

$$f(r_n) = \sum_{m=1}^{M} u(t_m) d_m(r_n),$$

where $u(t_m)$ denotes the induced voltage signal of the scanned object at time point t_m. Obviously, this method suffers from the drawback that a system matrix inversion is required. However, it offers the big advantage that the data acquisition and the image reconstruction can be performed simultaneously. This way, this methods offers the potential to update the image continuously during the measurement in interventional applications.

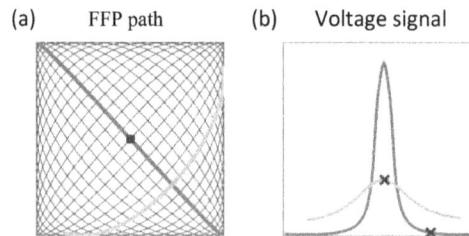

Figure 1: (a) Lissajous trajectory (b) Simulated voltage signals corresponding to the FFP paths marked in (a) for a delta sample placed in the origin of the FOV. The voltage signals at the intersection point of both FFP paths are marked with an x.

CONCLUSION Since the dependency of the induced voltage signal on the orientation of the FFP path results in a time-dependent PSF, the x-space reconstruction technique is not suitable for Lissajous-based data acquisition schemes. In order to solve this problem, a system matrix has to be determined. Based on the system matrix, time-dependent deconvolution kernels can be calculated, that allow for a simultaneous data acquisition and reconstruction.

REFERENCES
[1] P.W. Goodwill and S. M. Conolly. *IEEE Trans Med Imaging*, 30(9):1581-90,2011. doi: 10.1109/TMI.2011.2125982.
[2] P.W. Goodwill and S. M. Conolly. *IEEE Trans Med Imaging*, 29(11):1851-9, 2010. doi: 10.1109/TMI.2010.2052284
[3] P.W. Goodwill, K. Lu, B. Zheng and S. M. Conolly. *Rev. Sci. Intrum*, 83:033708,2012. doi: 10.1063/1.3694534.
[4] W. Erb, C. Kaethner, M. Ahlborg, T.M. Buzug. Numerische Mathematik, 1-21,2015. doi: 10.1007/s00211-015-0762-1
[5] P.W. Goodwill, Dissertation, University of California, Berkeley, 2010
[6] M. Grüttner, T. Knopp, J. Franke, M. Heidenreich, J. Rahmer, A. Halkola, C. Kaethner, J. Borgert, T. M. Buzug, *Biomed Tech*, 58(6):583-91. doi: 10.1515/bmt-2012-0063

Flexible reconstruction method for Traveling Wave MPI

T. Kampf [a,*], P. Vogel [a], M.A. Rückert [a], V.C. Behr [a]

[a] Department of Experimental Physics 5 (Biophysics), University of Würzburg, Würzburg
[*] Corresponding author, email: Thomas.Kampf@physik.uni-wuerzburg.de

INTRODUCTION Traveling wave MPI [1] reconstructs images in several steps [2] similar to x-space imaging [3]. Both are employing a gridding process and a 2D deconvolution. However, in TWMPI one major issue is geometry distortion of the reconstructed image caused by field inhomogeneities of the dynamic linear gradient array (dLGA).

In this abstract a flexible reconstruction based on an inverse method is presented, which can correct the geometry distortion.

MATERIAL AND METHODS In Fig. 1 (a) the original TWMPI reconstruction process is shown: starting with the signal from the digitizer the signal is corrected for distortions from the receive chain and filtered. Then it is gridded on a 2D plane according to the excitation frequencies prior to applying a 2D reconstruction using a Wiener deconvolution filter. Using morphing algorithms the geometry can be corrected prior to further post-processing.

Instead of performing a deconvolution and a geometry correction the new TWMPI reconstruction uses a system matrix after the gridding process to reconstruct the image (Fig. 1 (b)). The system matrix is also used for the reconstruction process in other MPI scanners (Fig. 1 (c)) [4, 5]. The major difference is the gridding process. While reconstruction proposed by Gleich and Weizenecker uses specific frequency ratios and sampling rates to avoid Fourier-bleeding before the peak-picking process to build the system matrix, the new TWMPI reconstruction uses an additional gridding process.

This additional step allows reconstructing datasets with different excitation frequencies and sampling rates with the same system matrix as long as the hardware parameters defining the field remain the same.

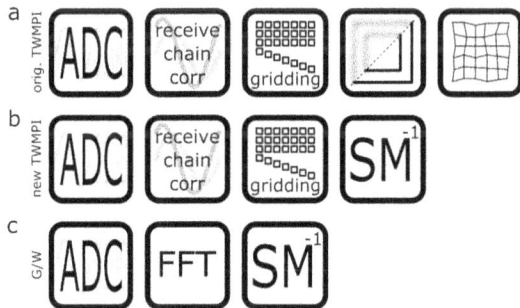

Figure 1: Schematic overview of the different reconstruction methods. **(a)** shows the old and **(b)** the new reconstruction for TWMPI while **(c)** sketches the Gleich/Weizenecker reconstruction process.

Furthermore, this flexibility allows implementing different system matrices, e.g. choosing image based data points or Fourier parameters (see Fig. 2). The system matrix has the dimensions $m \times n$, where n is the number of pixels of the image to be reconstructed and m is the number of gridded data used for reconstruction. The matrix is filled subsequently with the signal datasets from a point-like sample at each of the n positions of the image pixels. These datasets can be directly measured by scanning a point sample or can be calculated by a simulation.

The image-based approach also allows varying the size of the system matrix by changing the chosen cutout of the field of view. This has the advantage of decreasing the reconstruction time and offers the possibility of looking at single patches with higher resolution in the FOV, which avoids the issue of displaying samples with high dynamic ranges in one image.

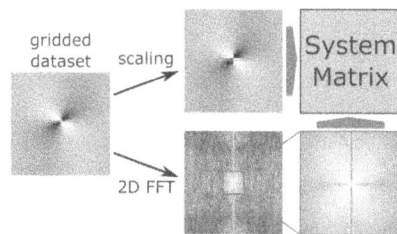

Figure 2: After the gridding process there are two different possibilities for picking the parameters for the system matrix: Either performing a 2D FFT and taking the inner area of the Fourier image or scaling the raw image down and using this information.

RESULTS A sample was scanned using a TWMPI system. Fig. 3 shows the results of the different reconstruction methods: the original TWMPI reconstruction cannot correct the geometry distortion. The reconstruction using the system matrix yields much better results with using either the image-based or the 2D FFT-based system matrix.

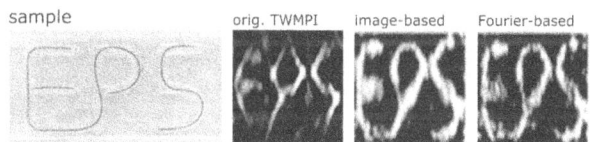

Figure 3: Comparison of the reconstruction methods: the original TWMPI, the reconstruction using the image-based and the Fourier-based system matrix.

CONCLUSION This abstract features a versatile reconstruction, which allows to correct for geometry distortions arising in TWMPI scanners. Furthermore, the input datasets are completely independent in terms of excitation frequencies and sampling rates because of the additional gridding process. This allows a simplification in the hardware of TWMPI scanners.

ACKNOWLEDGEMENTS This work was partially funded by the DFG (BE-5293/1-1).

REFERENCES

[1] P. Vogel, et al., *IEEE TMI*, 33(2):400-407, 2013. doi: 10.1109/TMI.2013.2285472.
[2] P. Vogel, et al., IEEE TMI, 33(10):1954-1959, 2014. doi: 10.1109/TMI:2014.2327515.
[3] P. Goodwill, et al., *IEEE TMI*, 29(11):1851-1859, 2010. doi: 10.1109/TMI.2010.2052284.
[4] B. Gleich and J. Weizenecker, *Nature*, 435:1214-1217, 2005. doi: 10.1038/NATURE03808.
[5] J. Rahmer, et al., *BMC Med Imag, 9:4*, 2009. doi: 10.1186/1471-2342-9-4.

Reconstruction of Experimental 2D MPI Data using a Hybrid System Matrix

Matthias Graeser[a,*], Anselm von Gladiss[a], Patryk Szwargulski[a,c], Mandy Ahlborg[a], Tobias Knopp[b,c], Thorsten M. Buzug[a]

[a]Institute of Medical Engineering, Universität zu Lübeck
[b] Section for Biomedical Imaging, University Medical Center Hamburg-Eppendorf
[c] Institute for Biomedical Imaging, Hamburg University of Technology
* Corresponding author, email: {graeser,buzug}@imt.uni-luebeck.de

INTRODUCTION In Magnetic Particle Imaging (MPI), a spatial particle distribution is visualized by x-space or frequency space reconstruction [1, 2]. One drawback of frequency space reconstruction is the long calibration process that has to be performed for each particle system that is used for imaging. This calibration process is time-demanding and occupies the scanner. Therefore, it is a serious cost factor for clinical use. To reduce the occupation time, different approaches like compressed sensing [3] or the usage of a system calibration unit can be applied [4]. A different approach is to use a magnetic particle spectrometer (MPS) to measure the particle response to a field sequence that equals the field sequence at a spatial position in the imaging device. Thus, the imaging device itself is not needed for the time-consuming calibration process. Prerequisites for this method are both a good description of the selection field and a known or estimated transfer function of the imaging device [5, 6]. In this work, a two dimensional MPS [7] is used to measure the particle response of Resovist® (Bayer-Schering Pharma AG) and reconstruct images of a multiple dots phantom recorded in a preclinical MPI scanner (Bruker/Philips).

MATERIAL AND METHODS The imaging and calibration processes have been performed with 2D excitation with excitation frequencies of $f_1 = 24.51$ kHz and $f_2 = 26.04$ kHz and an amplitude of 14 mT in both directions. The selection field has been set to 1.25 T/m in both directions. The system matrices (SM) have been acquired by a classical robot approach, referred to as robot based, and by a hybrid approach. To match the spatial discretization, the robot based SM has been interpolated onto the same grid as the hybrid SM. The imaged phantom consists of 5 dots with decreasing diameter from 5 mm to 1 mm and decreasing distance from 5 mm to 2 mm. The image data has been averaged 20,000 times for a good SNR. The system matrices have been averaged 1,500 times (robot based) and 200 times (hybrid approach). Due to limitations in SNR, the robot based system matrix has a grid of 22 x 22 spatial points while the hybrid grid has a size of 84 x 84 points. To match the measured data, the hybrid system function had to be corrected for the transfer function of the MPI scanner. This has been done by linear regression between the robot and the hybrid SM. The reconstruction of the phantom has been performed with two sets of reconstruction parameters using only frequency components with sufficient SNR in both system matrices. Parameter set 1 consisted of $\lambda = 0.1$, $\text{SNR}_{\min} = 1.5$, $f_{\min} = 60$ kHz, while parameter set 2 consisted of $\lambda = 0.01$, $\text{SNR}_{\min} = 1.5$, $f_{\min} = 80$ kHz.

RESULTS The resulting images show that the image can be reconstructed without loss of image quality using the hybrid approach. Some parts of the reconstructed image show a better separation of the dots in the hybrid approach while others result in a deformation of the dots depending on the reconstruction parameters.

CONCLUSION It has been shown, that the hybrid approach can be used for image reconstruction of multi-dimensional MPI data. As the hybrid SM has a much better SNR compared to the robot SM, less noise is added during the reconstruction process. Further improvements may be achieved for the hybrid approach by using additional frequency components depending on the SNR of the image data.

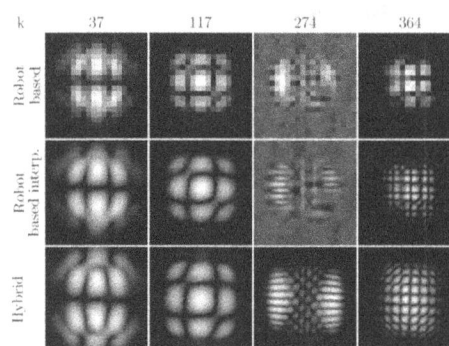

Figure 1: Comparison of system matrix components.

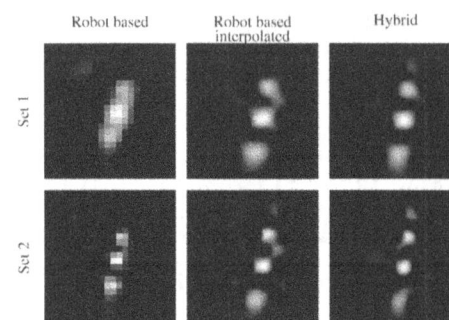

Figure 2: Reconstruction with two parameter sets.

ACKNOWLEDGEMENTS This work was supported by the Federal Ministry of Education and Research, Germany (BMBF) under Grant 13GW0069A.

REFERENCES
[1] J. Rahmer et al., BMC Medical Imaging, 2009, doi: 10.1186/1471-2342-9-4.
[2] P. W. Goodwill et al., IEEE Trans. Med. Imaging, 2012, doi: 10.1109/TMI.2012.218524.
[3] A. von Gladiss et al., IEEE Transactions on Magnetics, 2015, doi: 10.1109/TMAG.2014.2326432.
[4] A. Halkola et al.,IWMPI, 2013, doi: 10.1109/IWMPI.2013.6528344.
[5] M. Graeser et al., SPPY 2012, p. 59-64, Springer Berlin.
[6] M. Gruettner et al., NSS MIC, 2011, doi: 10.1109/NSSMIC.2011.6152687.
[7] M. Graeser et al., IWMPI 2015, doi: 10.1109/IWMPI.2015.7107078

130

Artefact Suppression in Time-resolved Magnetic Particle Imaging

Alexander Weber[a,b], Jochen Franke[a,c], Heinrich Lehr[a], Wolfgang Ruhm[a], Michael Heidenreich[a], Thorsten M. Buzug[b], Ulrich Heinen[a,*]

[a] Bruker BioSpin MRI GmbH, Rudolf-Plank-Straße 23, D-76275 Ettlingen, Germany
[b] Institut für Medizintechnik, Universität zu Lübeck, Ratzeburger Alle 160, D-23562 Lübeck, Germany
[c] Physics of Molecular Imaging Systems, University RWTH Aachen, Germany
[*] Corresponding author, email: ulrich.heinen@bruker.com

INTRODUCTION Magnetic Particle Imaging (MPI) is a new imaging modality that is capable of tracking magnetic nanoparticles (MNPs) *in vivo* with good spatial resolution and excellent temporal resolution [1,2]. However, as in any other imaging modality, MPI images are generated from data captured over a finite acquisition interval. Whenever the distribution of MNPs changes during this period, image artefacts have to be expected. In this contribution, we introduce a formal description of the temporal signal modulation by dynamic processes to establish mechanisms of artefact generation. We further propose a windowing procedure that substantially reduces dynamic artefacts. It is shown that this procedure allows generating a sequence of images which corresponds well to the true dynamic process.

THEORY In a time-dependent MPI experiment, the acquisition data corresponding to consecutive field cycles of the excitation system are reconstructed into an image sequence. In the system-matrix based reconstruction approach, a discrete Fourier transform (DFT) first converts the raw data into MPI spectra. Due to the periodic excitation, all harmonics generated by a static MNP distribution exactly correspond to multiples of the cycle rate of the scanner and are represented exactly by the frequency slots of the discretized spectrum, if the acquisition and excitation cycles match. This is no longer true if the local concentrations are modulated in time, by e.g. flow or cardiac movement. This effect corresponds to a convolution of the discrete harmonic spectrum by the Fourier transform of the modulation function. In the simple-most case of a sinusoidal signal modulation, each harmonic is split into side-bands separated from the main harmonic by the modulation frequency. These harmonics are no longer multiples of the cycle rate and their frequency position can no longer be exactly represented by the DFT of a single acquisition cycle. Therefore, they are subjected to considerable broadening by spectral leakage and cause harmonic cross-talk which manifests itself by reconstruction errors. At the same time, the modulation depth at the static harmonic frequency declines with increasing modulation frequency.

Extension of the observation period in combination with a windowing procedure [3] leads to a reduction of the leakage effect without affecting harmonic amplitudes corresponding to stationary signals. Using a specifically adapted flat-top window function [4] the full modulation depth at the stationary harmonic frequencies can be restored for practically relevant modulation speeds. Thus, a reconstruction with a system matrix containing spectra of a stationary reference sample is appropriate.

MATERIAL AND METHODS The harmonic cross-talk caused by dynamic processes was studied by a simulation of an asymmetric 8-point phantom, which was subjected to a uniform intensity modulation with 50% modulation depth and a modulation frequency corresponding to 10% of the cycle rate. The parameters of the 2D MPI simulation were Selection Field=1T/m in x- and y-direction, Drive Field $DF_x=DF_y=16mT$, frequencies $f_x=24.5kHz$, $f_y=25.3kHz$, sampling bandwidth 1.25MHz. The system matrix covers an area of $32\times32mm^2$ with a grid size of 32×32 pixels. To only assess the artefacts due to modulation, a noise-free simulation was used. The reconstruction was performed using Singular Value Decomposition with Tikhonov regularization ($\lambda=10^{-3}$). The windowing function was optimized for modulation frequencies up to 33% of the cycle rate.

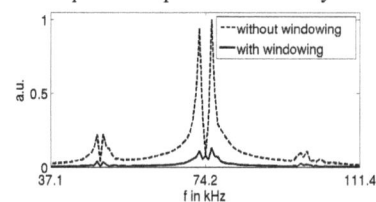

Figure 1: Modulation-induced spectral errors due to harmonic cross-talk on simulation data with and without windowing procedure. Such errors are not present for a static phantom.

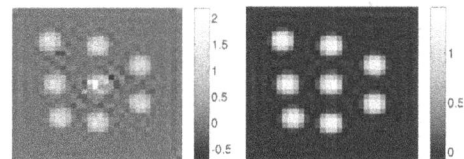

Figure 2: Reconstruction results of the modulated 8-point phantom without (left) and with (right) windowing procedure.

RESULTS Figure 1 shows the modulation-induced harmonic cross-talk spectrum of the simulation. With the help of the windowing procedure, this cross-talk effect is substantially reduced. In Figure 2, the reconstruction results of the modulated 8-point phantom are presented. It is evident that the windowing procedure substantially reduces the artefacts induced by the concentration modulation. This allows a further reduction of the regularization parameter, which corresponds to a higher achievable resolution.

CONCLUSION Dynamic processes cause quantifiable modulations of the MNP harmonics and introduce artefacts by harmonic cross-talk. The use of a suitable windowing procedure can reduce these artefacts and restores the image modulation depth for physiologically relevant processes.

REFERENCES
[1] B. Gleich, J. Weizenecker. *Nature*, 435(7046):1217—1217(2005)
[2] J. Weizenecker et al., *Phys. Med. Biol.*, 54, L1-L10(2009)
[3] F.J. Harris, *Proceedings of the IEEE*, 66(1), 51-83 (1978)
[4] L. Salvatore, A. Trotta, *IEE Proceedings B*, 135, 346-361 (1988)

Fused Lasso Regularization for Magnetic Particle Imaging

Martin Storath[a,*], Christina Brandt[b,*], Martin Hofmann[c], Tobias Knopp[c], Alexander Weber[d], Andreas Weinmann[e]

[a] Biomedical Imaging Group, École Polytechnique Fédérale de Lausanne, Switzerland
[b] Department of Applied Physics, University of Eastern Finland, Kuopio, Finland
[c] Section for Biomedical Imaging, University Medical Center Hamburg-Eppendorf, and Institute for Biomedical Imaging, Hamburg University of Technology, Germany
[d] Bruker Biospin, Ettlingen, Germany, and Institute of Medical Engineering, University of Lübeck, Germany
[e] Helmholtz Zentrum München, Germany
[*] Corresponding author, email: martin.storath@epfl.ch, christina.brandt@uef.fi

INTRODUCTION The image reconstruction in magnetic particle imaging (MPI) is an ill-posed inverse problem which, therefore, needs regularization. Current reconstruction methods for MPI are based on classical Tikhonov regularization. That is, the system equation $Ax = y$ is solved in a least squares sense while penalizing the squared L_2-norm of the resulting image x. Frequently, non-negativity constraints are imposed. The main advantage of Tikhonov regularization is its simplicity as it boils down to solving a linear system of equations. However, this comes with a drawback regarding the reconstruction quality. Loosely speaking, the squared L_2-penalty just prevents the solution from "blowing up" but it does not take into account any a priori knowledge on the structure of the underlying data, in particular no spatial neighborhood relations of the pixels. The shortcomings of this regularization model become very clear when viewed from a statistical perspective: Tikhonov regularization is the maximum a posteriori (MAP) estimator using a Gaussian noise model combined with the prior assumption that the pixels follow Gaussian distributions. While the Gaussian noise model is a reasonable assumption, the Gaussian prior is not well matched with the typical statistics of magnetic particle images.

MATERIAL AND METHODS We propose to regularize the MPI reconstruction by sparsity promoting priors which is motivated by the following observations: the particles often concentrate in a few locations within the field of view. In consequence, the number of non-zero pixels is often much smaller than the total number of pixels. Furthermore, we often encounter sparsity with respect to the edges of the imaged objects. As a particular example the reader may think of tracking a balloon catheter during a cardiovascular intervention; when filled with magnetic particles the density is homogeneous within the catheter and zero outside. A further reasonable prior is that the particle density is non-negative. In combination, we thus propose the regularization model

$$x^* = \arg\min_{x \geq 0} \alpha \, \mathrm{TV}(x) + \beta \|x\|_1 + \frac{1}{2} \|Ax - y\|_2^2$$

which is known as *fused lasso* in statistics [1]. It uses a total variation and an L_1-norm penalty. The priors promote sparsity with respect to the gradient and the value of the intensity, respectively. For its discretization, we propose a near-isotropic finite difference splitting scheme [2,3] which we adapted to the acquisition geometry. For finding a minimizer of the above variational problem, we derive a splitting method tailored to MPI. It is based on a generalized forward-backward splitting [4], thresholding [5], and the taut string algorithm [6,7].

RESULTS Figure 1 shows a reconstruction from simulated noisy data corrupted with 5 percent of white noise using the proposed method. For comparison, the classical Tikhonov-solution with non-negativity constraint is plotted. For both methods, the regularization parameters are tuned w.r.t. optimal PSNR.

Figure 1: Stenosis phantom (left), classical Tikhonov regularized reconstruction (middle, PSNR=18.5) and result obtained by the non-negative fused lasso regularization (right, PSNR=29.5).

CONCLUSION We have motivated that the non-negative fused lasso reconstruction is suited for typical magnetic particle images. We have proposed an efficient splitting algorithm scheme for the fused lasso which is particularly suited to MPI. The results illustrate that the proposed method improves the reconstruction results in the sense that noise is suppressed while edges are preserved. First results on real experimental data confirm the advantage of the algorithm and will be published in a forthcoming paper [8].

ACKNOWLEDGEMENTS This work was supported by the German Research Foundation (project ER777/1-1). M. Storath was supported by the European Research Council under the European Union's Seventh Framework Programme (FP7/2007-2013)/ERC grant agreement no. 267439. C. Brandt was supported by the Academy of Finland (project 286964). M. Hofmann and T. Knopp acknowledge funding and support of the German Research Foundation (DFG, grant no. AD125/5-1). A. Weber acknowledges the financial support by the German Federal Ministry of Education and Research (FKZ 13N11088). A. Weinmann was supported by the Helmholtz Association (VH-NG-526). The authors would like to thank L. Condat for making his implementation of the taut string algorithm publicly available.

REFERENCES
[1] R. Tibshirani, M. Saunders, S. Rosset, J. Zhu, and K. Knight. Sparsity and smoothness via the fused lasso. Journal of the Royal Statistical Society: Series B, 67 (1):91–108, 2005.
[2] M. Storath and A. Weinmann. Fast partitioning of vector-valued images. SIAM Journal on Imaging Sciences, 7(3):1826–1852, 2014.
[3] A. Chambolle. Finite-differences discretizations of the Mumford-Shah functional. ESAIM: Mathematical Modelling and Numerical Analysis, 33(02):261–288, 1999.
[4] H. Raguet, J. Fadili, and G. Peyré. A generalized forward-backward splitting. SIAM Journal on Imaging Sciences, 6(3): 1199–1226, 2013.
[5] J. Friedman, T. Hastie, H. Höfling, and R. Tibshirani. Pathwise coordinate optimization. Annals of Appl. Statistics, 1(2): 302–332, 2007.
[6] P. Davies and A. Kovac. Local extremes, runs, strings and multiresolution. Annals of Statistics, pages 1–48, 2001.
[7] L. Condat. A direct algorithm for 1-D total variation denoising. IEEE Signal Processing Letters, 20(11):1054–1057, 2013.
[8] M. Storath, C. Brandt, M. Hofmann, T. Knopp, A. Weber, A. Weinmann. Edge preserving image formation for 3D+time magnetic particle imaging. In preparation

Session 4:

Instrumentation 1

Keynote:

Safety Limits in MPI and Implications for Image Quality

Dr. Emine Ulku Saritas
Bilkent University, Ankara, Turkey

ABSTRACT Understanding the potential safety hazards of all the applied magnetic fields is crucial for fast imaging in Magnetic Particle Imaging (MPI), especially when translating MPI to clinics. The frequency and magnitude of the magnetic fields used in MPI are constrained by two safety limits: specific absorption rate (SAR) limit and magnetostimulation limit. Due to its frequency of operation (typically around 25 kHz and lately being extended to 150 kHz), the drive field in MPI is mainly constrained by magnetostimulation limits.

According to recent studies, the allowable drive field amplitude gets lower at higher frequencies, with longer magnetic pulses having considerably lower thresholds. Interestingly, recent developments in the field have also shown that the resolution in MPI improves when operating with lower drive field amplitudes, which would justify the migration to higher frequencies (e.g., 150 kHz). However, the safety limits of the slow shifting focus field, as well as the delayed response of the nanoparticles, also need to be taken into account when choosing the optimal parameters. This talk will examine the safety limits of the magnetic fields used in MPI, and their impact on image quality and the choice of scan parameters.

Signal path for a 10 kHz and 25 kHz mobility MPI System

Christian Kuhlmann[a,*], Sebastian Draack[a], Thilo Viereck[a], Frank Ludwig[a], Meinhard Schilling[a]

[a] Institut für Elektrische Messtechnik und Grundlagen der Elektrotechnik, TU Braunschweig, 38106 Braunschweig, Germany
* Corresponding author, email: c.kuhlmann@tu-bs.de

INTRODUCTION Magnetic particle imaging (MPI) has the capability of providing functional information either through the use of several particle types [1], [2] or by accessing the relaxation process of the nanoparticles [3]. One possibility to implement the latter method is to measure the sample at two frequencies, allowing for the separation of concentration and mobility information. In this work, we present the signal path of a dual frequency mobility MPI scanner.

MATERIAL AND METHODS
Transmit filters Only passive filters are capable of handling the power, dynamic range and linearity requirements for the drive field current. LC-ladder filters are appropriate for this role [4] and are used in our setup in a 3rd order band-pass configuration with a Bessel characteristic. The resulting asymmetric values for the filter elements are suited to incorporate the transmit coils into the filter topology, even though their inductance is several times larger. The slow transition into the stop band for this filter response can be offset by adding band-stop elements for the 3rd and 5th harmonic.
Receive coil A two-axis differential detection coil is used to mitigate the requirements on the detection filters.
For the x-direction (aligned with the bore) a differential solenoid coil is used to detect the particle signal, while feed-through from the drive field is rejected by compensation coils outside the field of view (FOV). For the y direction differential saddle coils are employed with the compensation coils shifted along the bore to reside outside the FOV.
Receive filters Fourth-order band-stop filters are used to suppress residual fundamental feed-through. Due to the differential detection coil setup, requirements on the filter elements considering linearity and power handling capability are greatly reduced. The filter impedance is chosen as 50 Ohms, resulting in optimal SNR for the chosen detection coil inductance. Additionally, this simplifies impedance matched cabling of the wideband detection path.
RESULTS The following results present the 25 kHz signal path of the system. Components for the 10 kHz drive field benefit from reduced parasitics and are therefore not shown.
Transmit filters The transmit filters for the x channel show high attenuation of 60 dB up to 300 kHz and maintain reasonable suppression (30 dB) up to 10 MHz. The y-axis filters maintain 50 dB attenuation up to 1 MHz (cf. Fig. 1). Attenuation in the pass-band is below 1.63 dB.

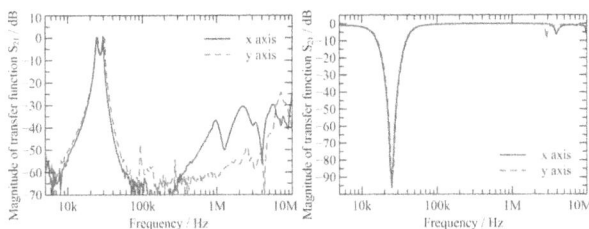

Figure 1: *left:* 25 kHz transmit filter response (including drive coil) *right:* 25 kHz detection filter response

Receive coil Figure 2 depicts a CAD image of the constructed receive coil. In the x-direction, a compensation of 32 dB is achieved, compensation in the y-axis is 20 dB.

Figure 2: Two-axis differential detection coil

Receive filters The constructed detection filters show an attenuation of 90 dB (cf. Fig. 1) at the fundamental and sharp transition into the pass-band (2nd harmonic at -3 dB) as well as low attenuation (0.05 dB) and a wide bandwidth (-3 dB at 3.6 MHz for x- and 2.75 MHz for the y-channel).
Measurements The system was successfully tested at 25 kHz and 10 kHz. Fig. 3 shows the first system matrices acquired with the scanner at 25 kHz.

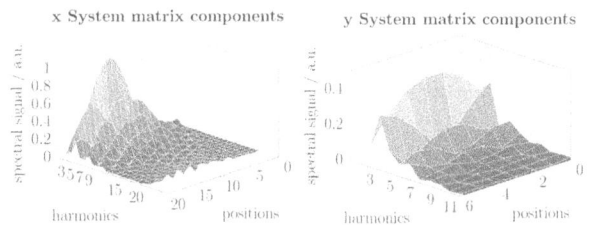

Figure 3: First acquired system matrices, using 1 T/m gradient; *left:* x-axis, using 22 mT$_{pk}$ drive field and a 2 mm Resovist sample along a 2 mm grid *right:* y-axis, using 12.5 mT$_{pk}$ drive field, along a 2 mm grid

CONCLUSION We have presented the signal path for a 2D mobility MPI system. Verification of individual components as well as first system tests have been successful.

ACKNOWLEDGEMENTS This work was financially supported by the DFG under grant no. LU800/5-1.

REFERENCES
[1] Rahmer, J. et al.. First experimental evidence of the feasibility of multi-color magnetic particle imaging. *Phys. Med. Biol.*, 60(5), pp.1775–91, 2015.
[2] Hensley, D. et al., Preliminary experimental X-space color MPI. In *Magnetic Particle Imaging, 2015 5th International Workshop on.*, 2015.
[3] Wawrzik, T. et al.. Debye-Based Frequency-Domain Magnetization Model for Magnetic Nanoparticles in Magnetic Particle Spectroscopy. *IEEE Transactions on Magnetics*, 51(2), pp.1–4, 2015.
[4] Casson, A. & Rodriguez-Villegas, E., A Review and Modern Approach to LC Ladder Synthesis. *JLPEA*, 1(3), pp.20–44, 2011

First Spectrum Measurements with a Rabbit-Sized FFL-Scanner

Jan Stelzner[a,*], Gael Bringout[a,*], Anselm von Gladiss[a], Hanne Medimagh[a], Mandy Ahlborg[a], Timo F. Sattel[b], Thorsten M. Buzug[a]

[a]Institute of Medical Engineering, University of Luebeck
[b]Philips Medical Systems DMC GmbH, Hamburg, Germany
([*] equally contributing authors), email: { stelzner, bringout, buzug}@imt.uni-luebeck.de

INTRODUCTION In Magnetic Particle Imaging (MPI), the signal quality depends on various aspects and elements within the whole signal chain. In 2015, the concept of a field-free-line (FFL) scanner with a bore diameter of 173 mm and the capability to accommodate objects in the size of rabbits was presented [1]. In this work, the results of the first commissioning of the drive-field (DF) signal chain are presented.

MATERIAL AND METHODS A scanner-device system including two pairs of coils for the drive-field excitation orthogonal to the bore axis as well as a set of surrounding coils for the spatial encoding has been set up. Additionally, the peripheral devices regarding the DF for the excitation in one direction have been implemented to complete the entire signal chain for one channel.

A sketch of the signal chain circuit is depicted in Fig. 1. It consists of transmission path (top row) and the receive path (bottom row). The transmission path contains a signal generator that provides a sinusoidal signal of 25 kHz, a power amplifier, a band-pass filter to attenuate the distortions caused by the power amplification and an impedance matching including a power factor correction. The impedance matching of the Drive-field coil (DFC) is implemented with a capacitive voltage divider (C_S and C_P). The receive path involves a decoupling circuit that consists of a serial oscillating circuit with C_{dec} and L_{dec}, which is resonant at 25 kHz. It decouples the receive chain from the fundamental frequency. The particle signal passes a band-stop filter to further attenuate the first harmonic and is amplified by a low noise amplifier (LNA). The amplitude spectrum of the output signal is either directly visualized by a signal analyzer or further processed on a PC.

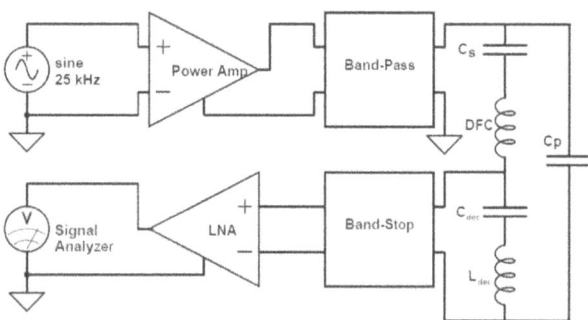

Figure 1: Simplified graph of the DF signal chain

Additionally, common mode chokes (CMC) and transformers for ground-loop suppression, which are not shown in Fig. 1, have been used. Furthermore, variometers [2] were used within the filters to adjust the resonant frequency.

RESULTS Measurements have been performed with the presented setup. The amplitude spectrum of an empty measurement has been recorded with 700 A_{PP} flowing through the DFC (Fig. 2). This results in a magnetic flux density of 30.5 mT_{PP} in the center of the bore. The fundamental and the fourth harmonic are tagged

by marker 1 und 4. Other harmonics are also clearly visible in the figure as well as some disturbances between.

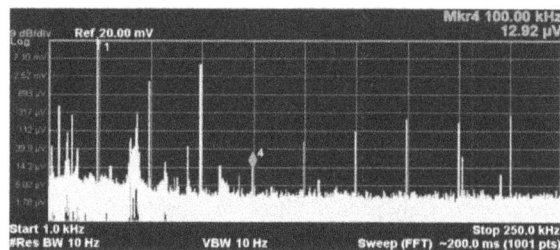

Figure 2: Amplitude spectrum of an unaveraged empty-measurement on a semi-logarithmic scale at 700 A_{PP} in the DFC.

In a second measurement, a particle signal has been recorded. In Fig. 3, the amplitudes of the 1st to the 8th harmonic are plotted over time, while different particle samples have been inserted into and removed from the center of the bore.

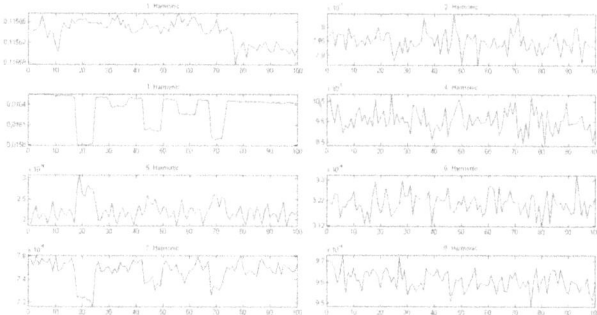

Figure 3: Time plot of the first eight harmonics. The x-axis shows the consecutive number of the 500 times averaged measurements and the y-axis the voltage amplitude in V_P.

CONCLUSION The performed measurements prove the capability of this setup to generate magnetic fields up to 30.5 mT_{PP} and to detect superparamagnetic nanoparticles. The empty measurement still shows room for improvements in terms of the amplitudes of the harmonics, disturbance and noise level.

ACKNOWLEDGEMENTS The authors acknowledge the financial support of the German Federal Ministry of Education and Research (BMBF) under grant number 13N11090 and of the European Union and the State Schleswig-Holstein (Programme for the Future – Economy) under grant number 122-10-004.

REFERENCES
[1] G. Bringout et al. *Book of Abstracts of the IWMPI 2015*, 49, 2015 doi: 10.1109/IWMPI.2015.7107032.
[2] J. Stelzner et al. *Book of Abstracts of the IWMPI 2015*, 92, 2015 doi: 10.1109/IWMPI.2015.7107074.

Micro Traveling Wave MPI – initial results with optimized tracer LS-008

P. Vogel[a,*], M.A. Rückert[a], S.J. Kemp[b], A.P. Khandhar[b], R.M. Ferguson[b], A. Vilter[a], P. Klauer[a], K.M. Krishnan[b,c], V.C. Behr[a]

[a] Department of Experimental Physics 5 (Biophysics), University of Würzburg, Würzburg, Germany
[b] Lodespin Labs, Seattle, USA
[c] Department of Materials Science & Engineering, University of Washington, Seattle, USA
* Corresponding author, email: Patrick.Vogel@physik.uni-wuerzburg.de

INTRODUCTION Different types of scanners have been presented since the first publication of Magnetic Particle Imaging (MPI) [1]. Most of them were designed for pre-clinical applications resulting in low spatial resolution of about 1 mm due to SAR and PNS limitations. Increasing the resolution for MPI implies increasing the gradient strength, which is necessary for spatial encoding in MPI. The presented µTWMPI system is an optimized setup for high gradient strength and high resolution.

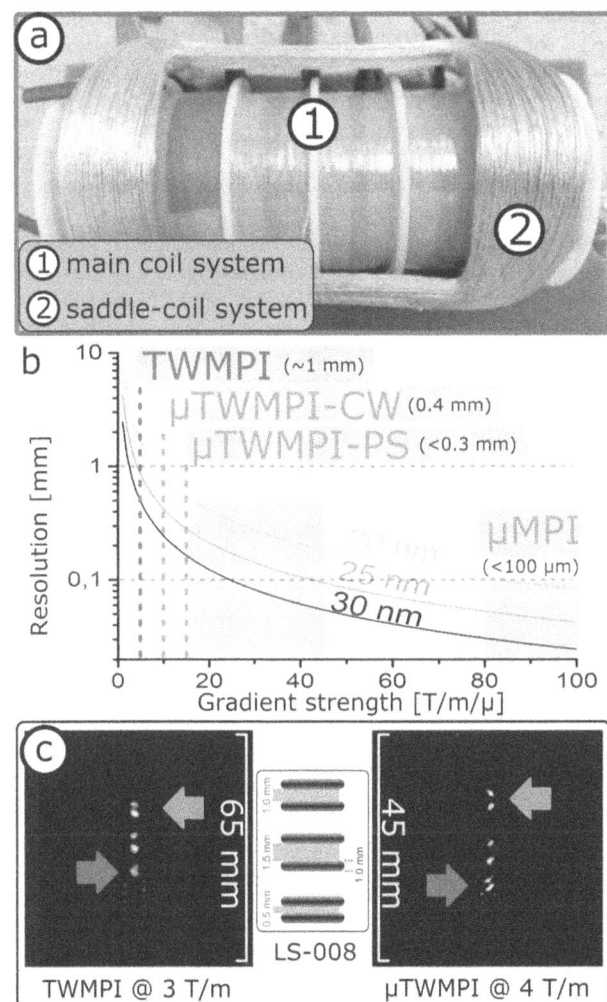

MATERIAL AND METHODS The µTWMPI system is a further development of the Traveling Wave MPI scanner [2]. For imaging a field-free point (FFP) with a strong gradient is generated by two coil pairs in Maxwell configuration, which can be driven individually. To achieve higher gradient strengths, the main coil system was optimized with respect to the ratio of power consumption versus field strength. The inner bore size was decreased to 23 mm, which greatly increases the achievable gradient, and the saddle-coil pairs for moving the FFP through the sample were placed on the outside of the system. The usable field of view (FOV) of the system is about 45 mm in length and 20 mm in diameter. Fig. 1. (a) shows the µTWMPI system consisting of the main coils (dynamic linear gradient array – dLGA) and one of the saddle-coil pairs.

For the µTWMPI system two different modes of operation are available. In the continuous-wave mode (CW-mode) the system can be driven by common audio amplifiers for several 100 ms. In the power-shot mode (PS-mode) a home-build power-amp drives the system in a pulse-mode. The PS-mode offers only a few milliseconds of acquisition time, but has the advantage of a distortion-free and noise-free excitation frequency.

RESULTS A gradient strength of about maximal 9 T/m was achieved in CW-mode. This yields a resolution of about 500 µm for Resovist (Fig. 1. (c)) [3]. First experiments with a phantom filled with LS-008 (Lodespin Labs, USA) was performed with the µTWMPI and compared with TWMPI results. Even at a gradient strength of 4 T/m it is possible to resolve the 500 µm parts of the phantom (see Fig. 1 (c)).

In initial tests using the PS-mode the gradient strength could be increased up to 15 T/m using an early version of the power-amp. This corresponds to a resolution of about 200 µm for Resovist and <100 µm using LS-008 (Fig. 1 (b)).

CONCLUSION The presented µTWMPI scanner system is a promising approach for ultra-high resolution MPI scanners. Compared to initial experiments using permanent magnets [4], the sample handling has improved significantly. With its good-sized FOV and a spatial resolution below 500 µm using an optimized MPI tracer (LS-008) the system enables new options for experiments in biology, material sciences and other areas [5]. For resolutions below 100 µm (~45 T/m) a modified power-amp is required.

ACKNOWLEDGEMENTS This work was partially funded by the DFG (BE-5293/1-1) and supported by the NIH grants 1R41EB013520-01 and 2R42EB013520-02A1.

Figure 1: **(a)** image of the µTWMPI system consisting of the main coils (dLGA) and one saddle-coil pair. **(b)** overview of the different scanner types and their achievable gradient strengths. **(c)** first results of the µTWMPI system: in comparison with the TWMPI results a resolution of 500 µm is achievable at a gradient of about 4 T/m using LS-008 (Lodespin Labs, USA).

REFERENCES
[1] N. Panagiotopoulos, et al., *Int J Nanomedicine*, 10:3097—3114, 2015. doi: 10.2147/IJN.S70488.
[2] P. Vogel, et al., *IEEE TMI*, 33(2):400-407, 2013. doi: 10.1109/TMI.2013.2285472.
[3] P. Goodwill, et al., *IEEE TMI*, 29(11):1851-1859, 2010. doi: 10.1109/TMI.2010.2052284.
[4] P. Vogel, et al., IEEE Trans. Magn., 51(2):6502104, 2015. doi: 10.1109/TMAG.2014.2329135.
[5] P. Vogel, et al., *diffusion-fundamentals.org*, 22(12):1-5, 2014.

M(H) dependence and size distribution of SPIONs measured by atomic magnetometry

Simone Colombo[a,*], Victor Lebedev[a], Zoran D. Grujić[a], Vladimir Dolgovskiy[a], Antoine Weis[a].

[a] Département de Physique, Université de Fribourg, Chemin du Musée 3, 1700 Fribourg, Switzerland
[*] Corresponding author, email: simone.colombo@unifr.ch

INTRODUCTION Most MNP applications call for a quantitative characterization and monitoring of the particle size distributions both prior to and after their administration into the biological tissue. Two imaging modalities for determining MNP distributions in biological tissues are being actively pursued, viz., magnetorelaxation (MRX) [1] and Magnetic Particle Imaging (MPI) [2]. High-sensitivity magnetic induction detection plays a key role in view of minimizing the administered MNP dose in biomedical applications. Established MNP characterization/detection methods mainly rely on detecting the oscillating induction $B(t) \propto M(t)$ induced by a harmonic excitation $H(t)$ by means of a magnetic pick-up (induction) coil.

Here we describe our successful attempt to replace the pick-up coil by an atomic optically-pumped magnetometer (OPM) which allows recording slow $B(t)$ variations in frequency ranges that are not accessible to induction coils.

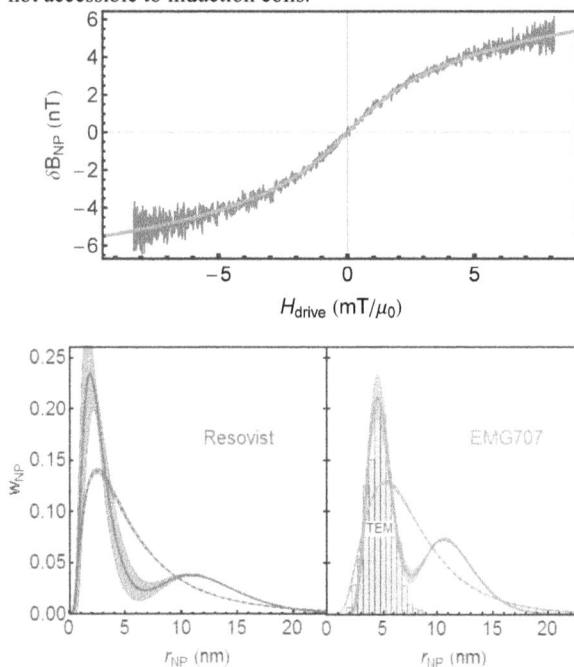

MATERIAL AND METHODS The OPM used in our apparatus is based on optically detected magnetic resonance in spin-polarized Cs vapor [3]. The magnetometer detects no signal without MNP sample down to its noise floor of ≈ 5 pT/Hz$^{1/2}$. Under typical experimental conditions the magnetometer can react to magnetic field changes with a bandwidth of ≈ 1 kHz, while keeping the mentioned sensitivity. The recordings were performed on dilution series of Ferrotec-EMG-707 and Resovist, water-suspended superparamagnetic iron oxide nanoparticles (SPIONs). We excite the sample by a field $H_{scan}(t)$ of amplitude up to 16 mT$_{pp}$/μ_0 that sinusoidally oscillates at a frequency f_{scan} of 10 Hz.

We record time series (sampled at a rate of 2000 S/s) of the drive current $I_{scan}(t)$ and the induced signals $B_{NP}(t)$ and combine those signals into $M(H)$ curves.

RESULTS We have fitted size-distribution-weighted Langevin functions to the experimental $M(H)$ data (Fig. 1, top). Fits were made assuming both mono-modal and bi-modal size distributions. The parameters extracted from the bi-modal distribution fits are in agreement with parameters found in the literature, thus validating our the method.

Freshly produced magnetic nanoparticle solutions are basically monomodal, but, because of cluster formation evolve during aging to a bimodal distribution, as described, e.g., in Ref. [4].

We report a current sensitivity limit of ~ 7 μg_{Fe}.

CONCLUSION We have demonstrated that an atomic magnetometer in an unshielded environment can be used for a direct quantitative measurement of magnetic nanoparticles' $M(H)$ curves using sub-kHz drive fields. The quality of the measurements allows the extraction of magnetic nanoparticles size-distributions. The method allows a quantitative monitoring of MNP cluster formation due to aging of MNP solutions.

ACKNOWLEDGEMENTS Work supported by Swiss National Science Foundation Grant No. 200021_149542.

REFERENCES
[1] M. Liebl, U. Steinhoff, F. Wiekhorst, J. Haueisen, and L. Trahms. *Phys. Med. Biol.*, 59(21):6607, 2014. doi: 10.1088/0031-9155/59/21/6607.
[2] N. Panagiotopoulos, R. L. Duschka, M. Ahlborg, G. Bringout, Ch. Debbeler, M. Graeser, Ch. Kaethner, K. Lüdtke-Buzug, H. Medimagh, J. Stelzner, et al. *Int. Journ.Nanomed.*, 10:3097, 2015. doi: 10.2147/IJN.S70488.
[3] G. Bison, N. Castagna, A. Hofer, P. Knowles, J. L. Schenker, M. Kasprzak, H. Saudan, and A. Weis. *Appl.Phys. Lett.*, 95:173701, 2009. doi: 10.1063/1.3255041
[4] D. Eberbeck, C. L. Dennis, N. F. Huls, K. L. Krycka, C. Gruttner, and F.Westphal. *IEEE Magn.*, 49(1):269–274, 2013. doi: 10.1109/TMAG.2012.2226438.

Figure 1: Top: $B_{NP}(H)$ response of a 0.5 ml Ferrotec EMG-707 sample containing 0.3 mg of iron (data, representing 50 averaged sinusoidal $H(t)$ cycles (5 s in total) and size-distribution-weighted fits. Bottom: Mono- and bi-modal MNP size distributions inferred from fits.

The Design of Magnetic Particle Imaging Gradient Magnetic Field Generator using Finite Element Method

Shiqiang Pi[a], Jingjing Cheng[a], Wenzhong Liu[a,*]

[a] School of Automation, Huazhong University of Science and Technology, Wuhan 430074, China
* Corresponding author, email: lwz7410@hust.edu.cn

INTRODUCTION Magnetic particle imaging (MPI) has been more and more attractive because of its potential to achieve real-time, nonionizing radiation and high spatial resolution 3-D imaging [1]. Under gradient and dynamic magnetic fields, MPI utilizes the nonlinear magnetization response to map the concentration spatial distribution of the magnetic particles *in vitro* and *in vivo* [1-3]. Gradient magnetic field generator (GMFG) is one of the most important components in MPI system making its design is of great significance. For given magnetic nanoparticles, the spatial resolution of MPI is mainly determined by the GMFG. The increase of magnetic field gradient improves the spatial resolution but meanwhile increases the power consumption and heating of GMFG. Two kinds of structural GMFGs (see Fig. 1) were analyzed through finite element method in this paper. The number of turns, power consumption and magnetic field gradient of the GMFGs were compared to each other.

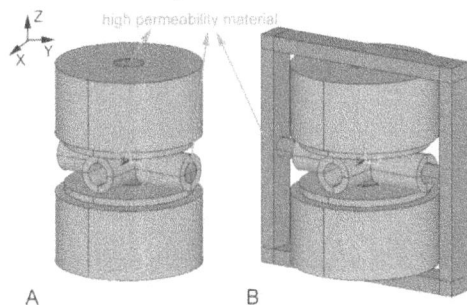

Figure 1: The geometrical structure of MPI systems.

MATERIAL AND METHODS The red components in Fig. 1 represent the magnetic coils whereas the blue components represent the iron cores and yoke steel, which are made of high permeability materials. The magnetic gradients of MPI systems mentioned above were analyzed by finite element method using ANSYS. To compare the magnetic gradients generated by the GMFGs, the same parameters (such as the number of turns (1550 turns), current (10 A) and geometrical dimensions) of the coils were used in the simulation. A phantom with two delta samples (radius of 2 mm and a distance of 7 mm) filled with magnetic nanoparticles (SHP-25, Ocean NanoTech, LLC) was employed in the MPI system with similar structure in Fig.1 B. The FOV of the MPI system is 30×28 mm^2 with magnetic gradients of about 1.8 and 5.5 T/m in x and z directions. Meanwhile, the exciting current of the GMFG was 8 A.

RESULTS The dashed lines and solid lines in Fig. 2 represent the absolute value of the magnetic field generated by GMFG with structure A and B, respectively. The magnetic gradients of structure A are about 1 and 3.5 T/m in x and z directions whereas those of structure B are about 2 and 6.6 T/m. The magnetic gradient of structure B is about twice as large as that of structure A. To generate the same magnetic gradient, double the power consumption and number of turns are demanded in the MPI

system with structure A. Fig. 3 shows the measurement of a phantom with two delta samples with 7 mm distance.

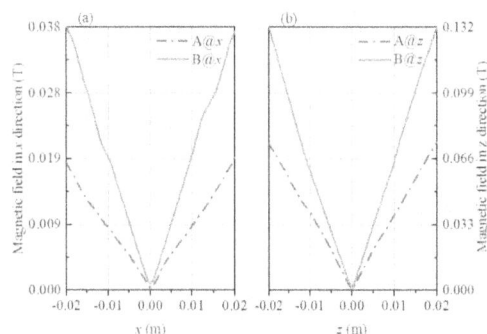

Figure 2: The absolute value of magnetic field in x and z directions.

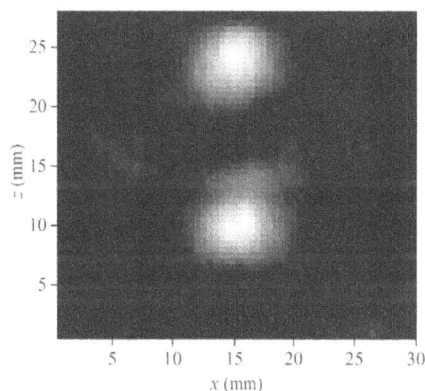

Figure 3: The measurement of a phantom with two delta samples.

CONCLUSION From the simulation results, we found that the yoke steel providing a high permeability magnetic circuit makes the GMFG easy to generate high gradient magnetic field and as well as to reduce power consumption. Although the image of the phantom has been reconstructed, our present MPI system needs to be improved in spatial resolution and SNR.

ACKNOWLEDGEMENTS This work was supported by 61571199 (NSFC) and Hubei Provincial project of 2014AEA048.

REFERENCES
[1] N. Panagiotopoulos, R. L. Duschka, et al., *Int. J. Nanomed.*, 10:3097—3114, 2015. doi: http://dx.doi.org/10.2147/ijn.s70488.
[2] B. Gleich and J. Weizenecker. *Nature*, 435(7046):1217—1217, 2005. doi: 10.1038/nature03808.
[3] P. W. Goodwill, E. U. Saritas, et al. *Adv.Mater.*, 24(28):3870—3877, 2012. doi: 10.1002/adma.201200221.

A 1.4 T/m Field Free Line Magnetic Particle Imaging Device

Matthias Weber[a,*], Klaas Bente[a], Steffen Bruns[a], Anselm von Gladiss[a], Matthias Graeser[a], Thorsten M. Buzug[a]

[a] Institute of Medical Engineering, Universität zu Lübeck, Germany
* Corresponding author, email: {weber,buzug}@imt.uni-luebeck.de

INTRODUCTION Several promising developments have been achieved with the field free line (FFL) technique in Magnetic Particle Imaging (MPI) [1, 2]. Besides high gradient strength featuring high resolution, real-time approaches have broadened the application spectrum. Nevertheless, the manufacturing of systems, especially of the field generating units, is challenging. In particular, the selection field generator is highly power-demanding if it has been implemented with electromagnetic coils to facilitate the rotation of the FFL. This work proposes a scanner setup that features an FFL generated by permanent magnets that is integrated in a rotating unit. Thus, the scanner requires a minimum of power.

Figure 1: Sectional view of the CAD construction. The system has a bore diameter of 30 mm.

MATERIAL AND METHODS The field generating units consist of a selection field, which is generated by permanent magnets [3] and a pair of Helmholtz coils that generate the homogeneous drive field. The selection field features an FFL and a gradient of 1.41 T/m. The drive field has an amplitude of approximately 11 mT. Both the selection and drive field units are mounted on a rotating unit. Since the drive field unit is wired, it limits arbitrary rotation. Therefore, the unit is restricted to rotations of 180° that are performed by a robot. Currently, a rotation of 180° is reached in approximately 5 seconds.

The homogeneous drive field has an excitation frequency of 25 kHz. It is amplified by a power amplifier (AE Techron 2105, Elkhart, IN 46516, USA). The signal is filtered with a third order band pass Butterworth filter. The drive field is impedance matched to 8 Ω.

A highly homogenous receive coil is aligned collinearly to the drive field coil to detect the particle signal [4]. A fourth order band stop Butterworth filter rejects the excitation signal. The particle signal is low-noise amplified (SRS560, Stanford Research Systems, Sunnyvale, CA, USA) and digitized.

The acquired data is x-space reconstructed. A necessary deconvolution algorithm is carried out to increase the resolution. This process has been explored in detail in [2].

RESULTS The reconstructed data is shown in Fig. 2 on the right. A field of view (FOV) of 16 mm x 16 mm is scanned whereat a total power of 200 W is consumed by the drive field. The particle samples can be clearly distinguished in the reconstructed image.

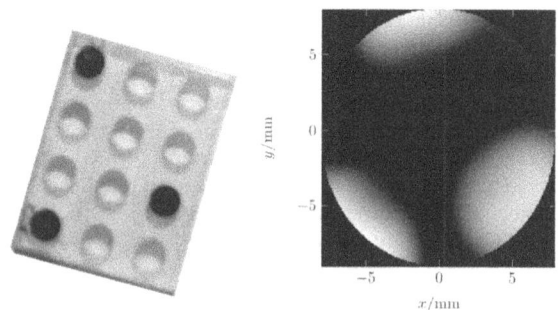

Figure 2: The phantom on the left features three holes with a diameter of 3 mm each. The holes are filled with undiluted Resovist. The reconstructed image is shown on the right.

CONCLUSION The concept of designing FFL-MPI scanners with a rotating permanent magnet unit has been proven in this work. The main advantage is a one-dimensional signal chain and consequently, a very low power consumption. Therefore, it is a concept for standalone MPI systems that may easily be installed in any lab environment.

ACKNOWLEDGEMENTS This work was supported by the LUMEN Research Group through the German Bundesministerium für Bildung und Forschung under Grant FKZ 13EZ1140A/B. LUMEN is a Joint Research Project of the Lübeck University of Applied Sciences, Lübeck, Germany, and the Universität zu Lübeck, Lübeck, and represents its own branch of the Graduate School for Computing in Medicine and Life Sciences, Universität zu Lübeck.

REFERENCES
[1] P. W. Goodwill, J. J. Konkle, B. Zheng, E. U. Saritas, and S. M. Conolly, "Projection X-space magnetic particle imaging," IEEE Trans. Med. Imaging, vol. 31, no. 5, pp. 1076–1085, 2012.
[2] K. Bente, M. Weber, M. Graeser, T. F. Sattel, M. Erbe, and T. M. Buzug, "Electronic Field Free Line Rotation and Relaxation Deconvolution in Magnetic Particle Imaging," IEEE Trans. Med. Imaging, vol. 34, no. 2, pp. 644–651, Feb. 2015.
[3] M. Weber, K. Bente, A. von Gladiss, and T. M. Buzug, "MPI with a mechanically rotated FFL," 2015 International Workshop on Magnetic Particle Imaging (IWMPI), 2015.
[4] M. Weber, K. Bente, M. Graeser, T. F. Sattel, and T. M. Buzug, "Implementation of a High-Precision 2-D Receiving Coil Set for Magnetic Particle Imaging," IEEE Trans. Mag., vol. 51, no. 2, pp. 1–4, Feb. 2015.

Session 5:

Application 2

Assessing flow dynamics in a 3D printed aneurysm model by magnetic particle imaging

Jan Sedlacik[1*], Andreas Frölich[1], Johanna Spallek[2], Nils D. Forkert[3], Tobias D. Faizy[1], Franziska Werner[4,5], Tobias Knopp[4,5], Dieter Krause[2], Jens Fiehler[1], Jan-Hendrik Buhk[1]

[1]Neuroradiology, University Medical Center Hamburg-Eppendorf, Hamburg, Germany.
[2] Product Development and Mechanical Engineering Design, Hamburg University of Technology, Hamburg, Germany.
[3] Department of Radiology and Hotchkiss Brain Institute, University of Calgary, Calgary, AB, Canada.
[4] Section for Biomedical Imaging, University Medical Center Hamburg-Eppendor, Hamburg, Germany.
[5] Institute for Biomedical Imaging, Hamburg University of Technology, Hamburg, Germany.
* Corresponding author, email: j.sedlacik@uke.de

INTRODUCTION Magnetic particle imaging (MPI) is capable of acquiring 3D datasets with high temporal resolution [1,2], which may be especially beneficial for in vivo hemodynamic imaging. The characterization of the hemodynamics of aneurysms is of particular interest [3], since treatment planning and follow-up diagnosis may benefit from this new imaging technique. The purpose of this work was to assess the flow dynamics in a realistic 3D printed aneurysm model [4] with MPI and to validate the measurements with magnetic resonance imaging (MRI), and dynamic subtraction angiography (DSA).

MATERIAL AND METHODS The 3D printed aneurysm model was derived from a static 3D subtraction angiography of a patient with an incidental Internal Carotid Artery (ICA) aneurysm of saccular morphology (ca. 5 mm diameter). The model was printed with 254 μm thick layers of acrylonitrile butadiene styrene at fused deposition modeling using the HP Designjet 3D printer and impregnated with Nano-Seal (Jeln Imprägnierung, Schwalmstedt, Germany) [5]. The aneurysm model was connected to a peristaltic pump, which was set to deliver a physiological flow and pulsation rate of about 250 mL/min and 70/s, respectively. However, due to the peristaltic nature of the pump, the pulsation profile is not comparable with physiology. 4D phase contrast flow quantification (4D pc-fq) and dynamic MRI, i.e. time-resolved contrast enhanced angiography with stochastic trajectories, was performed using a 7 T Bruker Clinscan small animal MRI. The 4D pc-fq was triggered with the pump pulsation and measured over 4 hours to obtain sufficient spatial resolution (500 μm isotropic) and signal to noise ratio. The dynamic MRI measurement was optimized for high temporal resolution (270 ms) while administering a bolus of 3 mL of 0.05 mol(Gd-DOTA)/L with a rate of 1 mL/s using a syringe pump and an angiographic catheter with 1 mm inner diameter. The tip of the catheter was placed close to the aneurysm model (ca. 5 cm upstream) to reduce bolus dispersion. The first commercially available pre-clinical MPI scanner (Bruker/Philips) was used to acquire 1 mm isotropic 3D data with 21.5 ms temporal resolution while administering a bolus with 50 mmol(Fe)/L (MM4, TOPASS GmbH, Berlin, Germany) similarly administered as for the dynamic MRI measurement. DSA was acquired using a Philips Allura FD20 with a temporal resolution of 33.3 ms during similar bolus injection of 150 mg(iodine)/mL (Imeron) as for the dynamic MRI and MPI measurements. Image post processing and visualization was done with in house written software using Matlab.

RESULTS Distinct pulsative flow patterns as well as a delayed contrast agent outflow from the aneurysm can be detected with MPI (Fig.1). Similar patterns were observed with dynamic MRI
.

and DSA (not shown). Furthermore, 4D pc-fq MRI showed distinct lower flow velocity and a vortex inside the aneurysm (not shown). Single MPI frames around a local signal peak are able to show the contrast agent passage through the model (Fig.2).

DISCUSSION AND CONCLUSION MPI was able to assess the flow dynamics of the 3D printed aneurysm model, which was verified by MRI and DSA. The delayed contrast agent outflow from the aneurysm implies important information about the intra-aneurysmal hemodynamics, i.e. the vortex. In addition, the lower broad signal maximum in between two high sharp signal peaks of the MPI signal depicts a secondary pulsation phase of the non-physiological pump pulsation, which was not detected by dynamic MRI or DSA. Future work needs to be done to improve spatial resolution of MPI to more better outperform the competing methods of MRI and DSA.

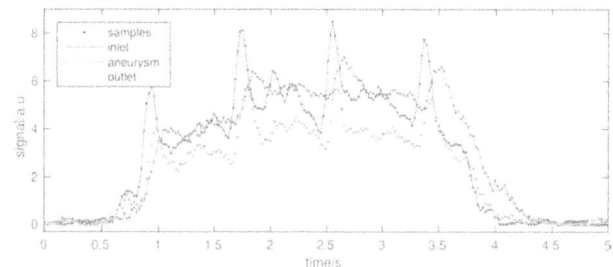

Figure 1: ROI analysis shows distinct pulsation and delayed contrast agent outflow from the aneurysm.

Figure 2: Single MPI frames around signal peak at t=3.5s depicting the contrast agent passage through the model.

ACKNOWLEDGEMENTS We wish to thank the German Research Foundation (DFG), grant no. AD 125 / 5-4, and the Forschungszentrum Medizintechnik Hamburg (fmthh) for financial support and Philips Healthcare for the support and realization of the "Hermann-Zeumer Research Laboratory" including a Philips AlluraClarity Angiography system.

REFERENCES
[1] B. Gleich and J. Weizenecker. *Nature*, 435(7046):1217—1217, 2005.
[2] T. Knopp and T. M. Buzug. Springer, Berlin/Heidelberg, 2012.
[3] Jeong W, Rhee K. Comput Math Methods Med. 2012;2012:782801.
[4] Anderson JR, et al. J Neurointerv Surg. 2015 Apr 10
[5] Frölich AM, et al. AJNR Am J Neuroradiol. 2015 Aug 20

Differential pick-up coils in magnetic particle spectrometry to detect low concentration SPIO nanoparticle tracers

Bharadwaj Muralidharan[a,*], Thomas E. Milner[b] and Chun Huh[c]

[a] Department of Electrical and Computer Engineering, The University of Texas at Austin, Austin TX 78712, USA.
[b] Department of Biomedical Engineering, The University of Texas at Austin, Austin TX 78712, USA.
[c] Center for Petroleum and Geosystems Engineering, The University of Texas at Austin, Austin TX 78712, USA.
[*] Corresponding author, email: bmuralidharan@utexas.edu

INTRODUCTION For the oil exploration and production, use of specially surface-coated magnetic nanoparticles (MNPs) to detect the presence and distribution of certain target fluid or mineral component in subsurface geologic formation is currently an active research area. For example, by injecting MNPs into an oil reservoir and making them at attach at the oil/water interfaces in rock pores, and then applying external magnetic field oscillation, the *in-situ* distribution of oil could potentially be discerned [1]. In such attempts, determining how much MNPs entered into which geologic layers is critically important. Typically, such determination is made by oil well logging, which is the practice of mapping the vertical distribution of certain petrophysical properties by inserting electromagnetic, NMR and/or sonic instruments into the well.

Since the core of the MNPs employed for such applications is super-paramagnetic, just as in medical tomographic applications, magnetic particle spectroscopy (MPS) is an attractive choice to determine the vertical distribution of the MNP concentration that had been injected into the oil well. Our work is the first such attempt to apply the MPS concept for oilfield applications. The lowest concentration detectable is limited by the noise floor introduced by the pick-up coil. We initially looked into traditional MPS coil design similar to the one by Biederer et al [2]. However, the noise limited detection to at least 10 mg/l concentration of small sized MNPs.

To decrease the noise level, we employ a differential pick-up coil design with which the noise can be removed with use of the control coil. Figure 1 shows the schematics of the system used for the experiment.

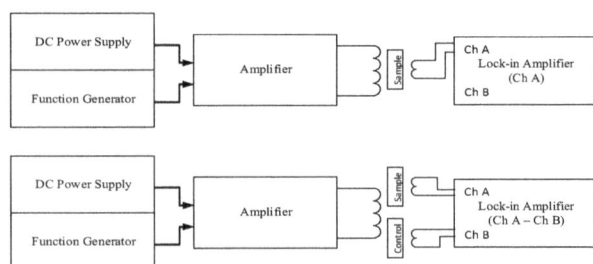

Figure 1: Schematic of standard (top) and differential (bottom) MPS system used for testing.

MATERIAL AND METHODS The MNPs used are EMG 605 from Ferrotec Corp. of ~10nm core size and similar polyacrylic acid-coated MNPs synthesized *in-house* as described in [3]. The volume of fluid used for testing is 1 ml. Various MNP concentrations were used to obtain the magnetization data shown in Figure 2. The control used in collecting data is 1 ml of de-ionized water.

The data was collected with multilayered solenoid coils. The excitation coil has 200 turns with inner diameter of 12.5 mm and length of 37.5mm, designed to provide a 50 mT field without damage excited by a 1kHz sinusoidal drive signal. The pickup coils were wound on a 10mm bobbin (3D printed with coils wound on it), having 300 turns with the length of 12.5 mm. The two pick-up coils were matched with maximum error of 0.1 Ω resistance and inductance of 0.1 mH.

RESULTS

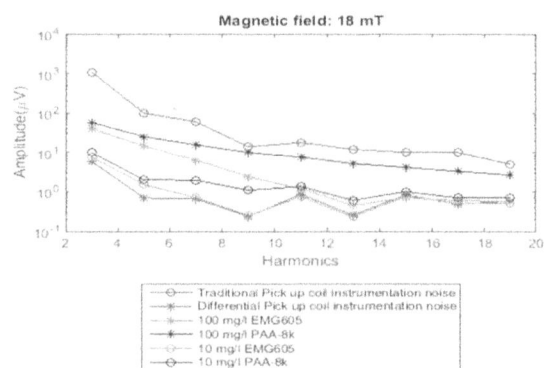

Figure 2: Different concentration sample data over odd harmonics (3 – 19) with differential pick – up coil design. The instrumentation noise of traditional pick-up coil is also plotted.

CONCLUSION The use of a differential pick-up coil reduces the noise floor as the noise is removed as a common mode signal. Owing to sample data collection at real time with noise removal, the sensitivity with respect to the detected particle concentration is improved by an order of 100.

ACKNOWLEDGEMENTS This research was supported by "Nanoparticles for Subsurface Engineering" IAP at University of Texas at Austin. Drs. Saebom Ko and Qing Wang kindly provided the MNPs with their characterization data. I would also like to thank Scott Jenney for teaching me the nuances in winding a coil.

REFERENCES

[1] Ryoo, S., Rahmani, A. R., Yoon, K. Y., Prodanovic, M., Kotsmar, C., Milner, T. E., Johnston, K. P., Bryant, S. L., and Huh, C., J. Petrol. Sci. Eng., 81, 129-144 (2012). doi: 10.1016/j.petrol.2011.11.008.
[2] S. Biederer, T. Knopp, T. F. Sattel, K. Lüdtke-Buzug, B. Gleich, J. Weizenecker, J. Borgert and T. M. Buzug. J. Phys. D: Appl. Phys. 42, 205007 (2009). doi: 10.1088/0022-3727/42/20/205007.
[3] Q. Wang, V. Prigiobbe, C. Huh, S. L. Bryant, M. V. Bennetzen and K. Mogensen. IPTC, December, 2014. doi: 10.2523/17901-MS Conference: 2014 International Petroleum Technology Conference

Devices for remote magnetic operation in an MPI scanner

Christian Stehning[a,*], Peter Mazurkewitz[a], Bernhard Gleich[a], Jürgen Rahmer[a]

[a] Philips GmbH Innovative Technologies, Research Laboratories, Hamburg
[*] Corresponding author, email: christian.stehning@philips.com

INTRODUCTION Beside its imaging capabilities, an MPI system [1] is a flexible and powerful field generator that allows localizing and applying significant forces on small magnetic devices [2]. A proof of principle for the operation of selected devices in a preclinical MPI system is shown in this study. Furthermore, a coarse theoretical assumption on the miniaturization scaling potential is presented.

MATERIAL AND METHODS Schematic plots of selected devices employed in this study are shown in Fig. 1 [a,b]:

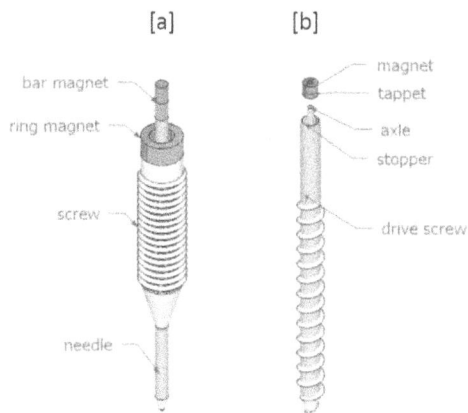

Figure 1: Devices for rotational/linear forces

A screw with an attached ring magnet [a] was operated via a rotating magnetic field about the cylinder axis. A gradient field exerted a force on a needle in the cylinder axis. Furthermore, larger dynamic torques using the centrifugal mass of the magnet were exerted using a drive screw illustrated in [b]. In this setup, a pivot-mounted magnet with a tappet is pre-wound by a rotating field, then performs a fast rotation against a stopper. Furthermore, multiple devices were positioned with a distance of 30-40mm, and operated independently using a combination of homogeneous and gradient fields. Selected device dimensions for e.g. a magnetic screw were 4mm × 20mm (diameter × length). A typically employed flux of the MPI system was $|\vec{B}|$=20mT.

For a theoretical assumption on the miniaturization scaling potential, the torque $|\vec{D}|$ required to rotate a screw of length l in (soft) tissue is determined by the friction surface times its distance r from the rotation axis:

$$|\vec{D}| \propto l \cdot 2\pi r \cdot r \qquad \text{(Eq.1)}$$

The exerted torque on a magnetic dipole $\vec{m}$ in the MPI focus field with magnetic flux $\vec{B}$ reads:

$$\vec{D}_{\vec{m}} = \vec{m} \times \vec{B} \qquad \text{(Eq. 2)}$$

For a screw fabricated from homogeneous magnetic material, the magnetic dipole moment $\vec{m}$ in Eq. 2 is proportional to the cylindrical volume:

$$|\vec{m}| \propto l \cdot 2\pi r^2 \qquad \text{(Eq. 3)}$$

These approximations suggest that the required torque on a screw in soft tissue (Eq.1), and that delivered by the MPI fields (Eqs. 2 and 3), both scale with the squared radius of the device. If sufficient flux B is provided by the MPI system, the device operability is independent of its dimensions.

RESULTS A selected device (magnetic screw) and one frame of a movie illustrating the operation of the device in a gelatine sample in an MPI system are shown in Fig. 2 [a,b], respectively.

Figure 2: Photograph of a magnetic screw [a], and frame of a movie illustrating the operation in a gelatine sample [b]

Using rotating magnetic fields provided by the MPI system, the device was screwed into the gelatine sample over a distance of 40mm, and back to its initial position.

CONCLUSION The powerful MPI field generators can deliver push/pull or bending forces, torque, or electric power to an in-body medical device. This may overcome the need for intravenous access or interventions, or address target regions that are difficult to access via conventional catheter interventions. Furthermore, devices may be designed with very small outer dimensions, and by the use of the selection field gradient, an individual control of multiple devices in the scanner bore is possible.

ACKNOWLEDGEMENTS This work was supported by the Bundesministerium für Bildung und Forschung (BMBF grant 13GW0069C).

REFERENCES
[1] B. Gleich and J. Weizenecker. Nature, 435(7046):1217—1217, 2005. doi: 10.1038/nature03808.
[2] N. Nothnagel et al, IWMPI Berkeley 2013

First Murine *in vivo* Cancer Imaging with MPI

Elaine Yu [a*], Mindy Bishop [a], Patrick W. Goodwill [a,c], Bo Zheng [a], Matt Ferguson [d], Kannan M. Krishnan [e], Steven M. Conolly [a,b]

[a] Department of Bioengineering , University of California, Berkeley, CA, USA
[b] Department of Electrical Engineering and Computer Sciences, University of California, Berkeley, CA, USA
[c] Magnetic Insight, Inc., Newark, CA, USA
[d] LodeSpin Labs, PO Box 95632, Seattle, WA, USA
[e] Department of Material Science and Engineering, University of Washington, Seattle, WA, USA
[*] Corresponding author, email: elaineyu@berkeley.edu

INTRODUCTION MPI [1, 2] is a high-contrast and quantitative new imaging modality with high tracer sensitivity—even 2 ng of Iron is detectable in an animal [3]. MPI also uses no ionizing radiation, and the SPIO tracers are safe even for patients with Chronic Kidney Disease (CKD). Here we demonstrate, for the first time, *in vivo* cancer imaging with MPI.

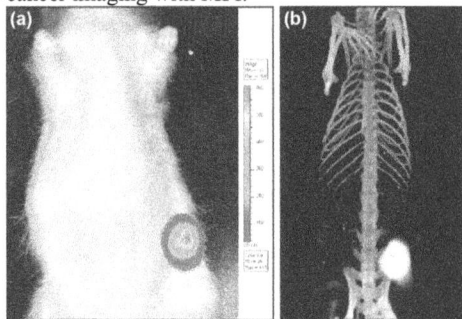

Figure 1: MPI images correlate well with a bioluminescence reference image. (a) Bioluminescence image of MDA-MB-231-luc tumor implant. (b) Projection of 3D MPI image of the tumor 24 hours after SPIO injection overlaid with CT reference. Unlike in BLI, the MPI signal shows no attenuation with depth.

MATERIAL AND METHODS Four athymic rats were implanted with 7 million MDA-MB-231-luc tumor cells in the mammary fat pad and monitored for 4 weeks *in vivo*. *Bioluminescence Imaging (BLI):* Rats were injected with luciferin prior to imaging. *MPI:* Long circulating Lodespin SPIO tracer (15 mg Fe/kg) were injected via the tail vein into anesthetized rats, which were scanned with respiratory gating at multiple time points post-injection: 10, 30 min, 1, 2, 4, 6, 24, 48 and 96 hrs. Images were acquired with a Field Free Point (FFP) MPI scanner with a drive field amplitude of 40 mTpp at 20.05 kHz (Resolution: 1.2 mm, FOV: 4 cm × 4 cm × 8 cm, Scan Time: 5 minutes). Images were reconstructed with x-space MPI reconstruction [2]. CT scans of two of the animals were acquired for anatomical reference (Resolution: 93μm, FOV: 4 cm × 4.7 cm × 16.5 cm, Scan Time: 25 minutes).

RESULTS BLI and MPI/CT images are shown in Fig. 1. BLI confirms the presence of live MDA-MB-231-luc cells. Slices from the 3D MPI/CT datasets taken at different time points are shown in Fig. 2. The tumor is well appreciated, with initial wash-in on the tumor rim, peak uptake at 24 hours, and washout beyond 48 hours, likely due to EPR. The superb contrast allows clear visualization of the tumor. In addition, MPI enables quantitative analysis of tracer dynamics.

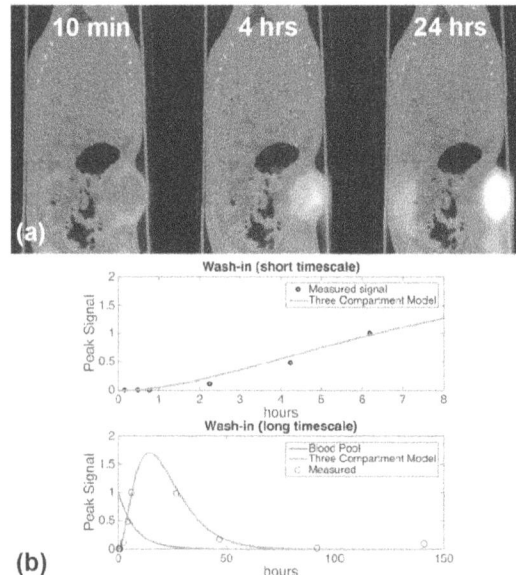

Figure 2: MPI visualizes and quantitates accumulation of SPIO tracer in the MBA-MB-231-luc tumor following systemic tracer administration. (a) Representative longitudinal MPI/CT image time points showing SPIO accumulation in the tumor. (b) The MPI image is quantitative and enables pharmacokinetic compartment modeling of the wash-in/wash-out curve.

CONCLUSION Here we present early results showing MPI's potential for sensitive and quantitative cancer imaging. Bioluminescence imaging requires genetic modification and suffers signal attenuation with depth. In contrast, MPI is quantitative, does not require genetic modifications, and has no signal attenuation with depth. MPI's high sensitivity and superb contrast produce images comparable with other molecular imaging modalities such as PET and SPECT. As we continue to develop the technique, we anticipate MPI will emerge as a robust imaging platform for evaluating cancer targeting moieties, detecting small metastases, and monitoring cell migration.

ACKNOWLEDGEMENTS We would like to acknowledge funding support from NSF GRFP, NIH R01 EB013689, CIRM RT2-01893, Keck Foundation 009323, NIH 1R24 MH106053, NIH 1R01 EB019458, and ACTG 037829.

REFERENCES

[1] Gleich, B., & Weizenecker, J. (2005). Nature, 435(7046).

[2] Goodwill, P. W., & Conolly, S. M. (2010). IEEE TMI, 29(11).

[3] Zheng, B., Vazin, T., Goodwill, P. W., Conway, A., Verma, A., Ulku Saritas, E., Conolly, S. M. (2015). Scientific Reports, 5, 14055.

In-vivo Measurements with UW-tracers in a harmonic 5.5 T/m MPI

Marcel Straub[a,*], Vera Päfgen[b], Eric Teeman[c], Kannan M. Krishnan[c], Fabian Kießling[b], Volkmar Schulz[a,d,*]

[a] Department of Physics of Molecular Imaging, Institute of Experimental Molecular Imaging, RWTH Aachen University, Aachen, Germany
[b] Institute of Experimental Molecular Imaging, RWTH Aachen University, Aachen, Germany
[c] Department of Materials Science, University of Washington, Seattle, WA 98195 USA
[d] Philips Research Europe, Germany
[*] Corresponding authors, email: marcel.straub@pmi.rwth-aachen.de, schulz@pmi.rwth-aachen.de

INTRODUCTION Magnetic Particle Imaging (MPI) is a tracer based imaging modality, and therefore, the properties of the used tracer strongly determine the performance of MPI. We investigate the performance of the UW1-209 particles [1] with our harmonic field free point MPI scanner [2]. We will show first *ex vivo* and in vivo results for this particular scanner tracer combination.

MATERIAL AND METHODS The MPI scanner is operated with a gradient of 5.5 T/m and drive field amplitudes of 20 mT. For image reconstruction a system matrix is recorded for the UW1 particles ($1.99\ mg_{Fe}/ml$) with a cylindrical calibration probe of 1mm diameter and 1mm length on a grid of $36 \times 33 \times 9$ voxels with an overall volume of $22 \times 20 \times 5.6\ mm^3$ corresponding to a voxel size of $0.6 \times 0.6 \times 0.7\ mm^3$. Comparison to Resovist ($27.9\ mg_{Fe}/ml$), from Bayer Schering, is done under identical conditions. The signal to noise ratio (SNR) for both tracers is calculated as described by J. Rahmer et.al. [3]. In order to compare the SNR of both tracers, the SNR is normalized by the different iron concentration.

Besides theses characterizations of UW1 particles, the tracer is administered during a running MPI acquisition via the tail vain to healthy female balb/c mice from Charles River with a weight between 19 and 22 g. The tracer is diluted in 0.9% NaCl and i.v. injected at a concentration of 40 $\mu mol_{Fe}/kg_{Body\ Mass}$. After MPI measurement, the animals were euthanized and scanned with a µCT to create an anatomical reference.

All frequency components of the system matrix with an SNR<3 are ignored for image reconstruction, because these do not show a reasonable quality during visual inspection. The images are reconstructed with an iterative solver, which constrains the solution to positive real values [4], with a regularization of $\lambda = 10^{10}$.

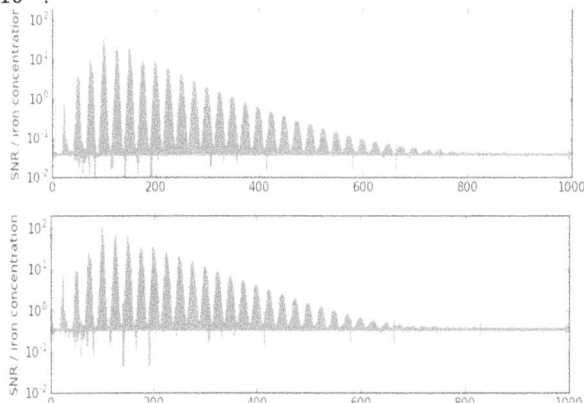

Figure 1: Signal to noise ratio (SNR) for two different MPI tracers. We normalize the SNR by iron concentration of the tracer. The SNR is determined from equal system matrix measurements. *(top)* Resovist. *(bottom)* UW1.

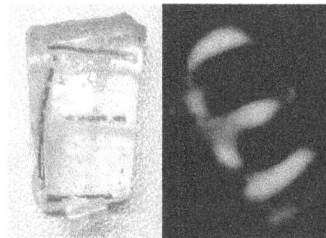

Figure 2: Phantom made of undiluted UW1 particles. *(left)* Photo of the phantom (12 mm × 7 mm). *(right)* Reconstructed MPI image.

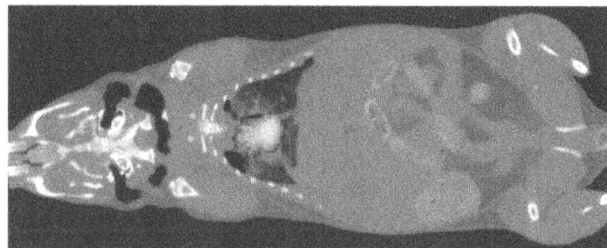

Figure 3: CT scan of a mouse after tracer injection with the MPI image of the heart region as overlay.

RESULTS Fig. 1 shows the SNR of the recorded UW1 system matrix for our scanner compared to a comparable Resovist system matrix. The iron concentration corrected SNR plots demonstrate the well-known fact, that the UW1 particles generate a higher signal than Resovist. Fig. 2 shows an exemplary reconstructed phantom, which is filled with undiluted UW1, reflecting the high spatial resolution.

Fig. 3 shows a CT scan of a mouse after tracer injection with the MPI image of the heart region as overlay. The MPI image was recorded shortly after the injection.

CONCLUSION We were able to show the MPI performance benefits of UW1 over Resovist. Furthermore, we have shown first *in vivo* imaging with our scanner and UW1 particles.

ACKNOWLEDGEMENTS The authors would like to thank Philips for their financial support of the Ph.D. position of Marcel Straub and for donating us their first MPI scanner. Work at UW was supported by NIH grant 2R42EB013520-02A1.

REFERENCES
[1] R. M. Ferguson, A. P. Khandhar, and K. M. Krishnan. *Journal of Applied Physics*, 111, 07B318, 2012. doi: 10.1063/1.3676053
[2] B. Gleich and J. Weizenecker. *Nature*, 435(7046):1217—1217, 2005. doi: 10.1038/nature03808.
[3] J. Rahmer, J. Weizenecker, B. Gleich, and J. Borgert. IEEE Transactions on Medical Imaging, 31(6): 1289 – 1299, 2012. doi: 10.1109/TMI.2012.2188639.
[4] A. Dax. *SIAM J. Sci. Comput.*, 14(3): 570–584, 1993. doi: 10.1137/0914036.

Multi-patch MPI allows whole body imaging of mice using a long circulating blood pool tracer

C. Jung[1,*], J. Salamon[1], P. Szwargulski[2,3], N. Gdaniec[2,3], M. Hofmann[2,3], M.G. Kaul[1], G. Adam[1], S.J. Kemp[4], R.M. Ferguson[4], A.P. Khandhar[4] and K.M. Krishnan[4,5],T. Knopp[1,2,3], H. Ittrich[1]

[1]Department of Diagnostic and Interventional Radiology, University Medical Center Hamburg-Eppendorf, Hamburg, Germany
[2]Section for Biomedical Imaging, University Medical Center Hamburg-Eppendorf, Hamburg, Germany
[3]Institute for Biomedical Imaging, Hamburg University of Technology (TUHH), Hamburg, Germany
[4]LodeSpin Labs, Seattle, Washington, USA
[5]Department of Materials Sciences & Engineering, University of Washington, Seattle, Washington, USA
* Corresponding author, email: cjung@uke.de

INTRODUCTION Magnetic particle imaging (MPI) is a new imaging technology that allows the direct quantitative mapping of the spatial distribution of superparamagnetic iron oxide nanoparticles (SPIO) [1]. Current SPIO contrast agents developed for MRI, when used off-the-shelf, are grossly inadequate for MPI – a mere 3% of nanoparticles in Resovist™ contribute to the MPI signal [2] – and simply do not translate very well for clinical applications. A variety of strategies has been proposed for the development of appropriate nanoparticular systems. One common approach for MPI is the synthesis of homogeneously distributed single-core SPIOs with a dedicated iron core diameter for ideal MPI characteristics [3]. Kandhar et al have already presented first results of a SPIO tracer which showed a 3-fold greater signal per unit mass, and 20% better spatial resolution compared to Resovist [4].

One further challenge regarding human applications is the limited field of view (FoV). The restriction in the size of the covered imaging area is needed because of two phenomena, the peripheral nerve stimulation (PNS) and tissue heating. The preclinical MPI scanner considered in this work has a maximum drive field strength of 14 mT offering a FoV of 22.4x22.4x11.2 mm^3 for a selection field strength of 2.5T/m. Accordingly whole body imaging even in small animals like mice is impossible.

In this study we used LS-008, a long circulating MPI tracer from LodeSpin Labs for *in vivo* whole body imaging in mice by implementing a multi-station approach [5].

MATERIAL AND METHODS MPI Scans of FVB mice (n=4) were carried out using a 3D imaging sequence (2.5 T/m gradient strength, 14mT drive-field strength, FoV 22.4x22.4x11.2 mm^3). Ten minutes after the injection of 60µl of LS-008 via the tailvein six different drive-field patches (two cranial, two median and two caudal) each taking 1.5 min were performed. As MPI delivers no anatomic information, MRI scans at 7T ClinScan (Bruker) were performed before MPI examination using a T2-weighted 2D turbo spin echo sequence. Fiducial markers were used to enable MRI/MPI image fusion.

Image reconstruction was performed offline using a custom reconstruction framework developed in the programming language Julia using the join formulation outlined in [5].

RESULTS The combined MRI/MPI measurements were carried out successfully. The reconstruction of the drive-field patches generated no artifacts at the margins resulting in a whole mice body MP imaging. Compared to previous experiments that we carried out using a gradient strength of 1 T/m [6] the multi-patch method with an increased gradient strength of 2.5 T/m resulted in a higher spatial resolution. Therefore we were able not only to visualize the inferior vena cava, the heart and the liver but also the cerebral vessels, the thoracic aorta and the kidneys (Fig. 1).

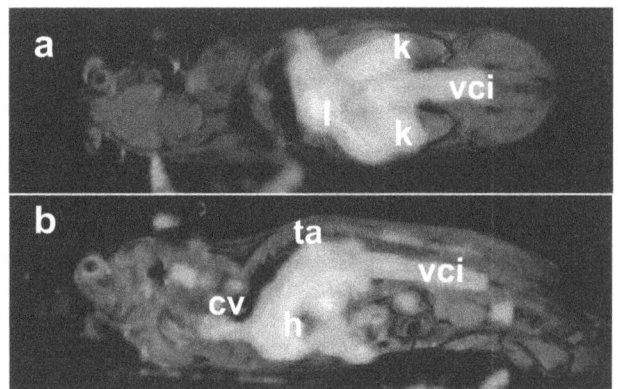

Figure 1: Coronal (a) and sagittal (b), MRI (grayscale)/MPI (yellow) fusion image of the whole body of a mouse. (vci = inferior vena cava, h = heart, l = liver, cv = cerebral vessels, ta thoracic aorta, k= kidney)

CONCLUSION *In vivo* whole body imaging of mice using multi-patch MPI is feasible. The long circulating blood tracer enabled us to visualize whole mice without motion artifacts that would occur using short half-life contrast agents with fast liver uptake. The presented technique may offer a strong tool for fast and radiation free whole body angiography for example in case of atherosclerotic disease.

ACKNOWLEDGEMENTS We thankfully acknowledge funding and support by the German Research Foundation (DFG, grant number AD 125 / 5-1) and the city of Hamburg. RMF, KMK, APK, SJK acknowledge NIH funding under 2R42EB013520-02A1.

REFERENCES
1.Gleich B, Weizenecker J. Tomographic imaging using the nonlinear response of magnetic particles. Nature 2005;435:1214-7.
2.Weizenecker J, Gleich B, Rahmer J, Dahnke H, Borgert J. Three-dimensional real-time in vivo magnetic particle imaging. Phys Med Biol 2009;54:L1-L10.
3.Panagiotopoulos N, Duschka RL, Ahlborg M, et al. Magnetic particle imaging: current developments and future directions. Int J Nanomedicine 2015;10:3097-114.
4.Ferguson RM, Khandhar AP, Kemp SJ, et al. Magnetic particle imaging with tailored iron oxide nanoparticle tracers. IEEE Trans Med Imaging 2015;34:1077-84.
5.Knopp T, Them K, Kaul M, Gdaniec N. Joint reconstruction of non-overlapping magnetic particle imaging focus-field data. Phys Med Biol 2015;60:L15-21.
6.Kaul MG, Weber O, Heinen U, et al. Combined Preclinical Magnetic Particle Imaging and Magnetic Resonance Imaging: Initial Results in Mice. Rofo 2015;187:347-52.

Preliminary results of a hybrid cardio vascular *in vivo* study using a highly integrated hybrid MPI-MRI system

Jochen Franke [a,b*], Nicoleta Baxan [c], Ulrich Heinen [a], Alexander Weber [a,d], Heinrich Lehr [a], Martin Ilg [a], Michael Heidenreich [a], Wolfgang Ruhm [a] and Volkmar Schulz [b]

[a] Bruker BioSpin MRI GmbH, Germany
[b] Physics of Molecular Imaging Systems, University RWTH Aachen, Germany
[c] Biomedical Imaging Centre, Imperial College London, United Kingdom
[d] Institute of Medical Engineering, University of Lübeck, Germany
* Corresponding author, email: jochen.franke@bruker.com

INTRODUCTION: In modern medicine, multimodal imaging is emerging to raise the diagnostic value by combining the complementary multimodal information. In 2005, the novel tracer-based imaging modality Magnetic Particle Imaging (MPI) was presented by Gleich and Weizenecker [1], allowing the electromagnetic detection of superparamagnetic iron oxide (SPIO) nanoparticle distributions in three dimensions (3D) with a high temporal resolution. While MPI is highly sensitive to the non-linear SPIO response, the acquired data possess a high signal-to-noise-ratio (SNR) with no background signal generated by e.g. biological tissue. This, however, requests for additional morphological information. This was achieved in [2] by two stand-alone MPI and Magnetic Resonance Imaging (MRI) systems, while in [3-6] hybrid MPI-MRI scanner designs with the capability of sequential data acquisition were presented. The latter systems allow reduced co-registration errors induced by object transportation and/or repositioning.

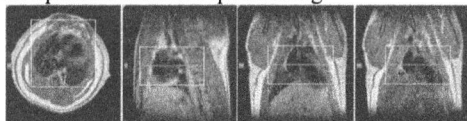

Figure 1: MPI FOV-planning virtually superimposed as white boxes to the high resolution MRI scans. LEFT TO RIGHT: pre injection axial, sagittal and coronal, post-injection coronal (note the darkening of the liver).

MATERIAL AND METHODS: For a cardio vascular *in vivo* study on one anesthetized rat (Lewis, 262 g body weight), a highly integrated hybrid MPI-MRI scanner [3,6] was used. The subject was scanned (*ParaVision6* software, Bruker BioSpin, Germany) sequentially as follows: **1) pre-injection MRI:** Low-resolution MRI scout scans (tri-planar FLASH sequence) allowed matching the desired imaging region to the predefined static 3D MPI field-of-view (FOV). 1st and 2nd order global shim were optimized in the final subject position using a Free-Induction-Decay based algorithm including a re-adjustment of the basic frequency. Three orthogonal high spatial resolution MRI datasets (MSME, TE/TR=8.6/300 ms, matrix=128×128×9, FOV=50×50×27, BW=25 kHz, AVG=25, scan time=16 min) were acquired. **2) MPI:** During MPI data acquisition ($A_{DFx,y,z}$=12 mT, G_{SFmax}=2.2 T/m, BW=625 kHz, repetitions=3000, scan time=65 s), one 20 µl bolus of undiluted Resovist (Bayer Schering Pharma AG, Germany) was administered manually into the tail vein. For the system function based image reconstruction, the bolus passage (repetition 130…350) was selected and background corrected (Kaczmarz [7], AVG=1, iterations=60, λ=8·10^{-3}, matrix=28×28×16, FOV=28×28×16 mm³). **3) post-injection MRI:** 1st and 2nd order shims were optimized as previously described to reduce susceptibility artifacts caused by the iron loaded liver. The coronal high spatial resolution MRI scan was repeated 40 min after tracer injection. **4) Data fusion and visualization:** The 3D coronal MRI pre-injection dataset and the 4D MPI dataset were loaded, fused, globally rigid translated according to pre-determined values (-2.5 mm, -1.5 mm, -1 mm, for x, y, and z) and visualized using PMOD 3.6 (PMOD Technologies Ltd, Switzerland).

RESULTS Fig. 1 shows the MRI-based MPI FOV-planning result as overlay to the high resolution MRI datasets. Fig. 2 depicts the fused coronal 3D MRI dataset (dark grayscale) with one time point of the time resolved 3D MPI dataset (light grayscale). The passage of the tracer bolus through the heart is visualized on the MPI image as it fills the right ventricle.

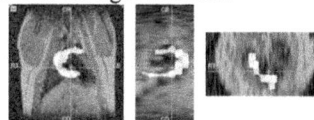

Figure 2: Orthogonal fused static 3D coronal MRI (dark grayscale) and one time point of the reformatted dynamic 3D MPI (light grayscale) dataset.

CONCLUSION: This study demonstrated the systems capability of sequential hybrid static 3D MRI and dynamic 3D MPI *in vivo* experiments with a maximal MPI temporal resolution of 21.45 ms/volume using a non-toxic dosage of Resovist. MRI based pre-injection MPI FOV planning proved to be a reliable method in the described setting. The rat was scanned sequentially within the same study with MRI, MPI and MRI without subject movement. The modality transition time was performed in less than 2.5 min in each direction. A global pre-determined rigid translation had to be used to compensate constant offset of the MPI and MRI coordinate system origins. For future experiments, this offset can be omitted by adjusting the MPI calibration robot coordinate system to the MRI gradient iso-center prior to a new system function acquisition.

IN VIVO PERMISSION & ACKNOWLEDGEMENTS: All animal experiments were conducted with the legal approval of the responsible Animal Ethics Committee (Regional Council of Karlsruhe, Germany; 35-9185.81/G-178/12). The authors thankfully acknowledge the financial support by the German Federal Ministry of Education and Research, FKZ 13N11088.

REFERENCES:

[1] B. Gleich and J. Weizenecker. Nature, 10.1038/nature03808.

[2] J. Weizenecker et al. Physics in Medicine and Biology, 10.1088/0031-9155/54/5/L01

[3] J. Franke et al. Proc. IWMPI 2013, 10.1109/IWMPI.2013.6528363

[4] J. Franke et al. Proc. IWMPI 2013, 10.1109/IWMPI.2013.6528367

[5] P. Vogel et al. IEEE TMI, 10.1109/TMI.2014.2327515.

[6] J. Franke et al. Proc. IWMPI 2015; 10.1109/IWMPI.2015.7106990

[7] S. Kaczmarz. Bull. Internat. Acad. Polon. Sci. Lett., 35:355357, 1937.

Systemic Real-time Cell Tracking with Magnetic Particle Imaging

Bo Zheng[a,*], Marc P. von See[b], Elaine Yu[a], Beliz Gunel[c], Kuan Lu[a], Tandis Vazin[d], David V. Schaffer[a,d], Patrick W. Goodwill[e], Steven M. Conolly[a,c]

[a] Department of Bioengineering, University of California at Berkeley, Berkeley, CA, USA
[b] Institute of Medical Technology, Hamburg University of Technology, Hamburg, Germany
[c] Department of Electrical Engineering, University of California at Berkeley, Berkeley, CA, USA
[d] Department of Molecular and Biochemical Engineering, University of California at Berkeley, Berkeley, CA, USA
[e] Magnetic Insight, Newark, CA, USA
* Corresponding author, email: bozheng@berkeley.edu

INTRODUCTION Magnetic Particle Imaging (MPI) offers robust, non-invasive, long-term, and quantitative stem cell tracking, making it suitable for clinical translation. We evaluated MPI for dynamic monitoring of systemically administered mesenchymal stem cells (MSCs). MSCs are of therapeutic interest since they can control inflammation and modify the proliferation and cytokine production of immune cells [1]. MSC-based therapies have shown promise for diseases such as stroke, myocardial infarction, and cancer [1, 2]. Intravenous injections have been used to deliver MSCs in both animal models and clinical trials [2-4]. However, it remains difficult to noninvasively monitor the delivery and biodistribution of administered cells into target organs (27, 29, 7). Recent studies have suggested that more than 80% of MSCs are entrapped in pulmonary vasculature following intravenous injection [2-4]. Here we demonstrate the first use of MPI to quantitatively track systemically administered MSCs.

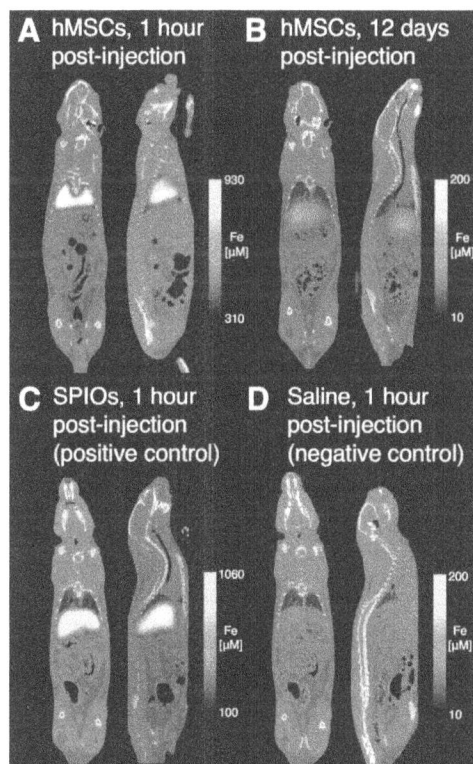

Figure 1. MPI-CT imaging of intravenously injected MSCs. MPI shows immediate entrapment of MSCs in the lungs (A) and eventual clearance to liver by day 12 (B). SPIO-only injections are immediately cleared (C), while saline injections show no signal (D).

MATERIAL AND METHODS Four groups of Fisher 344 rats (n = 3 each) received tail vein injections of 5×10^6 to 8×10^6 Resovist-labeled hMSCs, 100 µL Resovist, or 1 mL isotonic saline solution. The animals were imaged using MPI-CT for up to 12 days post-injection. MPI imaging for each time point (n = 4) was performed on a 3D FFP scanner with 7 T/m gradient 4× 3.75×10 cm FOV, 9 minute acquisition. CT imaging: 25 minute acquisition, 184 µm isotropic resolution.

RESULTS Similar to previous studies, our MPI studies indicated a rapid trafficking of intravenously injected MSCs to lung tissue, followed by clearance through the liver and spleen. MPI measurements of MSC liver clearance (4.6 day half-life) show agreement with previous studies [2-4]. These measurements are validated by gold-standard ICP data ($R^2 = 0.943$).

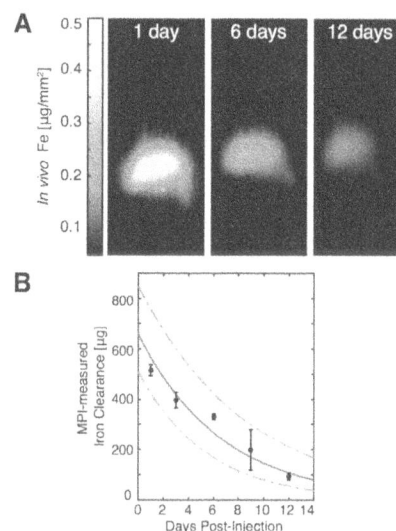

Figure 2. (A) MPI monitoring of MSC clearance in liver from 1 to 12 days post-injection. (B) *In vivo* MPI measurements of MSC clearance from liver ($\tau = 4.6$ days).

CONCLUSION These results indicate that MPI is highly useful for whole-body imaging measurements of the biodistribution and clearance of stem cell therapies in real time.

ACKNOWLEDGEMENTS We gratefully acknowledge funding from Siebel Scholars Foundation, NIH R01 EB013689, CIRM RT2-01893, Keck Foundation 009323, NIH 1R24 MH106053 and NIH 1R01 EB019458, and ACTG 037829.

REFERENCES
[1] E. Eggenhofer, F. Luk, M. H. Dahlke, M. J. Hoogduijn, Front. Immunol. 5, 1–6 (2014).
[2] M. T. Harting et al., J. Neurosurg. 110, 1189–1197 (2009).
[3] U.M. Fischer et al. Stem Cells Dev. 18:683–92 (2009).
[4] J Gao et al. Cells Tissues Organs. 169:12–20 (2001).

Session 6:

Tracer Materials 1

Keynote:

High Resolution Temperature Estimation by using Magnetic Nanoparticles

Prof. Wenzhong Liu

Huazhong University of Science and Technology, Wuhan, China

ABSTRACT Magnetic nanothermometry (MNTM) using magnetic nanoparticles (MNPs) has a unique property that allows temperature probing without the limitation of measuring depth. The MNTM using Langevin function and finite terms of Taylor expansion of Langevin function to model the magnetization and the inverse susceptibility of MNPs in dc magnetic field was proposed to measure temperature by our group.

The first experiments performed in MPMS SQUID VSM (QUANTUM Design, USA) were time-consuming with an accuracy of 0.57 K in the temperature range from 310 to 350 K. Under the calibration of Bloch's Law, maximum temperature estimation error of 0.022 K with a standard deviation of 0.017 K was achieved using the experiment data obtained in SQUID. In weak sinusoidal magnetic field, a real-time MNTM with maximum temperature estimation error of 0.67 K and standard deviation of 0.29 K in 1 s measurement was achieved, whereas the experiments shown the maximum temperature estimation error was 0.48 K with a standard deviation of 0.19 K in sinusoidal ac plus dc magnetic fields. The influence of particle size distribution on the accuracy of MNTM was also concerned in our recent study.

Localization of magnetic nanoparticles and its effect on magnetic relaxation evaluated by dynamic magnetization measurement for magnetic particle imaging

Yasushi Takemura* and Satoshi Ota

Department of Electrical and Computer Engineering, Yokohama National University, Japan
* Corresponding author, email: takemura@ynu.ac.j

EFFECTIVE RELAXATION TIME DERIVED FROM DYNAMIC HYSTERESIS

MEASUREMENT Understanding of dynamic response of magnetic nanoparticles (MNPs) is significant for optimizing material and applied field condition for magnetic particle imaging (MPI) [1]. AC susceptibility of MNPs has been studied both by experimentally and theoretically [2]. We have measured dynamic magnetization curves at the applied field frequency of 1-500 kHz [3]. This evaluation method is useful for research on MPI [4], and hyperthermia.

We measured AC hysteresis curves of water-based magnetite nanoparticles [3]. The average primary and hydrodynamic diameters were 11 nm and 52 nm, respectively. Figure 1 shows the intrinsic loss power, ILP of the samples, which is equivalent to the area of hysteresis curve. Although the magnetization of the particles is quickly rotated by the Néel relaxation with its relaxation time, τ_N, less than 10^{-6} s, the ILP was maximum at the frequency range of 1-30 kHz. The peak frequency corresponds to the Brownian relaxation time, τ_B, which may not be explained by an effective relaxation time, τ_{eff} as $1/\tau_{eff} = 1/\tau_B + 1/\tau_N$. The larger ILP of the fluid samples than that of the non-rotatable sample (shown as Fixed in the figure) at $f > 100$ kHz suggests the rotation of particles. This frequency range is much higher than frequency for the Brownian relaxation time.

LOCALIZATION OF MNPs AND DIPOLAR INTERACTION

Magnetic properties of intercellular MNPs were also studied. The prepared samples were MNPs dispersed in liquid, fixed in solid epoxy and added to cells. A MNP solution was added to HeLa cells. Figure 2 shows the microscopic image of detached HeLa cells [5]. Agglomerated MNPs were observed in the cytoplasm and on the membrane. We collected 7×10^7 cells and used as the cellular sample for dynamic hysteresis measurement. All the liquid, fixed and cellular samples exhibited a superparamagnetic feature without any remanence magnetization in their DC hysteresis curves. Figure 3 shows the AC hysteresis curves of the samples measured at 400 kHz. The magnetization under the applied field of 4 kA/m was highest in the liquid sample, because of the magnetization rotation due to Brownian relaxation. The magnetization of the cellular sample was reproducibly lower than that of the fixed sample, which can be explained by the dipolar interaction in the agglomerated MNPs. The dipolar interaction is also observed as a shift in the Brownian relaxation time depending on the density of fluid MNP samples in Fig. 1.

CONCLUSION

Measurements of dynamic hysteresis curves are valuable in understanding magnetic relaxation of MNPs for MPI.

ACKNOWLEDGEMENTS This work was partially supported by the JSPS KAKENHI Grant Number 15H05764 and 26289124.

REFERENCES

[1] B. Gleich, J. Weizenecker, *Nature*, **435**, 1214 (2005).
[2] T. Yoshida, K. Ogawa, K. Enpuku, N. Usuki, H. Kanzaki, *Jpn. J. Appl. Phys.*, **49**, 053001 (2010).
[3] S. Ota, T. Yamada, Y. Takemura, *J. Appl. Phys.*, **117**, 17D713 (2015).
[4] A. Tomitaka, R. M. Ferguson, A. P. Khandhar, S. J. Kemp, S. Ota, K. Nakamura, Y. Takemura, K. M. Krishnan, *IEEE Trans. Magn.*, **51**, 6100504 (2015).
[5] S. Ota, T. Yamada, Y. Takemura, *J. Nanomater.*, **2015**, 836761 (2015).

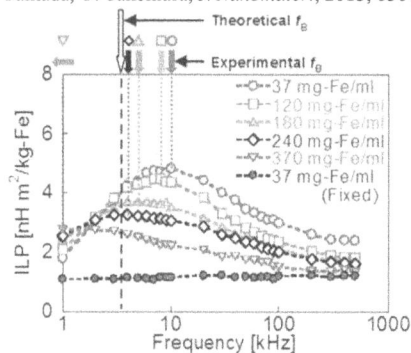

Fig. 1 Dependence of area of hysteresis loop (intrinsic loss power) on frequency for liquid and fixed MNPs.

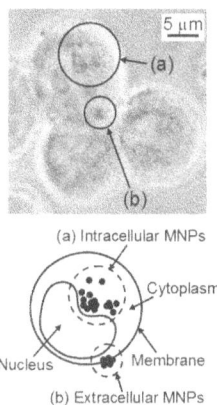

Fig. 2 (a) Intracellular and (b) extracellular MNPs

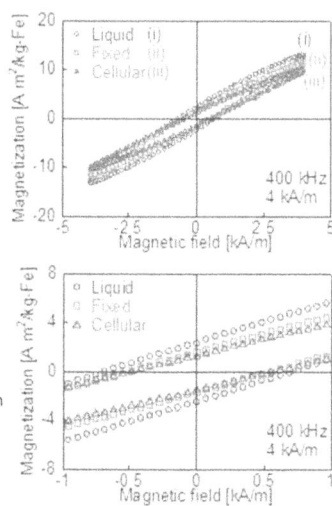

Fig. 3 Minor hysteresis curves of liquid, fixed, and cellular MNPs at 400 kHz.

In vivo velocity determination in the inferior vena cava in mice by Magnetic Particle Imaging and Magnetic Resonance Imaging

Michael G. Kaul[a,*], Tobias Mummert[a], Johannes Salamon[a], Martin Hofmann[a,b], Harald Ittrich[a], Gerhard Adam[a], Tobias Knopp[a,b], Caroline Jung[a]

[a] Department of Diagnstic and Interventional Radiology, University Medial Center Hamburg, Germany
[b] Hamburg University of Technology, Hamburg, Germany
* Corresponding author, email: mkaul@uke.uni-hamburg.de

INTRODUCTION The estimation of velocities of flowing blood in major vessels is of major importance in the diagnostics of cardio vascular diseases.

Magnetic particle imaging (MPI) is a new radiologic imaging modality providing a high temporal resolution. Angiography techniques with up to 30 fps have proven to be able to track contrast agent injections and to depict the velocity in flowing blood [1]. The goal of this study was to perform velocity measurements by bolus tracking in phantom and in vivo in mice in the inferior vena cava.

MATERIAL AND METHODS A tube with diameter of 1.3 mm was wrapped spirally around a cylinder to build a flow phantom. The inner diameter was chosen to have a similar mean diameter as a caval vein of a mouse. A drop of ferucarbotran (500 mM) was placed inside the tube surrounded by oil. The drop was pushed by an injection pump with velocities of 2, 5, 10, 14, 17, and 19 cm/s. The movement was scanned with a preclinical MPI scanner (Philips/Bruker) dynamically sampling the 3D volume within 21.5 ms. Each measurement was repeated five times.

In vivo measurements proved by a local animal care committee were carried out in ten healthy mice. For anatomic referencing MRI was applied [2]. Velocity encoded phase contrast MRI was performed in the hepatic and infrarenal part of the VCI. Mice were transferred to the MPI scanner. A bolus of 30μL ferucarbotran was injected during a dynamic MPI measurement by a syringe pump.

In MPI arrival times of the ferucarbotran drop as well as of the in vivo bolus were estimated. In the phantom the propagated distance was known, so that velocities could be calculated. Distances between the hepatic infrarenal region of the inferior vena cava were individually estimated from MPI data by two observers.

RESULTS Estimated velocities in the phantom were 1.8, 4.7, 9.8, 13.9, 17.0, 19.1 cm/s and in agreement to the reference (fig. 1).

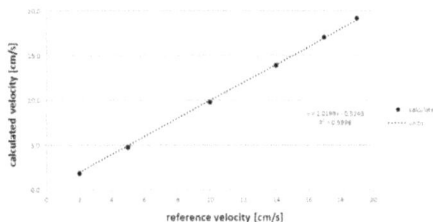

Figure 1: Scatter plot of reference velocities given by flow rates adjusted at the pump and estimated values by bolus tracking.

The animal experiments were performed successfully. Image quality was sufficient to reconstruct arrival time maps (fig.2).

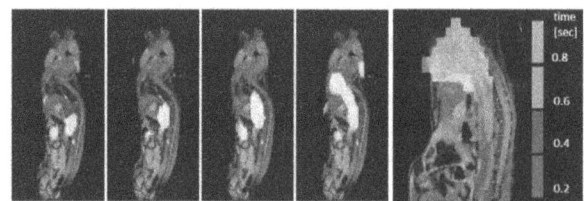

Figure 2: Bolus arrival in the inferior vena cava direction heat. Arrival time of the bolus were determined provided as parametric maps and co-registered with T2-weighted MRI data.

In vivo analysis of MPI revealed velocities between 3.8 (hepatic) and 7.4 cm/s (infrarenal). This was in good agreement to MRI with 2.7 and 7.6 cm/s (fig.3).

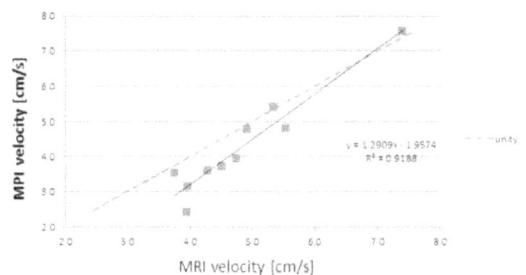

Figure 3: Velocity estimations by MRI and MPI in the inferior vena cava. Hepatic and infrarenal MRI values were averaged.

CONCLUSION MPI facilitates velocity measurements in the inferior vena cava. Phantom measurements approve that MPI is fast enough to detect velocities up to 19 cm/s.

ACKNOWLEDGEMENTS We gratefully acknowledge funding and support of the German Research Foundation. (DFG, grant number AD 125/5-1)

REFERENCES
[1] Simon D. Shpilfoygel, Robert A. Close, Daniel J. Valentino, and Gary R. Duckwiler . X-ray videodensitometric methods for blood flow and velocity measurement: A critical review of literature. *Medical Physics* 27(9), 2008-2023, 2000
[2] Michael G Kaul, Oliver Weber, Ulrich Heinen, Aline Reitmeier, Tobias Mummert, Caroline Jung, Nina Raabe, Tobias Knopp, Harald Ittrich, Gerhard.Adam. Combined Preclinical Magnetic Particle Imaging and Magnetic Resonance Imaging: Initial Results in Mice. *Fortschr Röntgenstr* 187: 347–352, 2015

Imaging brain cancer xenografts *in vivo* using tailored nanoparticles functionalized for glioma tumor targeting and MPI-NIRF contrast

Hamed Arami[a], Eric Teeman[a], Alyssa Troksa[a], Haydin Bradshaw[a], Denny Liggitt[b], and Kannan M. Krishnan[a]

[a] Department of Materials Sciences & Engineering, University of Washington, Seattle, Washington, USA
[b] Department of Comparative Medicine, University of Washington, Seattle, Washington, USA
* Corresponding author, email: kannanmk@uw.edu

INTRODUCTION We have developed a general PMAO-PEG coating platform for functionalizing the surfaces of our chemically synthesized iron-oxide nanoparticles[1] tailored for MPI [2]. We showed the potential of this coating platform for enhancing *in vivo* circulation time [3], multimodal imaging [4], and by conjugating lactoferrin, demonstrated their MPS efficacy in specific internalization of glioma cells *in vitro* [5]. Here, we demonstrate the targeting efficiency of lactoferrin-Cy5.5 conjugated NPs in a rodent model with brain cancer xenografts, using *in vivo* near infra red fluorescent (NIRF) imaging (IVIS system) and *ex vivo* magnetic particle spectroscopy (MPS).

MATERIAL AND METHODS Iron oxide NPs with median core size of 27nm were synthesized, thiolated lactoferrin was conjugated to their PMAO-PEG coatings and then purified using PD-10 columns. Cy5.5-NHS NIRF molecules were also conjugated to the lactoferrin molecules using methods reported earlier[4,5]. The right side flanks of athymic CD-1 nude mice (Charles River,; female, 8 weeks old) were injected with C6 glioma cells (1,000,000 cells in 100μL of DMEM-10%FBS cell culture media and 100μL matrigel). Tumor growth was monitored daily and when the tumor size reached ~10% of the body weight (after about 3-4 weeks) lactoferrin-conjugated tracers were injected via the tail vein. In vivo evaluation of the NPs uptake in tumors, 30 minutes and 2 hours after injection, was carried out using an IVIS system; then the animals were sacrificed and the magnetic and fluorescence signals of excised tumors were measured by IVIS and MPS (25 kHz). Mice were injected in four different groups with: (1) Lactoferrin-conjugated NPs and external magnet to enhance the targeting, (2) Lactoferrin without magnetic targeting, (3) NPs without lactoferrin on their surface to evaluate EPR perfusion, and (4) phosphate buffered solution (PBS) as a control.

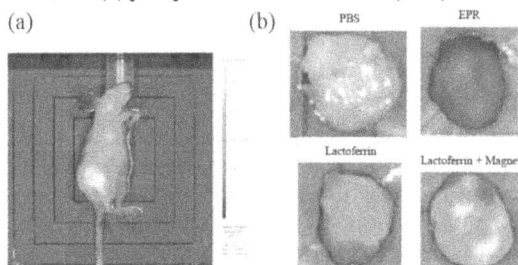

Figure 1: NIRF images of (a) brain cancer xenograft in nude mice 2 hours after injection and (b) excised tumors for the four different groups, where brightness is proportional to the uptake.

RESULTS Overall, our *in vivo* and *ex vivo* results showed that nanoparticles were accumulated in tumors based on three mechanisms: 1) enhanced permeation and retention (EPR) effect, based on the diffusion of particles through the tumors leaky vasculature (enhanced permeation) and subsequent accumulation in tumors (retention); 2) ligand (lactoferrin) assisted targeting and 3) magnetic targeting. These three mechanisms depend on

nanoparticles hydrodynamic size, surface coating and charge [6]. *In vivo* NIRF imaging showed a very small uptake in the tumors due to the EPR effect; however, the lactoferrin conjugated NPs were readily internalized into xenografts, with further enhanced uptake when we placed a magnet adjacent to the tumors (Figure 1a). NIRF images of the excised tumors (Figure 1b) showed that the tumor uptake, based on these mechanisms is cumulative, with the combination of magnetic and lactoferrin-assisted targeting showing the most increased uptake. These results were also confirmed by magnetic particle spectroscopy (MPS) analyses of the tumor tissues from the excised xenografts (Fig 2).

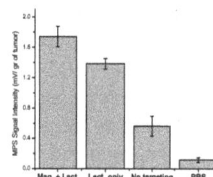

Figure 2: MPS signal intensity per tumor mass for the four different groups.

CONCLUSION Our functionalized platform, with glioma-targeting lactoferrin conjugated to the PMAO-PEG surface coatings of the optimized MPI tracers, enhances uptake by tumors. Magnetic targeting further improves this phenomenon. High tracer mass sensitivity and fast image processing of NIRF, in combination with MPS analyses (as a first measure of the MPI performance of the NPs) enabled us to monitor the NPs uptake by tumors and do preliminary investigation of the feasibility of *in vivo* MPI cancer diagnosis. NPs uptake by the tumors can be improved by increasing blood circulation and decreasing hydrodynamic size [6]. Our current focus is on improving tracer mass sensitivity which is required for future applications of MPI in detecting solid tumors at early stages.

ACKNOWLEDGEMENTS This research was supported by NIH grant 2R42EB013520-02A1. Animal studies were approved by UW IACUC.

REFERENCES

[1] R. Hufschmid *et al*, *Nanoscale*, **7** (2015) 11142-11154.
[2] R.M. Ferguson *et al*, *IEEE Trans. Med. Imag.* **34** (2015), 1077
[3] A.P. Khandhar *et al*, *IEEE Trans. Mag.* **51** (2015), 5300304
[4] H. Arami *et al*, *Biomaterials*, **52** (2015) 251-261
[5] Asahi Tomitaka-Kami, Sonu Gandhi, Hamed Arami and Kannan M. Krishnan, *Nanoscale*, **7** (2015), 16890-16898
[6] H. Arami, A. Khandhar, D. Liggitt, K.M. Krishnan, *Chem. Soc. Rev.*, **44**(2015) 8576 – 8607

Study on the *in vivo* survival of murine Ferucarbotran-loaded RBCs for their use as new MPI contrast agents

Antonella Antonelli[a], Carla Sfara[a], Ulrich Pison[b], Oliver Weber[c] and Mauro Magnani[a*]

[a] Department of Biomolecular Sciences, University of Urbino "Carlo Bo", Via Saffi 2, 61029 Urbino (PU), Italy
[b] Charité-Universitätsmedizin Berlin, CC7, Augustenburger Platz 1, 13353 Berlin, Germany
[c] Philips Medical Systems DMC GmbH, Röntgenstraße 24-26, D-22315 Hamburg, Germany
* Corresponding author, email: mauro.magnani@uniurb.it

INTRODUCTION Recently, the potential of red blood cells (RBCs) loaded with SPIO nanoparticles as a tracer material for magnetic particle imaging (MPI) to realize a blood-pool tracer agent with longer blood retention time for imaging of the circulatory system has been investigated [1, 2]. We have demonstrated that it is possible to encapsulate magnetic nanoparticles in human and murine RBCs without changing the main features of the natural cells using a procedure that permits a transient opening of cell membrane pores through a controlled hypotonic dialysis and successive isotonic resealing and reannealing of cells. Several SPIO nanoparticles, either commercially available or newly synthesized, have been evaluated for encapsulation into RBCs, and the results have shown that not all iron oxide nanoparticles can be efficiently encapsulated into RBCs, depending on several factors such as dispersant agent nature and nanoparticle size [3]. Therefore, it is very important to identify those nanoparticles potentially eligible for loading into red blood cells in order to produce SPIO-RBCs carriers available as intravascular contrast agents in the MPI application. Here, we report the results obtained applying the loading procedure of a new Ferucarbotran nanoparticle suspension (from TOPASS Clinical Solution) to murine RBCs. Although previous studies have showed a lower MPS spectral response of human Ferucarbotran-Loaded RBCs compared to bulk suspension [4], the reduced signal could be counterbalanced by superior in vivo stability of loaded RBCs compared to free nanoparticles. The in vivo survival of murine Ferucarbotran-Loaded RBCs in the blood circulation of mouse and iron biodistribution after their intravenous administration were also evaluated.

MATERIAL AND METHODS The encapsulation into murine RBCs of Ferucarbotran contrast agent, consisting of superparamagnetic iron oxide nanoparticles (total iron, 59.2 mg/ml) coated with carboxydextran and with a particle size in the range of 45-65 nm, was performed by same procedure of hypotonic dialysis, isotonic resealing and reannealing reported in ref.1. The loading procedure was performed using Ferucarbotran amounts ranging from 6 to 9 mg Fe added to 1 ml of dialysed RBCs. The concentrations of Ferucarbotran encapsulated in the murine RBCs were evaluated by an Avance-400 NMR Bruker spectrometer, as reported in ref.1 using a longitudinal relaxivity (r1) value of 1.353 $s^{-1}mM^{-1}$. In vivo experiments on mice (Female six-week-old ICR-CD-1® mice from Harlan) were performed to evaluate the presence of magnetic Ferucarbotran-Loaded-RBCs (L-RBCs) in the vascular system. Through a single bolus injection into mice retro orbital sinus 3 ± 0.4 Fe µmoles were administered by L-RBCs at 44% hematocrit or by bulk Ferucarbotran. At pre-set periods of time after injection, blood samples were withdrawn from ocular arteria of treated mice to study the pharmacokinetic through NMR T1 value measurements. The studies on organ biodistribution was determined by injection of 4.8 µmoles Fe by L-RBCs or free

Ferucarbotran suspension. Iron content in liver, spleen and kidney specimens of mice, euthanized at 13 days after treatment, was analysed (by ICP-OES 8000, PerkinElmer).

RESULTS NMR T1 value of murine L-RBCs ranged from 284 to 144 ms (versus 2370 ms of control RBCs) corresponding to an encapsulation of iron in the range of 2.5-5 mM with a cell recovery ranging from 35 to 50%. The survival of the L-RBCs in the bloodstream was monitored through T1 NMR measurements on blood samples taken at different times from treated mice (Fig.1A). Free Ferucarbotran injected in mice induces immediately a decrease in blood T1 value ($730ms\pm200$) still measurable at 1 hour ($1705ms\pm171$ corresponding to 0.4 Fe µmoles) but this effect disappears within 3 hours (2124 ms±132) from injection; on the contrary, the half-life of L-RBCs in bloodstream is around 48 hours when 1.3 ± 0.1 Fe µmoles are still present. In fact, the decrease of blood T1 value after L-RBCs injection is detectable and significantly different up to 9 days, after which the T1 values return to control values (2354 ms at 13 days). At this time, ICP-OES analyses on mice organs have showed that liver is the organ with highest Fe concentration with amounts that are higher in liver of L-RBCs treated mice (70%±1) respect to free Ferucarbotran treated mice (14%±9), Fig. 1B.

Figure 1: A) In vivo pharmacokinetic of murine Ferucarbotran-L-RBCs in mice. B) Fe organ biodistribution expressed as percentage of injected dose.

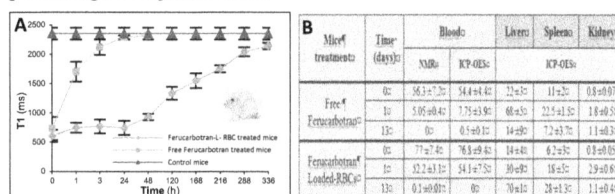

CONCLUSION Our investigations showed that it is possible to encapsulate the new ferucarbotran nanoparticles into murine RBCs similarly to human cells [4]. Moreover, the blood retention time of murine Ferucarbotran-L-RBCs is longer (~14 days) than the free contrast agent (~1 h) also evidenced by the slower Fe clearance from liver and spleen organs; these new constructs have a different *in vivo* kinetic removal respect to Ferucarbotran but are eliminated by similar routes.

ACKNOWLEDGEMENTS The authors would like to thank TOPASS Clinical Solution for providing the Ferucarbotran suspension.

REFERENCES
[1] A. Antonelli, C. Sfara, et al., Plos One 8, e78542, 2013.
[2] Rahmer J, Antonelli A, et al., *Phys Med Biol*, 58: 3965-3977, 2013.
[3] A. Antonelli, C. Sfara, et al., *Nanomedicine*, 6(2), 211-223, 2011.
[4] Antonelli A, Weber O, et al. In *Magnetic Particle Imaging (IWMPI)*, 5th International Workshop, Istanbul, Turkey, March 23-24, 2015.

Session 7:

Tracer Materials 2

Blood half-life of a long-circulating MPI tracer (LS-008)

Amit P Khandhar[a*], Paul Keselman[b], Scott J Kemp[a], R Matthew Ferguson[a], Bo Zheng[b], Patrick W Goodwill[c], Steven M Conolly[b] and Kannan M Krishnan[a,d]

[a] LodeSpin Labs, PO Box 95632, Seattle WA; [b] University of California, Department of Bioengineering, Berkeley CA; [c] Magnetic Insight, Inc., Newark CA; [d] University of Washington, Department of Materials Science & Engineering, Seattle WA.
[*] Corresponding author, email: amit@lodespin.com

INTRODUCTION Knowledge of pharmacokinetic (PK) properties of superparamagnetic iron oxide (SPIO) tracers is critical to understanding their capabilities for *in vivo* MPI. For applications requiring intravenous administration (e.g. cardiovascular, stroke and molecular imaging), blood half-life is an important PK parameter that defines the time it takes for the tracer to lose half its initial concentration in blood. For MPI, it is also important to evaluate the loss in signal, which correlates directly to imaging capability and may or may not follow the same loss profile as concentration. In this study, we evaluate the blood half-life of LS-008 – a long-circulating high resolution MPI tracer – in mice using both, *ex vivo* magnetic particle spectroscopy (MPS) measurements and *in vivo* imaging in a MPI scanner.

MATERIAL AND METHODS LS-008 tracer was synthesized using the thermolysis of iron oleate [1] and coated with poly(maleic anhydride-alt-1-octadecene) polymer grafted with 20 kDa methoxy-PEG-amine. Core diameter (d_C) of SPIO cores was measured using transmission electron microscopy, and hydrodynamic diameter (d_H) and zeta potential (ζ) in water were measured using dynamic light scattering. CD-1 female mice were used for all blood half-life experiments. LS-008 in 1X PBS was injected as a bolus at 5 mgFe/kg dose in one of the lateral tail veins. For half-life evaluation with MPS, we used 3 mice (6-8 weeks old) per time point (9 mice total to generate 6 time points). Blood collected retro-orbitally was analyzed in our MPS (25 kHz, 20 mT/μ_0). For half-life evaluation using imaging, a total of 3 mice (12-15 weeks old) were used. Mice were imaged with respiratory gating in a 3.5 T/m x 3.5 T/m x 7 T/m imager, with 40mT$_{p-p}$ drive field in z and a 3.5cm x 3.5cm x 8.5cm FOV. Scans (~7 min. duration) were done at 30 minute intervals with the longest time point at 6 hours after injection. We selected the neck region for signal analysis due to the presence of major blood vessels and its relative isolation from perfused tissues in the abdomen and head regions. Animal experiments were performed according to approved protocols at UW and UCB.

RESULTS Physical properties of LS-008 are provided in *Table 1*, including signal intensity and full width at half maximum (*FWHM*) of the differential susceptibility (χ_d) measured in MPS.

Table 1: Physical properties of LS-008.

	d_C [nm] (σ)[a]	d_H [nm][b] (PDI)	ζ [mV]	χ_d [m^3/gFe]	*FWHM* [mT/μ0]
LS-008	24.3 (0.1)	78.3 (0.1)	− 4.5	2.74E-05	6.0

a – σ is the standard deviation assuming a lognormal size distribution

b – d_H is the Z-average diameter based on the intensity measurement in DLS

MPS signal from blood samples was converted to concentration (μgFe/ml) using LS-008's known signal per unit mass, and plotted as a function of time in *Figure 1a*. Since MPS measures the scanner-independent signal from the tracer, it provides a crucial snapshot of the physicochemical state of SPIO in blood. The χ_d at different time points was compared with the native

response in water (data not shown here) to ensure tracer performance was preserved during circulation. Data from MPS showed excellent agreement with a first-order elimination profile (R^2 = 0.99). Blood half-life ($t_{1/2}$) from *Figure 1a* was 105 ± 8 minutes, calculated from: $t_{1/2} = \frac{0.693}{\lambda}$, where λ is the elimination rate constant determined from the fit. In the *in vivo* imaging experiment, $t_{1/2}$ values from 3 mice were averaged. There are several advantages to this method: first, each mouse provides the entire profile, *removing subject-to-subject variability within a curve* but *highlighting variability between subjects*. Secondly, no blood is collected, thus the number of time points acquired per mouse is only limited by imaging frequency. *Figure 1b* shows SPIO elimination curves (in μgFe) from all 3 mice. The average $t_{1/2}$, calculated assuming first-order elimination kinetics, was 103 ± 46 minutes from imaging. This agrees well with the MPS method but there is high variability between mice. Thus, more animals would be needed to achieve good statistical significance from the imaging experiment.

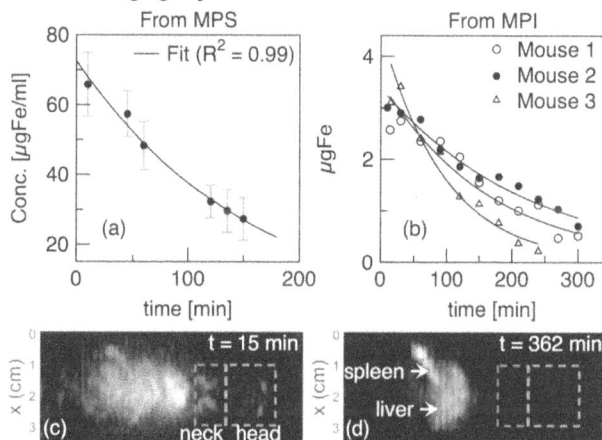

Figure 1: LS-008 elimination profiles from (a) MPS and (b) imaging in a MPI scanner. Coronal view of *mouse 3* showing maximum intensity projections at (c) 15 and (d) 362 minutes. In (c), the brain boundary and vessels in the neck region are clearly visible. In (d), 362 minutes after injection LS-008 is mostly distributed in the liver and spleen.

CONCLUSION The development of long-circulating MPI tracers is critical to demonstrating new MPI applications. We evaluated the blood half-life of LS-008, a long circulating PEG-ylated MPI tracer, using *ex vivo* MPS and *in vivo* MPI imaging. Both methods are complementary and help correlate tracer behavior with imaging capability. LS-008, injected intravenously, showed $t_{1/2}$ of over 100 minutes in mice.

ACKNOWLEDGEMENTS Work at LSL and UW supported by NIH/NIBIB R42EB013520. Work at UCB supported by NIH 1R01EB013689-01, NIH 1R24MH106053-01, NIH 1R01EB019458-01.

REFERENCES

[1] S. J. Kemp, R. M. Ferguson, A.P. Khandhar and K.M. Krishnan. *IEEE Xplore – 2015 IWMPI* doi: 10.1109/IWMPI.2015.7107016

Concentration Dependent MPI Tracer Performance

Norbert Löwa*, Patricia Radon, Olaf Kosch, Frank Wiekhorst

Physikalisch-Technische Bundesanstalt, Abbestr. 2-12, 10587 Berlin, Germany
* Corresponding author, email: norbert.loewa@ptb.de

INTRODUCTION Magnetic Particle Imaging (MPI) is a promising method capable for quantitative visualization of magnetic nanoparticles (MNP) exploiting their nonlinear dynamic magnetization curve. Obtaining the spatial MNP distribution from the measured MPI signals generally requires image reconstruction. Therefore, current commercially available systems use a calibration scan of a point-like reference at every possible position [1]. Similarly, Magnetic Particle Spectroscopy (MPS), which is ideally suitable to quantify MNP in different environments, requires the signal of a reference of known MNP content [2]. However, all reference sample based approaches presume that the dynamic magnetization behavior of MNP does not change, e.g. caused by viscosity changes [3], aggregation [2] or concentration dependent dipole-dipole interactions [4].

Here, we investigated Resovist® (Bayer HealthCare, GER) and its precursor Ferucarbotran (Meito Sangyo, JPN) with respect to concentration dependent MPS signal change. Both MNP types have been thoroughly investigated for their magnetic properties [4][5] and were intensively studied in the field of MPI [1][6].

MATERIAL AND METHODS To determine the effective magnetic domain size of the MNP, we measured the quasistatic magnetization $M(H)$ as a function of applied field strength (up to 5 T) of the samples by means of a commercial DC-susceptometer (MPMS XL, Quantum Design, USA) and fitted the magnetization data using the Langevin model. Furthermore, the MNP samples were measured by dynamic light scattering (DLS) (Zetasizer NanoZS, Malvern, UK) to determine the volume weighted hydrodynamic size distributions. Using a magnetic particle spectrometer (Bruker Biospin, GER) MPS measurements (B_{excit}=5 mT up to 25 mT, f_{excit}=25 kHz) on liquid FER and Resovist® samples at different iron concentrations (1:1 down to 1:10) were performed to study the concentration dependent MPS signal change.

RESULTS We found that Resovist® and its magnetically active ingredient FER are only slightly different according to size and magnetic properties in higher dilutions as measured by DLS and DC-susceptometry, respectively. The similarity was also found for MPS signals of diluted FER and Resovist®. Below 100 mmol/L no significant differences in the moment spectrum and small differences in the phase spectrum could be observed. Remarkably, Resovist® shows an unexpectedly strong concentration dependence of the MPS signal (Fig. 1). Above an iron concentration of about 150 mmol/L the moment and phase spectra exhibit a wave-like shape developing with increasing iron concentration. As MNP chain formation could be excluded this effect can be attributed to probable dipole-dipole interactions at high concentrations which was already reported before [4]. Interestingly, this effect is not present in FER even at a two-fold higher concentration.

CONCLUSION Our experimental results indicate that the dynamic magnetic behaviour of an MPI tracer may be concentration dependent and should be studied prior to MPI experiments. Particularly, for MPI experiments using Resovist® it may be necessary to record a system function using a reference sample at an iron concentration below 150 mmol/L to guarantee a valid image reconstruction of a sample having much lower MNP concentrations. Moreover, a system function acquired with highly concentrated RES may be only valid for the image reconstruction of a small range of tracer concentrations. In contrast, for FER we found no concentration dependence of the dynamic magnetic behavior even though a two-fold higher initial concentration was used. FER should thus be preferred for phantom experiments where full in vivo compatibility is not mandatory. MPI experiments investigating the concentration dependence of MNP under 3D excitation are important to validate the relevance of our spectroscopic measurement results.

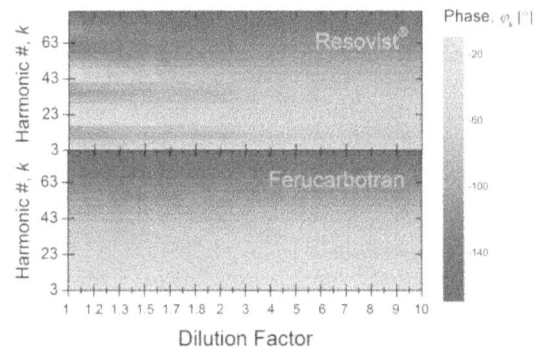

Figure 1: 2D color scale map of MPS phase signal j depending on dilution factor and harmonic number k. As it can be seen the phase of FER did not change with dilution whereas for Resovist® the wave-like shape of the phase disappears at a dilution factor of about 3.

ACKNOWLEDGEMENTS The research was supported by German Research Foundation, through DFG Research Unit FOR917 (Nanoguide) and German Ministry of Education and Research under Grant FKZ 13N11092 (MAPIT).

REFERENCES
[1] M.G. Kaul, H. Ittrich, O. Weber, U. Heinen, A. Reitmeier, T. Mummert, C. Jung, N. Raabe, T. Knopp, and G. Adam. *RoFo: Fortschritte auf dem Gebiete der Röntgenstrahlen und der Nuklearmedizin*, 187(5): 347—352, 2015. doi: 10.1055/s-0034-1399344.
[2] W.C. Poller, N. Löwa, F. Wiekhorst, M. Taupitz, S. Wagner, K. Möller, G. Baumann, V. Stangl, L. Trahms, and A. Ludwig. *J. Biomed. Nanotech.*, 12(2): 337—346, 2016. doi: 10.1166/jbn.2016.2204.
[3] T. Wawrzik, T. Yoshida, M. Schilling, and F. Ludwig. *IEEE Trans. Magn.*, 51(2): 1—4, 2015. doi: 10.1109/TMAG.2014.2332371.
[4] D. Eberbeck and L. Trahms. *J. Magn. Magn. Mater.*, 323(10): 1228—1232, 2011. doi: 10.1016/j.jmmm.2010.11.011.
[5] A.F. Thünemann, S. Rolf, P. Knappe, and S. Weidner. *Anal. Chem.*, 81(1): 296—301, 2009. doi: 10.1021/ac802009q.
[6] J. Rahmer, A. Halkola, B. Gleich, I. Schmale, and J. Borgert. *Phys. Med. Biol.*, 60(5): 1775—1791, 2015. doi: 10.1088/0031-9155/60/5/1775.

MPI Analysis of Metal Doped and Anisotropic Nanoparticles

Lisa M. Bauer[a], Shu F. Situ[b], Mark A. Griswold[a,c], Anna Cristina S. Samia[b,*]

[a] Department of Physics, Case Western Reserve University, Cleveland, Ohio, USA
[b] Department of Chemistry, Case Western Reserve University, Cleveland, Ohio, USA
[c] Department of Radiology, Case Western Reserve University and University Hospitals of Cleveland, Cleveland, Ohio, USA
[*] Corresponding author, email: anna.samia@case.edu

INTRODUCTION A major challenge to the growth of MPI is the need for tracers. While recent developments are promising, it would be advantageous to have a large collection of tracers that enable high-resolution imaging with large SNR.[1,2] Historically, MPI tracer development has focused on optimizing spherical magnetite nanoparticles, and little effort has been devoted to studying the effects of metal doping or shape anisotropy.

In this study, we examined the effect of metal doping on the MPI signal, as well as comparing spherical nanoparticles with cubic nanoparticles. As modifying nanoparticle properties also modifies relaxation, we used a magnetic nanocomposite platform in our studies, to minimize the effect of Brownian relaxation.[3]

MATERIAL AND METHODS Magnetite and Zn-doped magnetite nanoparticles were synthesized using a modified thermal decomposition method.[4-6] The magnetic nanocomposite films were fabricated using a liquid-solid compounding method; dispersed suspensions of the magnetic nanoparticles and UHMWPE were dried in vacuum and compression molded into films with 0.2 inch thickness at 200° C and 10 metric tons pressure. The final incorporation of magnetic nanoparticles was evaluated through thermal gravimetric analysis. [3]

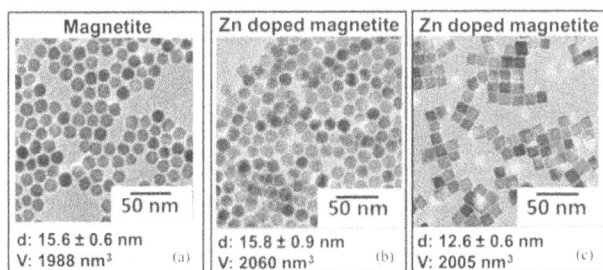

FIGURE 1. TEMs of (a) magnetite spheres, (b) doped magnetite spheres, and (c) doped magnetite cubes.

RESULTS Doped magnetite nanospheres (DS15) displayed a 17.6-fold increase in peak signal as compared to un-doped magnetite nanospheres (S15), as well as a decrease in FWHM. Doped magnetite nanocubes of both sizes (DC12 and DC50) outperformed magnetite nanospheres, in both SNR and FWHM. DC12 displayed a 27.75-fold increase in signal as compared to S15, and a 1.58-fold increase in signal over DS15. The SNR of DC50 was similar to that of DS15, though with larger FWHM. These results are summarized in Table 1 and in Figure 2.

CONCLUSION In developing new MPI tracers, it will be useful to consider the effects of metal doping and shape anisotropy. We believe that the enhanced MPI performance of doped nanospheres compared to magnetite spheres is due to the

TABLE 1. FWHM and SNRx summary. SNRx is the SNR normalized to the SNR of nanocomposites prepared with 12 nm doped nanocubes (DC12).

Name	Description	FWHM(mT)	SNRx
S15	Magnetite Spheres–15nm	13.3	27.75
DS15	Doped Magnetite Spheres–15nm	9.5	1.58

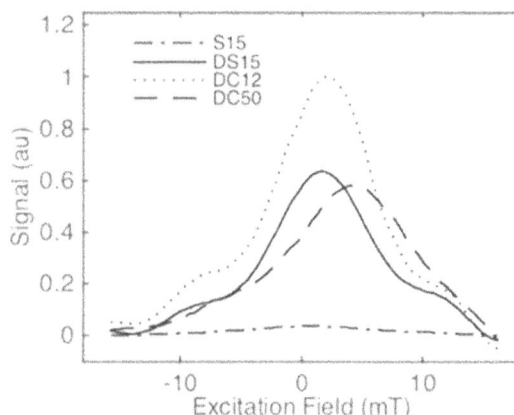

FIGURE 2. PSFs normalized to DC12. PSFs were measured using CWRU's relaxometer, with f_0=16.8kHz and B_{pp}=40mT.

increased saturation magnetization of Zn-doped magnetite, and that the enhanced performance of the nanocubes is due to the surface spin orientation.

ACKNOWLEDGEMENTS This work was supported in part by the National Cancer Institute's Training Programs in Cancer Pharmacology (5R25CA148052), grant number 1R24MH106053-01 from the National Institutes of Health, and an NSF-CAREER Grant (DMR-1253358) from the Solid State and Materials Chemistry Program.

REFERENCES
[1] L.M Bauer, S.F. Situ, M.A. Griswold, A.C.S. Samia. *J Phys Chem Lett*, 6(13):2509-2517, 2015. doi:10.1021/acs.jpclett.5b00610.
[2] R.M. Ferguson, et al. *IEEE Trans Med Imag*, 34(5):1077-1084, 2015. doi:10.1109/TMI.2014.2375065.
[3] M.H. Pablico-Lansigan, S.F. Situ, A.C.S. Samia. *Nanoscale*, 5(10):4040-4055, 2013. doi:10.1039/c3nr00544e.
[4] J. Jang, et al. *Angew Chemie Int. Ed.*, 48(7): 1234-1238, 2009. doi:10.1002/anie.200805149.
[5] D. Kim, et al. *J Am Chem Soc*, 131(2):454-455, 2009. doi:10.1021/ja8086906.
[6] S. Noh, et al. *Nano Lett*, 12(7):3716-3721, 2012. doi:10.1021/nl301499u

Oncogenic protease detection using magnetic particle spectrometry

Sonu Gandhi[#], Hamed Arami and Kannan M. Krishnan[*]

Department of Materials Sciences & Engineering, University of Washington, Seattle, Washington, USA
[*]Corresponding author, email: kannanmk@uw.edu
[#]Present Address: Amity Institute of Biotechnology, Amity University, Noida, India

INTRODUCTION Malignant phenotypes of solid tumors express specific proteases; the latter are conventionally detected using fluorescent molecules or radioisotopes, with associated major drawbacks of fluorescent quenching and health hazards caused by radionuclides. Here, we describe a new method for detecting proteases by sensitively monitoring the magnetic relaxation of monodisperse iron oxide nanoparticles (IONPs) using magnetic particle spectrometery (MPS)[1]. The key to this assay is the design of specific and tailored linker peptides that function as activatable nanosensors linking iron oxide nanoparticles. Further, these peptides are designed to possess selective sites that are recognizeable and cleaveable by specific proteases to be detected. When these linker peptides, labeled with biotin at N– and C– terminals, are added to the neutravidin functionalized IONPs, nanoparticles aggregate, resulting in well-defined changes in the MPS signal. However, as designed, in the presence of proteases these peptides are cleaved at predetermined sites, redispersing IONPs, and returning the MPS signal(s) close to its pre-aggregation state. These changes observed in all aspects of the MPS signal (peak intensity, its position as function of field amplitude, and full width at half-maximun – when combined, these three also eliminate false positives), help to detect specific proteases, relying only on the changes in magnetic relaxation characteristics of the functionalized IONPs (figure1).

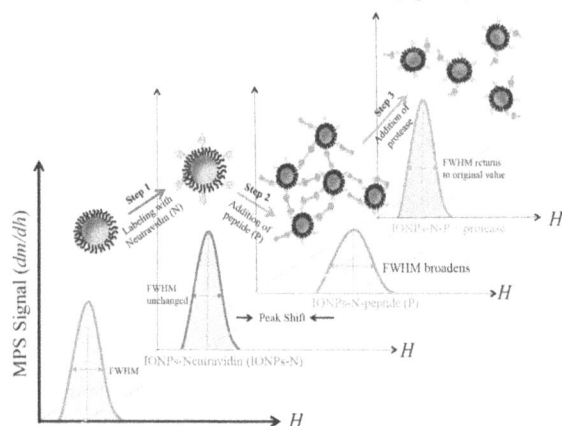

Figure 1: Basic principles of the MPS assay for protease detection. Specific linker peptides cause nanoparticle agglomeration. The presence of proteases which specifically cleave these linker peptides redisperse the nanoparticles. These changes affect the magnetic relaxation characteristics of the nanoparticles which can be reproducibly detected in the MPS signal as shown.

MATERIAL AND METHODS Highly monodisperse (dia = 22 ±1.94 nm) IONPs were prepared by thermal decomposition of iron oxyhydroxide (FeOOH) precursor in the presence of oleic acid surfactant[2]. IONPs were coated with PMAO polymer followed by hydrolysis of maleic anhydride groups of PMAO and generation of free carboxyl groups (– COOH) on the surface of the IONPs.

The, neutravidin was covalently conjugated to carboxyl functionalized IONPs, using carbodiimide chemistry. Two specific linker peptides — Biotin- GPARLAI-K-Biotin (GK-8) for trypsin and Biotin-GGPLGVRGK-Biotin (GK-9) for MMP-2 — were designed with specific protease cleavage sites, flanked with biotin at both ends to facilitate binding to neutravidin. All steps of the subsequent assay, i.e. agglomeration, peptide cleavage by proteases and redispersion of the IONPs were monitored by a home-built magnetic particle spectrometer (MPS).

RESULTS The MPS signal — peak height, half-width and position —is highly dependent on the relaxation mechanism (*i.e.* Néel, Brownian or hysteretic reversal) of the nanoparticles, which in turn, for a fixed applied frequency, is a strong function of their size, inter-particle interactions, surface functionalization, and the surrounding environment[3]. Addition of peptides (either GK-8 and GK-9) to IONPs-N causes immediate aggregation of the nanoparticles. The MPS data, $dm(H)/dH$, showed the clearest indication of particle binding and agglomeration, with peak heights decreasing to 5 and 3 mV from 13mV, and FWHMs increasing to 20 and 30 $mT\mu_0^{-1}$ from 10 $mT\mu_0^{-1}$. Addition of trypsin and MMP-2 proteases to the aggregated IONPs complexes cleaved the specific linker peptides, redispersing the aggregated IONPs, decreasing the FWHMs to 14 and 18 $mT\mu_0^{-1}$ and increasing MPS signal intensities to 12 and 10 mV . Finally, the *in-vitro* efficacy of this technique for detection of proteases secreted from two representative cancer cell lines (*i.e.* pancreatic carcinoma for trypsin, and fibrosarcoma cells for MMP-2) was independently confirmed. Details will be presented.

CONCLUSION We demonstrated the general utility of this rapid and low cost MPS-based assay by detecting one each from the two general classes of proteases, trypsin (a digestive serine protease, involved in various cancers, promoting proliferation, invasion and metastasis) and matrix metalloproteinase (MMP-2, observed through metastasis and tumor angiogenesis), using specifically designed linker peptides.

ACKNOWLEDGEMENTS This research was supported by grants NIH 2R42EB013520-02A1 and 1RO1EB013689-01/NIBIB

REFERENCES
[1] Gandhi S, Arami H, andand Krishnan KM, *Proc. Nat. Acad. Sci.* (submitted)
[2] Hufschmid R, et al *Nanoscale* 2015; 7(25):11142-11154.
[3] Arami H, Ferguson RM, Khandhar AP, Krishnan KM (2013) *Med Phys* 40(7): 071904

New Synthetic Route to Magnetic Multicore Particles for Magnetic Particle Imaging and other Biomedical Applications

Harald Kratz[a],*, Dietmar Eberbeck[b], Olaf Kosch[b], Jochen Franke[c], Ines Gemeinhardt[a], Susanne Wagner[a], Matthias Taupitz[a], Lutz Trahms[b] and Jörg Schnorr[a]

[a] Department of Radiology, Charité – Universitätsmedizin Berlin, Berlin, Germany
[b] Physikalische Technische Bundesanstalt, Berlin, Germany
[c] Bruker BioSpin MRI GmbH, Ettlingen, Germany
* Corresponding author, email: harald.kratz.@charite.d

INTRODUCTION Magnetic nanoparticles for use as in vivo magnetic particle imaging (MPI) tracers have to meet some criteria beyond good magnetic particle spectroscopy (MPS) and MPI performance. These criteria include biocompatibility, biodegradability, and stability of particle suspensions at physiologic pH [1]. Therefore, the substances and materials used for synthesis require careful selection with regard to the intended nanoparticle use. In addition to thermal decomposition, the coprecipitation method is widely used for nanoparticle synthesis because of it is simple and requires no organic solvents or high temperatures. Furthermore, no phase transfer to water of particles is necessary after synthesis.

MATERIAL AND METHODS Magnetic multicore particles (MCP) were synthesized using an aqueous synthesis route with green rust as intermediate stage. After oxidation by hydrogen peroxide, the maghemite/magnetite particles were coated with carboxymethyl dextran (CMD) and annealed at 90°C for several hours. After synthesis the MCP were washed by ultra filtration and fractionated by repeated magnetic separation [2]. MCP size and structure were examined by transmission electron microscope (TEM). Hydrodynamic diameters were measured by dynamic light scattering (DLS), and the particles were characterized by M(H) measurement and MPS. Initial phantom and in vivo investigations (in rats) were performed (see Fig. 1). Tracer degradation in the liver was assessed by magnetic resonance imaging (MRI). MCP tolerance and tracer MPI performance were tested.

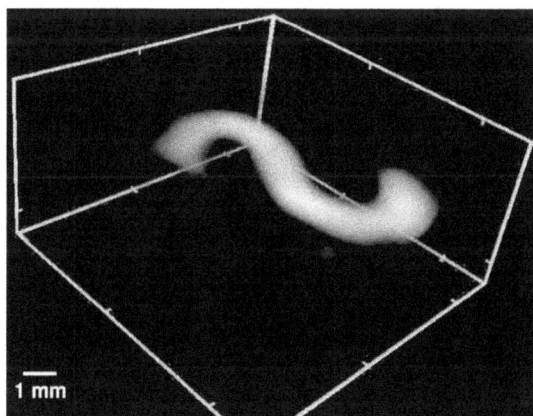

Figure 1: MPI image of a phantom (0.139 mM Fe/ml MCP, Silicon hose pipe (Ø = 0.9 mm))

RESULTS The synthetic route presented here yields multicore particles with improved tracer performance compared to Resovist® (see Fig. 2).

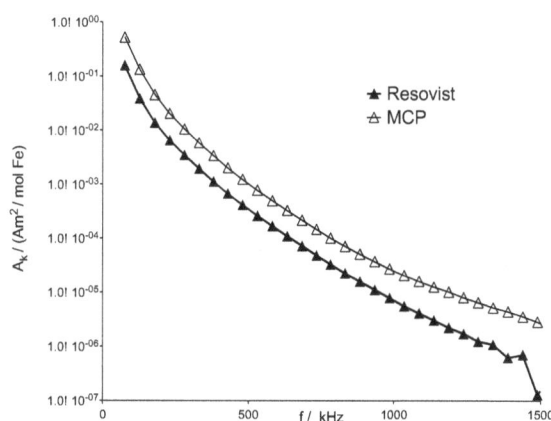

Figure 2: MPS (10 mT, 25 kHz) of multicore particles (MPC) in comparison with Resovist

TEM investigations revealed two different types of particles with mean core diameters of approx. 30 nm. The hydrodynamic size distribution (DLS) was in the range of 28-59 nm by volume. The in vivo rat experiments showed good compatibility, resulting in an No Observed Adverse Effect Level (NOAEL) > 3 mmol Fe/kg body weight. MRI demonstrated degradation within 6 weeks at dosages up to 0.1 mmol Fe/kg body weight. In vivo circulation times measured by MRI were comparable to Resovist® and first in vivo MPI results in rats suggest that the MCP are highly suitable for MPI.

CONCLUSION Besides thermal decomposition, aqueous routes are a promising alternative for MPI tracer development. The notorious problem of particle size control is compensated for by simplicity of synthesis in combination with subsequent magnetic separation. The MPI performance of the resulting MCP should be further improved by optimization of synthesis and separation techniques.

ACKNOWLEDGEMENTS We thank Bettina Herwig for language editing and Sören Selve for TEM investigations at ZELMI (TU Berlin). The research was supported by the German Ministry for Education and Research (BMBF), Grant Nos. FKZ 13N11091, 13N11092 and 13N11088 and by grants from the European Fund for Regional Development (EFRE), Investitionsbank Berlin (IBB), Grant No. 10146995 and the German Research Foundation DFG (Förderkennzeichen TA 166/9-1)

REFERENCES
[1] H. Kratz et al., *Biomedizinische Technik. Biomedical engineering*, 2013, 58(6):509-515, 2013. doi:10.1515/bmt-2012-0057.
[2] H.Kratz, et al., WO2013150118 A1 (October 10, 2013)

Session 8:

Instrumentation 2 / Methodology 3

Imaging and Localized Nanoparticle Heating with MPI

Daniel Hensley[a,*], Patrick Goodwill[b], Rohan Dhavalikar[c], Zhi Wei Tay[a], Bo Zheng[a], Carlos Rinaldi[c,d], Steven Conolly[a]

[a] Departments of Bioengineering and EECS, University of California, Berkeley
[b] Magnetic Insight, Inc.
[c] Department of Chemical Engineering, University of Florida, Gainesville
[d] J. Crayton Pruitt Family Department of Biomedical Engineering, University of Florida, Gainesville
[*] Corresponding author, email: dwhensley@berkeley.edu

INTRODUCTION Magnetic fluid hyperthermia (MFH) is a promising approach for drug delivery and the treatment of cancer [1, 2]. Previously, groups have separately explored spatial selectivity in MFH [3] and MPI for MFH pre-planning [4]. Here we show results combining MFH and MPI for both imaging and spatial selectivity. A typical MPI scan does not result in detectable heating. However, the gradient field used in MPI can be leveraged to spatially localize or focus the specific absorption rate (SAR) induced by particles in a typical MFH excitation regime (Fig. 1). We require modifications to the MPI transmit system [5, 6] to allow for an MFH mode. Here we show experimental results demonstrating spatial selectivity of nanoparticle heating (heating rate: 2 degrees C per minute) using a prototype MPI field-free line (FFL) magnet system and subsequent imaging of the same phantom with an MPI field-free point (FFP) scanner (Fig. 3).

MATERIAL AND METHODS We built a prototype MPI-based heating system using a 2.4 T/m FFL magnet and a resonant transmit coil with a 305 kHz center frequency and 15 mT maximum field (Fig. 1). Three 70 μL samples of MFH-optimized superparamagnetic iron oxide (SPIO) particles (7.24 mg/mL) were placed in vials 1.2 cm apart as shown in Fig. 2. Optical-based thermal probes were placed in each vial along with a reference probe in the bore. In heating trials, each vial was separately aligned with the FFL and baseline data was gathered for one minute before energizing the transmit coil for 30 seconds. The temperature data was normalized to the baseline and the reference probe temperature was subtracted to remove systemic thermal interference. The phantom was separately imaged with our Berkeley FFP scanner (4 x 4 x 6 cm FOV, 5.12 minutes total scan time).

RESULTS As shown in Fig. 2 we demonstrated the ability to quantitatively image the phantom and selectively heat only one vial at a time. The first two vials were heated at an approximately linear rate of 2 degrees C per minute. The third probe increased temperature at a reduced rate, likely due to imperfect positioning of the SPIO vial with respect to the FFL.

CONCLUSION We showed preliminary results for the first dual MPI-MFH system capable of image guided, spatially-selective nanoparticle heating. These data suggest that simultaneous imaging and SPIO heating may be accomplished using modified MPI hardware and standard MFH SPIO tracers.

ACKNOWLEDGEMENTS We gratefully acknowledge support from NIH 5R01EB013689-03, CIRM RT2-01893, Keck Foundation 034317, NIH 1R24MH106053-01, NIH 1R01EB019458-01, ACTG: 037829, and NIH 1R21EB018453.

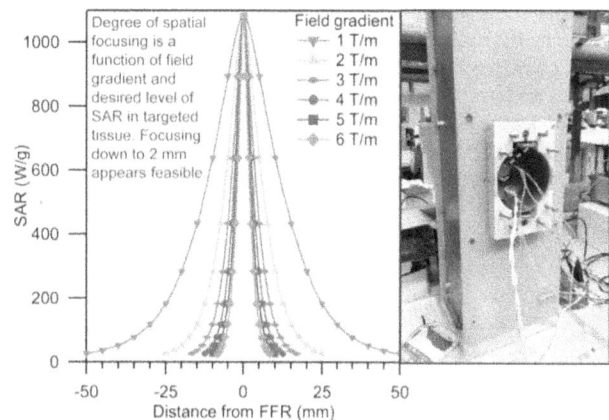

Figure 1: MPI for MFH. (Left) Spatial localization of SAR simulated by a heating model extended to include the MPI spatial gradient [7]. (Right) Berkeley FFL MFH heating device.

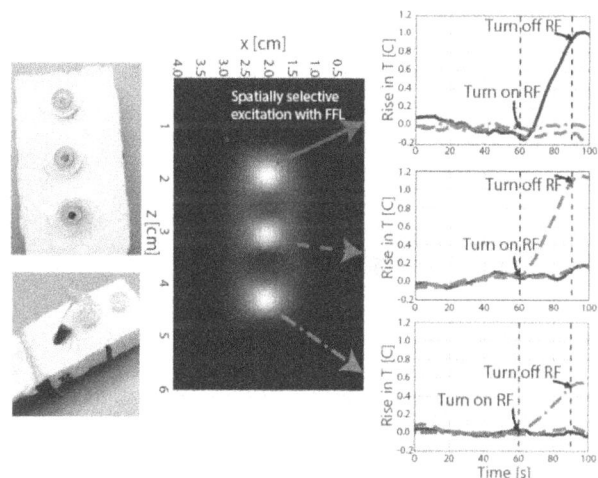

Figure 2: Combined MPI-MFH experimental data. (Left) Simple phantom composed of vials filled with MFH SPIOs. (Center) Phantom image obtained using the Berkeley FFP scanner. (Right) Temperature data from separate heating experiments demonstrating selective heating of each vial independently.

REFERENCES

[1] A. Jordan et al., *J Magn. and Magn. Mater.*, 201: 413-419, 1999.
[2] R. E. Rosensweig, *J Magn. and Magn. Mater.*, 252: 370-374, 2002.
[3] K. Murase et al., *Physica Medica*, 29: 624-630, 2013.
[4] K. Murase et al., *Open J Med. Img.*, 5: 85-99, 2015.
[5] T. Buzug et al. *Med Phys*, 22: 323-334, 2012.
[6] P. W. Goodwill et al., *Rev Sci. Instr.*, 83, 2012.
[7] D. Soto-Aquino et al., *J. Magn. Magn. Mater.*, 393: 46-55, 2015.

Device manipulation in an MPI-Scanner

Daniel Wirtz*, Claas Bontus, Jürgen Rahmer, Peter Mazurkewitz, Christian Stehning and Bernhard Gleich

Philips Technologie GmbH Innovative Technologies, Research Laboratories, Röntgenstraße 24-26, 22335 Hamburg, Germany
* Corresponding author, email: daniel.wirtz@philips.com

INTRODUCTION In addition to their imaging capabilities MPI systems can be utilized as powerful field generators, especially regarding the selection and focus fields. Compared to an MRI scanner the available gradient strengths in MPI systems are superior by roughly an order of magnitude. Thus steering of devices employing magnetic forces resulting in deflection of a device and/or the application of gradients resulting in a torque on the device are feasible and have been shown in [1] and [2]. In this work initial results from steering experiments with prototype catheter models are presented.

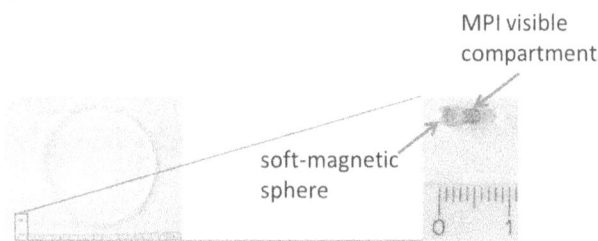

Figure 1: Catheter prototype with soft-magnetic sphere and MPI visible compartment containing MM4 powder.

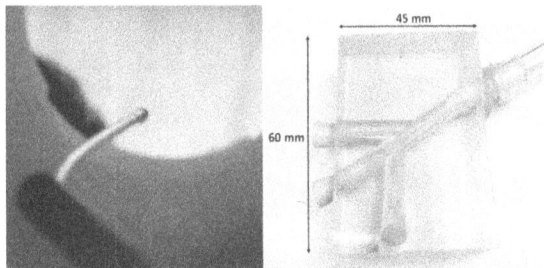

Figure 2: Catheter prototype positioned in the bore of the PCD scanner during manipulation experiments (left) and 3D vascular phantom. The canals are 6 mm in diameter and filled with a Resovist solution for MPI-visibility. Catheters enter the phantom through a piece of tube (right).

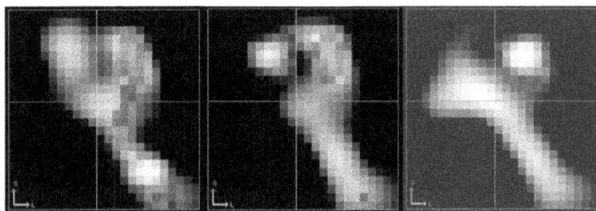

Figure 3: Real-time Color-MPI images during catheter advancement: The device enters the phantom (left) and is advanced into the central branch (center). Applying a homogeneous magnetic field in left-right direction during advancement, the device enters the right branch (right).

MATERIAL AND METHODS. Prototype catheters were constructed using six french (6F) polyurethane tubing (2 mm diameter). The devices have been equipped with up to four soft-magnetic beads of 2 mm diameter at the tip as well as an MPI-visible compartment of ~2 mm length containing MM4-powder, cf. Fig. 1.

For bench tests, the same tubing material has also been equipped with permanent magnetic beads of 2 mm diameter. Furthermore, several vascular phantoms were constructed from polycarbonate material. Therefore, canals of 5 and 6 mm diameter resembling typical vessel geometries have been drilled into the material and filled with a 1:200 Resovist-solution (0.5 mmol/l) in order to provide MPI-visibility, cf. Fig. 2.

RESULTS In bench tests both catheter types could be steered through the vessel phantom using an external permanent magnet. Especially the catheter containing the permanent magnet was easily steerable but would generate significant artifacts when used within the MPI scanner.

Thus in an initial step, the z-gradient of the MPI-scanner was used for manipulation of devices containing soft-magnetic beads. The available deflection force on the soft-magnetic sphere(s) was large enough to select between the paths in the bifurcation of the phantom while advancing the catheter. During the process, the catheter position was observed both, visually from the outside as well as in the real time Color-MPI signal, cf. Fig. 3.

While the tip position of the catheter was clearly visible through the displaced contrast fluid the tip sphere still created a significant artifact. Efforts on shielding the soft-magnetic beads for reduced artifact level are underway.

CONCLUSION It was shown that magnetic manipulation of a catheter device equipped with soft-magnetic material as well as permanent magnets using the selection field of an MPI-scanner is feasible. While position control using MPI visible compartments is possible, the artifact levels created by the soft-magnetic material as well as the permanent magnets have to be reduced for clinical applications by appropriate shielding.

Real-time control of the manipulations performed was possible using a real-time reconstruction algorithm and two-color visualization.

ACKNOWLEDGEMENTS

REFERENCES
[1] N. Nothnagel, „Lokalisierung und Manipulation ferromagnetischer Objekte mit einem MPI Scanner", Master Thesis, Univ. Lübeck, 2012
[2] N. Nothnagel et al., „Steering of Magnetic Devices with a Magnetic Particle Imaging System", to be published

Magnetic particle detection based on non-linear response to magnetic susecptibilty changes

Florian Fidler[a,*], Karl-Heinz Hiller[a], Peter M. Jakob[a]

[a] Research Center Magnetic-Resonance-Bavaria (MRB), Am Hubland, D-97074 Würzburg, Germany
* Corresponding author, email: fidler@mr-bavaria.de

INTRODUCTION Magnetic Particle Imaging (MPI) [1] and Spectroscopy (MPS) as introduced in general is based on the measurement of an induced voltage generated from the magnetic moments of suitable magnetic nanoparticles in an alternating magnetic excitation field. The non-linear response of the magnetization of such particles generate a receive signal on higher harmonics of the excitation frequency if no sufficient magnetic offset field is present to saturate the magnetization of the particles [2]. The magnetic susceptibility behaves for most cases quite similar since it is altered in a comparable way in a non-linear matter by applying an alternating magnetic field. The main difference in both physical properties is that they lead to different detection strategies [3]. Still the non-linear behavior can be detected as signal on higher harmonic frequencies. In this work the magnetic susceptibility is measured by time resolved measurement of the inductivity of the excitation field coil. This strategy reduces the efforts for the magnetic particle detection to a minimized single coil system sufficient to generate the excitation field and readout the answer from the magnetic particle system. Despite this the presented solution offers many more advantages, as the possibility to suppress any desired and unwanted signal enabling the optimized use of the dynamic range of the receiver digitizer. Exemplarily a single coil surface MPS system is shown and a volume detector, which can be combined with a gradient system for field-free-point or field-free-line generation. Additionally the presented setup is able to detect the first harmonic signal as well [4].

MATERIAL AND METHODS The basic setup presented here is based on a measurement of the magnetic susceptibility using a Wheatstone bridge like circuit. The Voltage in the coil is compared to an impedance network Zr, in our case this network was purely resistive. The coil is tuned to the excitation frequency to reduce necessary voltage. Any change in the inductivity will lead to a change of the electrical potential at the connection of both networks. Two digital transmitters T1 and T2 with digital adjustable amplitude and phase generate the voltage for both networks, amplified by amplifier A1 and A2. This allows to set the voltage digitally to zero for any desired situation, e.g. for no probe inside the scanner, in terms of frequencies, in general the excitation frequency, and of time, e.g. to remove some unwanted signal parts from a probe inside the scanner. The receive signal can optional be modeled by additional circuits. It is included here to compensate for the fact that a tuned coil will enhance the signal from the first harmonic. The optional bandstop is tuned to a frequency lower than the excitation frequency; this will reduce the voltage of lower harmonics and their noise levels adequate to use the dynamic range of the receiver for detection in an optimized way and represents a basic circuit to achieve this. The basic schematic of the scanner electronics is shown in fig. 1. A variety of surface coils and volume detectors was build; figure 2 shows exemplarily a surface MPS scanner with 30 mm diameter and an excitation field up to 180 mT.

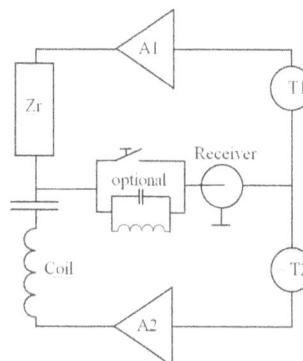

Figure 1: Basic schematic of the setup. T1 and T2 are digital transmitters, A1 and A2 power amplifiers, Zr an impedance network. The shown bandstop filter is optional.

Figure 2: Surface MPS scanner with 30 mm diameter and an excitation field of up to 180 mT at the surface.

RESULTS The presented technique is able to detect the presence of magnetic nanoparticles based on magnetic susceptibility changes. The effort to achieve this is drastically reduced. The presented system is fully digital adjustable.

CONCLUSION The novel technique based on the detection of non-linear response of the magnetic susceptibility of magnetic nanoparticles presented here offers a great potential in magnetic particle imaging and spetcroscopy.

ACKNOWLEDGEMENTS This work was supported by the EU FP7 HEALTH program IDEA – "Identification, homing and monitoring of therapeutic cells for regenerative medicine – Identify, Enrich, Accelerate" under grant agreement no 279288.

REFERENCES
[1] B. Gleich and J. Weizenecker. *Nature*, 435(7046):1217—1217, 2005. doi: 10.1038/nature03808.
[2] N. Panagiotopoulos et al. . *Int. J. Nanomed.*, 2015(10):3097—3114, 2015. doi: 10.2147IJN.S70488.
[3] Patent Application WO00215028569.
[4] F. Fidler et al. P28, Proc. IWMPI, 2015.

MPI system matrix reconstruction: making assumptions on the imaging device rather than on the tracer spatial distribution

Gael Bringout[a,*], Ksenija Gräfe[a], Thorsten M. Buzug[a]

[a] Institue of Medical Engineering, Universität zu Lübeck, Germany
* Corresponding author, email: bringout@imt.uni-luebeck.de

INTRODUCTION To reconstruct a first approximation of an unknown spatial tracer distribution, a preliminary solution of the inconsistent system of linear equations $\hat{S}c = \hat{u}$ (1) can be calculated, using a regularized weighted normal equation of the first kind formulated as $(\hat{S}^H WS + \lambda I)c = \hat{S}^H W\hat{u}$ (2), with $\hat{S}$ the system matrix (SM) and $\hat{S}^H$ its complex conjugate transpose, c the tracer distribution, $\hat{u}$ the acquired signal in Fourier space, W a weighting matrix, and λ the regularisation parameter together with the identity matrix I for Tikhonov regularization [1].

Furthermore, it is common to truncate the measurements according to the SNR, using a correspondingly truncated SM, $\hat{S}^\dagger$, and measurements $\hat{u}^\dagger$ in (2). However, the regularization scheme is chosen to enforce certain properties of the tracer distribution, which are in general unknown.

The presented technique does not make any assumptions on the spatial tracer distribution, besides that it can be approximated by solving directly $\hat{S}^\dagger c = \hat{u}^\dagger$ (3). Information only coming from the scanner design, imaging sequence, and acquired signal is used to truncate the matrices. The truncation thresholds can then be used for several measurements.

MATERIAL AND METHODS A single-sided scanner encoding a 2D plane by moving a low field volume along a 2D Lissajous curve is used [2, 3]. The measurements are used in form of the power spectrum of the acquired signal. The SM is measured on $M = 225$ positions $\vec{r}$ equidistantly spaced on a 2D grid.

In a first truncation step, data are rejected according to an SNR measurement defined for each frequency component k as

$$\text{SNR}(k) = \frac{\|\hat{u}_k\|}{std(Empty_k)},$$

with $std(Empty_k)$ the standard deviation determined from several air measurements. Here, a hard threshold of SNR > 10 is applied on the data coming from each receive channel.

Then, the energy $\tilde{w}_k$, defined by

$$\tilde{w}_k = \sum_{p=1}^{M} \left\| \hat{S}_k(\vec{r}_p) \right\|^2,$$

is used to further truncate the measurements, retaining the frequency components with the highest energy [4]. A hard threshold of $\tilde{w}_k > 0.01$ is applied on each channel. Arbitrary units are used due to the missing calibration of the receive channels.

Finally, the orthogonality map [1, 4] or Gramian matrix [5], calculated between the frequency components i and j of the SM as

$$G_{ij} = \langle \hat{s}_i, \hat{s}_j \rangle = \left\| \sum_{p=1}^{M} \frac{\hat{s}_i(\vec{r}_p)}{\|\hat{s}_i\|} \frac{conj(\hat{s}_j(\vec{r}_p))}{\|\hat{s}_j\|} \right\|,$$

is evaluated and used to derive another two hard thresholds. They are based on the standard deviation and the mean value of the Gramian matrix for a given i and any j. They have to be smaller than 0.06 and 0.07, respectively.

A modified Kaczmarz method is used to solve (3), imposing a real and non-negativity constraint on the solution, which leads to an approximation c^p of c after p iterations.

RESULTS Fig. 1 compares the reconstruction of two phantoms (Fig. 1, left) solving either equations (2) (Fig. 1, middle) or (3) (Fig. 1, right). The latter distributions are obtained after a calculation time of 164 ms and 143 ms on an Intel i5-760. The reconstructed distributions look similar or better for the presented approach. See imt.uni-luebeck.de or [6] for the corresponding scripts and data.

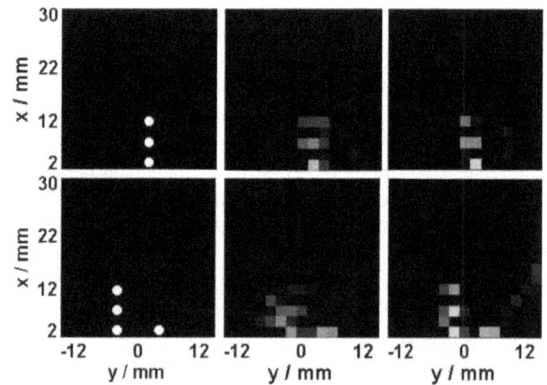

Figure 1: From left to right: model of the used phantoms. Reconstructions with a weighted regularized least squares approach with $\lambda \approx 4.65 \cdot 10^{-6}$ and $\lambda \approx 5.86 \cdot 10^{-6}$ and stopped after 5 iterations. Reconstruction with the presented method stopped after 50 iterations. All images use the same colour scale.

CONCLUSION The presented reconstruction technique solved the inconsistent system only by truncating the acquired data and by early stopping the iterative solver. Doing so, no assumption on the tracer distribution is made. The used thresholds can all be related to quantities, which can be linked to the scanner properties. Moreover, this technique does not seem to reduce the quality of the reconstruction. This strategy shows a way to the development of automatic techniques designed to obtain first and reliable reconstructions of the tracer distribution.

ACKNOWLEDGEMENTS The authors thank Christina Brandt and Hanne Medimagh for the fruitful discussion and gratefully acknowledge the financial support of the German Federal Ministry of Education and Research (BMBF, grant numbers 13N11090 and 01EZ0912) and of the European Union and the State Schleswig-Holstein (Programme for the Future - Economy, grant number 122-10-004).

REFERENCES
[1] T. Knopp and T. M. Buzug, Springer-Verlag, Berlin Heidelberg, 2012. ISBN 978-3-642-04199-0.
[2] T. F. Sattel et al., *Journal of Physics D: Applied Physics*, 42(2):1-5, 2009. doi: 10.1088/0022-3727/42/2/02200110.1109.
[3] K. Gräfe et al., 2015 5th International Workshop on Magnetic Particle Imaging (IWMPI). doi: 10.1109/IWMPI.2015.7107024.
[4] G. Bringout, Universität zu Lübeck, 2015, submitted. Ph.D. Thesis.
[5] N. Traulsen, Infinite Science Publishing, Lübeck. 2015.
[6] Data available under github.com/KsenijaGraefe/SingleSidedData.

The Influence of Trajectory and System Matrix Overlap on Image Reconstruction Results in Magnetic Particle Imaging

M. Ahlborg[a,*], C. Kaethner[a], T. Knopp[b,c], P. Szwargulski[b,c] and T.M. Buzug[a]

[a] Institute of Medical Engineering, Universität zu Lübeck, Lübeck, Germany
[b] Section for Biomedical Imaging, University Medical Center Hamburg-Eppendorf, Hamburg, Germany
[c] Institute for Biomedical Imaging, Hamburg University of Technology, Hamburg, Germany
[*] Corresponding author, email: {ahlborg, buzug}@imt.uni-luebeck.de

INTRODUCTION In Magnetic Particle Imaging (MPI) the field of view (FOV) size defined by the drive field amplitude and the gradient strength is limited by safety considerations. A possible way to increase the scan area is to use FOV patches [1]. The position of the patches can be chosen application specific and with respect to the resulting image reconstruction quality. In this contribution, two parameters are analyzed for a Lissajous sequence: a trajectory overlap and a system matrix overlap of patches. In a first simulation, both effects are studied thoroughly with focus on truncation artifact reduction [2] by gaining information redundancy. Subsequently, a validation on experimental data is performed.

MATERIAL AND METHODS Four 2D patches arranged on a grid are considered to examine the influence of both overlaps. The trajectory overlap is investigated by a successive shifting of the FOV patches towards each other. Thereby, the density of the trajectory is kept constant while increasing the overlap. The system matrix overlap is realized by a pixel-wise increase of the system matrix sampling area to compare different overlap configurations. In a simulation study, both overlaps are studied separately. Following, the findings are validated on experimental data and a continuative analysis of a combination is performed.

Any kind of overlap leads to information redundancy, which has to be handled. This can be done by reconstructing the individual patches separately with a consecutive post-processing of the redundantly reconstructed pixels [1]. In this contribution, four post-processing steps are considered: a cut-off, an average, a linear weighting and a $\sin^2$ weighting of the respective pixels. An alternative way to handle the information redundancy is to set up a joint system of equations of the individual patches [3,4] that inherently enables a processing of the multiple sampled pixels.

First, in a simulation a separate and detailed evaluation of both overlap types is performed. Second, a validation on selected experimental data obtained with a preclinical MPI scanner (Bruker/Philips) is done. Additionally, the experimental data is used to analyze the combination of both overlaps. The imaging plane is positioned in the xy-plane of the scanner. The gradient in both directions is 1.25 T/m. The drive-field strength amounts 14 mT. Image reconstruction is realized by solving the least squares problem of the system of equations Sc = u with the Kaczmarz algorithm and an additional Tikhonov regularization.

RESULTS Selected results of the simulation study are shown in Fig. 1. It is visible that a almost artifact-free reconstruction of an individual system matrix is possible. The best results can be obtained by a cut-off of redundant pixels where the patches overlap. With a trajectory overlap more information is included in the actual sampling of the FOV. This is clearly visible in the

reconstructed images that show a good correlation to the used phantom with little artifacts around the object.

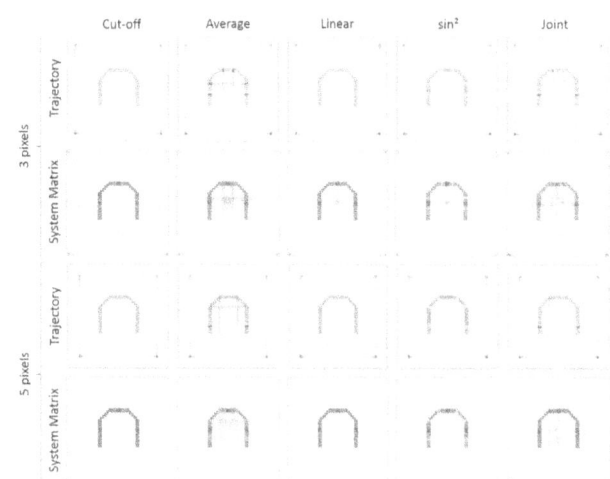

Figure 1: Simulation results of separate trajectory and system matrix overlaps of each three and five pixels.

CONCLUSION The use of trajectory and system matrix overlap is an important factor for artifact-free image reconstruction using FOV patches in MPI. The results show that a combination is commendable to reduce truncation artifacts at the patch edges.

ACKNOWLEDGEMENTS MA, CK, TMB like to thank the Federal Ministry of Education and Research (BMBF, 13N11090) and the European Union and the State Schleswig-Holstein (Programme for the Future – Economy, 122-10-004). TK and PS gratefully acknowledge funding and support of the German Research Foundation (DFG, AD125/5-1).

REFERENCES
[1] J. Rahmer et al. Proceedings of the International Society for Magnetic Resonance in Medicine, 19:629, 2011.
[2] M. Grüttner et al. International Workshop on Magnetic Particle Imaging (IWMPI) 2013, IEEE Xplore Digital Library, 2013, 10.1109/IWMPI.2013.6528335.
[3] M. Ahlborg et al. International Workshop on Magnetic Particle Imaging (IWMPI) 2015, IEEE Xplore Digital Library, 2015, 10.1109/IWMPI.2015.7107017.
[4] T. Knopp et al. Phys. Med. Biol. 60(8):L15, 2015, 10.1088/0031-9155/60/8/L15.

Fast Implicit Reconstruction of Focus Field Data in MPI

P. Szwargulski[a,b*], M. Hofmann[a,b], N. Gdaniec[a,b], and T. Knopp[a,b*]

[a] Section for Biomedical Imaging, University Medical Center Hamburg-Eppendorf, Hamburg, Germany
[b] Institute for Biomedical Imaging, Hamburg University of Technology, Hamburg, Germany
[*] Corresponding author, email: {p.szwargulski, t.knopp}@uke.de

INTRODUCTION Magnetic Particle Imaging (MPI) is a highly sensitive tomographic imaging modality, which is based on the response of magnetic material to dynamic magnetic fields [1]. The size of the field of view (FOV) is restricted due to limits on the strength of the applied oscillating magnetic fields. Additional focus fields [2] enable spatial translation of the FOV, while complying with physiological constraints. A larger FOV is realized by sequentially acquiring a series of shifted sampling areas [3], which is called multi-patch approach. For image reconstruction of multi-patch MPI data two different strategies have been proposed: The separate reconstruction of each measured patch followed by image post-processing [3, 4] and the joint reconstruction in a single step [5].

At present, the joint reconstruction requires the acquisition of a dedicated system matrix for each patch [5]. Furthermore, time and memory demands restrict its application to only a small number of patches. In order to overcome the limitations of the explicit join multi-patch reconstruction approach, in this work an efficient, implicit reconstruction scheme for the joint multi-patch imaging equation is derived.

MATERIAL AND METHODS For the joint reconstruction of the image c from the L focus field measurements $\tilde{u}^l$ an explicit constructed system matrix S is required [5]. In a continuous form the imaging equation is given by

$$\int_{\mathbb{R}^3} S_k^l(r)c(r)dr^3 = \tilde{u}_k^l. \tag{1}$$

Under the assumption of a linear gradient field a single calibration measurement S_k^0 is spatially shifted to each patch and can be used for reconstruction

$$\int_{\mathbb{R}^3} S_k^l(r)c(r)dr^3 = \int_{\mathbb{R}^3} S_k^0(T^l(r))c(r)dr^3 = \tilde{u}_k^l. \tag{2}$$

Using Eq. (2) an algorithm based on indirect indexing (see Fig. 1) is derived for reconstruction. Similar to sparse matrix formats indices for accessing the calibration matrix (Ind_S) and the image (Ind_c) are precomputed and stored for each drive-field patch.

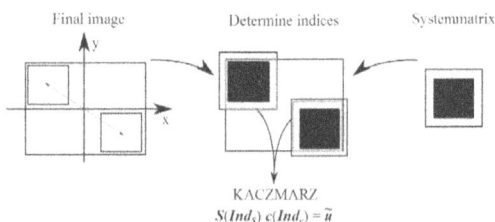

Figure 1: The principle of indirect indexing for an efficient joint reconstruction.

The time and memory requirements of the implicit and explicit [5] reconstruction are compared using a phantom dataset measured with a preclinical MPI scanner (Bruker/Philips) at eight non-overlapping focus field positions along the x-axis and a 3D calibration matrix with a grid size of 25×25×25.

RESULTS The time and memory requirements of the explicit and implicit reconstruction in dependence of the number of patches are shown in Fig. 2. For the explicit method a quadratic increase in the time and memory requirement is observed. Contrary, the reconstruction times of the implicit reconstruction scale linearly and the memory requirements are independent of the number of patches.

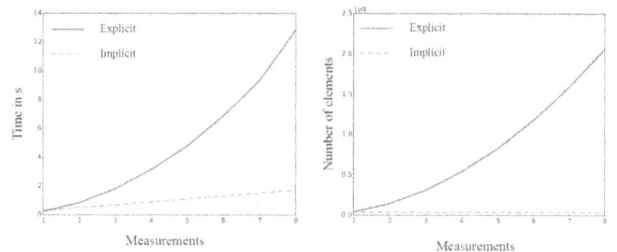

Figure 2: Comparison of the explicit and implicit reconstruction with respect to the time and memory requirement.

CONCLUSION In this work, an efficient method for the reconstruction of focus field data was introduced. It was shown that the reconstruction time scales linearly with the number of patches used, which is essential for future human application that will require a large number of patches.

ACKNOWLEDGEMENTS The authors gratefully acknowledge funding and support of the German Research Foundation (DFG, ADI 125/5-1).

REFERENCES
[1] B. Gleich and J. Weizenecker. Nature, 435(7046):1217—1217, 2005. doi: 10.1038/nature03808.
[2] B. Gleich et al. Proceedings of the International Society for Magnetic Resonance in Medicine, 18: 218, 2010.
[3] J. Rahmer et al. Proceedings of the International Society for Magnetic Resonance in Medicine, 19:629, 2011.
[4] M. Ahlborg et al. International Workshop on Magnetic Particle Imaging 2015, IEEE Xplore Digital Library, 2015.
[5] T. Knopp et al. Physics in Medicine and Biology, 60: L15, 2015

Interactive Positioning and Sizing of the Imaging Volume in Real-Time Magnetic Particle Imaging

Jürgen Rahmer[a,*], Claas Bontus[a], Jörn Borgert[a]

[a] Tomographic Imaging Systems, Philips Research Hamburg, Germany
[*] Corresponding author, email: juergen.rahmer@philips.com

INTRODUCTION The use of coils instead of permanent magnets for the generation of the MPI selection field enables interactive control of the size of the MPI imaging volume. Focus fields add the possibility of shifting the imaging volume in space [1]. If combined with online reconstruction, this technology allows flexible image-based 3D scan planning, giving the user the freedom to shift and resize the imaging volume during imaging. This contribution reports on the implementation of an interactive console that enables manipulation of selection and focus fields during an imaging experiment with visual control via online image reconstruction. Experimental results were obtained on a preclinical MPI demonstrator system as shown in Fig. 1.

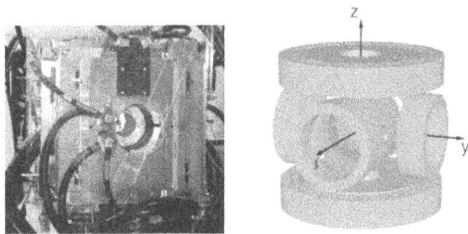

Figure 1: Pre-clinical demonstrator system [1]: Photograph and sketch of field generator unit.

MATERIAL AND METHODS The pre-clinical demonstrator system was operated using 3D drive field excitation at frequencies 24.5, 26.0, and 25.3 kHz with an amplitude of 16 mT. A variable selection field gradient between about 1.25 and 3.0 T/m was applied in the strong gradient direction (vertical or z axis). Focus fields were applied to shift the imaging volume up to a few centimeters in the desired direction. The system software was extended to enable interactive control of the selection and focus fields during imaging via a command line interface. For image-based scan planning, an online reconstruction and visualization was implemented. To this end, the standard iterative reconstruction approach [2] was accelerated by reducing the number of frequency components and the number of iterations at the cost of image quality and resolution. The resulting update rate for a 3D volume with $28 \times 28 \times 24$ voxels (à $1.5 \times 1.5 \times 1.0$ mm³) was about 5 Hz, with a latency between measurement and image display of roughly 1 s. The custom-built visualization tool enabled fast display updates with asynchronous pipelining to decouple reconstruction and processing from visualization. For background correction, noise spectra were acquired at the beginning of the imaging sequence by switching the selection field to a homogeneous field of about 150 mT to saturate all object signal.

RESULTS Fig. 2 shows online-reconstructed imaging volumes of a P-shaped phantom for different selection field gradients and focus field offsets. A maximum intensity projection through the 3D data set is calculated along the vertical axis. In the top row, the gradient is set to values of roughly 1.25, 2.0, 2.5, and 3.0 T/m, respectively. One finds that at lower gradients, a larger imaging volume is covered, whereas at higher gradients, a smaller volume is imaged at higher resolution. In the bottom row, the imaging volume is translated in R-L- and S-I-direction using respective focus field amplitudes at a gradient of 2.5 T/m. For all reconstructions, the same system function acquired at 2.5 T/m at the isocenter of the scanner has been used. Due to field inhomogeneities, at other gradient values and positions slight distortions occur that are not yet compensated for.

CONCLUSION The demonstrated approach offers highly flexible interactive sizing and positioning of the imaging volume during imaging. It thus enables zooming out of or into a region of interest. The approach can also form the basis for automated online tracking of a catheter or an injected bolus by centering the imaging volume on the position of the respective signal response.

ACKNOWLEDGEMENTS This work was supported by the German Federal Ministry of Education and Research (BMBF grants FKZ 13N9079, 13N11086, and 13GW0069C).

REFERENCES
[1] B. Gleich et al., "Fast MPI Demonstrator with Enlarged Field of View." In Proc. ISMRM, 18:218. Stockholm: ISMRM, 2010.
[2] J. Weizenecker et al., "Three-Dimensional Real-Time in Vivo MPI" Phys Med Biol 54, 5 (2009): L1–10.

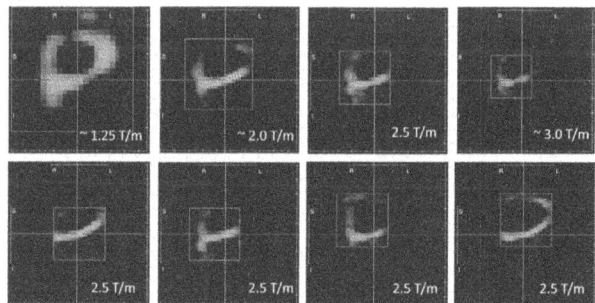

Figure 2: Top row: Maximum intensity projection (MIP) orthoviews for gradients ~1.25, ~2.0, ~2.5, and ~3.0 T/m. Bottom row: different focus field offsets in R-L- and S-I-direction at 2.5 T/m.

Infinite Science Publishing is a University Press and Academic Printing Incorporation. It provides a publication platform for excellent theses as well as scientific monographies and conference proceedings for reasonable costs.

These publications enable scientists and research organizations to reach the maximum attention for their results.

The service of Infinite Science Publishing comprises the entire range from the publication of print-ready documents up to cover design as well as copy-editing of single articles.

Infinite Science Publishing is an imprint of the Infinite Science GmbH, a University of Lübeck spin-off and service partner of the BioMedTec Science Campus.

www.infinite-science.de/publishing

Infinite Science GmbH
MFC 1 | BioMedTec Wissenschaftscampus
Maria-Goeppert-Str. 1, 23562 Lübeck
book@infinite-science.de

Infinite Science
Publishing

www.ingramcontent.com/pod-product-compliance
Lightning Source LLC
Chambersburg PA
CBHW082308210326
41598CB00029B/4479